LONDON MATHEMATICAL SOCIETY LECTURE NOTE SERIES

Managing Editor: Professor J.W.S. Cassels, Department of Pure Mathematics and Mathematical Statistics, University of Cambridge, 16 Mill Lane, Cambridge CB2 1SB, England

The titles below are available from booksellers, or, in case of difficulty, from Cambridge University Press.

34 Representation theory of Lie groups, M.F. ATIYAH *et al*
36 Homological group theory, C.T.C. WALL (ed)
39 Affine sets and affine groups, D.G. NORTHCOTT
46 p-adic analysis: a short course on recent work, N. KOBLITZ
50 Commutator calculus and groups of homotopy classes, H.J. BAUES
59 Applicable differential geometry, M. CRAMPIN & F.A.E. PIRANI
66 Several complex variables and complex manifolds II, M.J. FIELD
69 Representation theory, I.M. GELFAND *et al*
76 Spectral theory of linear differential operators and comparison algebras, H.O. CORDES
77 Isolated singular points on complete intersections, E.J.N. LOOIJENGA
83 Homogeneous structures on Riemannian manifolds, F. TRICERRI & L. VANHECKE
86 Topological topics, I.M. JAMES (ed)
87 Surveys in set theory, A.R.D. MATHIAS (ed)
88 FPF ring theory, C. FAITH & S. PAGE
89 An F-space sampler, N.J. KALTON, N.T. PECK & J.W. ROBERTS
90 Polytopes and symmetry, S.A. ROBERTSON
92 Representation of rings over skew fields, A.H. SCHOFIELD
93 Aspects of topology, I.M. JAMES & E.H. KRONHEIMER (eds)
94 Representations of general linear groups, G.D. JAMES
95 Low-dimensional topology 1982, R.A. FENN (ed)
96 Diophantine equations over function fields, R.C. MASON
97 Varieties of constructive mathematics, D.S. BRIDGES & F. RICHMAN
98 Localization in Noetherian rings, A.V. JATEGAONKAR
99 Methods of differential geometry in algebraic topology, M. KAROUBI & C. LERUSTE
100 Stopping time techniques for analysts and probabilists, L. EGGHE
101 Groups and geometry, ROGER C. LYNDON
103 Surveys in combinatorics 1985, I. ANDERSON (ed)
104 Elliptic structures on 3-manifolds, C.B. THOMAS
105 A local spectral theory for closed operators, I. ERDELYI & WANG SHENGWANG
106 Syzygies, E.G. EVANS & P. GRIFFITH
107 Compactification of Siegel moduli schemes, C-L. CHAI
108 Some topics in graph theory, H.P. YAP
109 Diophantine analysis, J. LOXTON & A. VAN DER POORTEN (eds)
110 An introduction to surreal numbers, H. GONSHOR
113 Lectures on the asymptotic theory of ideals, D. REES
114 Lectures on Bochner-Riesz means, K.M. DAVIS & Y-C. CHANG
115 An introduction to independence for analysts, H.G. DALES & W.H. WOODIN
116 Representations of algebras, P.J. WEBB (ed)
117 Homotopy theory, E. REES & J.D.S. JONES (eds)
118 Skew linear groups, M. SHIRVANI & B. WEHRFRITZ
119 Triangulated categories in the representation theory of finite-dimensional algebras, D. HAPPEL
121 Proceedings of *Groups - St Andrews 1985*, E. ROBERTSON & C. CAMPBELL (eds)
122 Non-classical continuum mechanics, R.J. KNOPS & A.A. LACEY (eds)
125 Commutator theory for congruence modular varieties, R. FREESE & R. MCKENZIE
126 Van der Corput's method of exponential sums, S.W. GRAHAM & G. KOLESNIK
127 New directions in dynamical systems, T.J. BEDFORD & J.W. SWIFT (eds)
128 Descriptive set theory and the structure of sets of uniqueness, A.S. KECHRIS & A. LOUVEAU
129 The subgroup structure of the finite classical groups, P.B. KLEIDMAN & M.W.LIEBECK
130 Model theory and modules, M. PREST
131 Algebraic, extremal & metric combinatorics, M-M. DEZA, P. FRANKL & I.G. ROSENBERG (eds)
132 Whitehead groups of finite groups, ROBERT OLIVER
133 Linear algebraic monoids, MOHAN S. PUTCHA
134 Number theory and dynamical systems, M. DODSON & J. VICKERS (eds)
135 Operator algebras and applications, 1, D. EVANS & M. TAKESAKI (eds)
136 Operator algebras and applications, 2, D. EVANS & M. TAKESAKI (eds)
137 Analysis at Urbana, I, E. BERKSON, T. PECK, & J. UHL (eds)
138 Analysis at Urbana, II, E. BERKSON, T. PECK, & J. UHL (eds)
139 Advances in homotopy theory, S. SALAMON, B. STEER & W. SUTHERLAND (eds)
140 Geometric aspects of Banach spaces, E.M. PEINADOR & A. RODES (eds)
141 Surveys in combinatorics 1989, J. SIEMONS (ed)

142 The geometry of jet bundles, D.J. SAUNDERS
143 The ergodic theory of discrete groups, PETER J. NICHOLLS
144 Introduction to uniform spaces, I.M. JAMES
145 Homological questions in local algebra, JAN R. STROOKER
146 Cohen-Macaulay modules over Cohen-Macaulay rings, Y. YOSHINO
147 Continuous and discrete modules, S.H. MOHAMED & B.J. MÜLLER
148 Helices and vector bundles, A.N. RUDAKOV *et al*
149 Solitons, nonlinear evolution equations and inverse scattering, M. ABLOWITZ & P. CLARKSON
150 Geometry of low-dimensional manifolds 1, S. DONALDSON & C.B. THOMAS (eds)
151 Geometry of low-dimensional manifolds 2, S. DONALDSON & C.B. THOMAS (eds)
152 Oligomorphic permutation groups, P. CAMERON
153 L-functions and arithmetic, J. COATES & M.J. TAYLOR (eds)
154 Number theory and cryptography, J. LOXTON (ed)
155 Classification theories of polarized varieties, TAKAO FUJITA
156 Twistors in mathematics and physics, T.N. BAILEY & R.J. BASTON (eds)
157 Analytic pro-*p* groups, J.D. DIXON, M.P.F. DU SAUTOY, A. MANN & D. SEGAL
158 Geometry of Banach spaces, P.F.X. MÜLLER & W. SCHACHERMAYER (eds)
159 Groups St Andrews 1989 volume 1, C.M. CAMPBELL & E.F. ROBERTSON (eds)
160 Groups St Andrews 1989 volume 2, C.M. CAMPBELL & E.F. ROBERTSON (eds)
161 Lectures on block theory, BURKHARD KÜLSHAMMER
162 Harmonic analysis and representation theory, A. FIGA-TALAMANCA & C. NEBBIA
163 Topics in varieties of group representations, S.M. VOVSI
164 Quasi-symmetric designs, M.S. SHRIKANDE & S.S. SANE
165 Groups, combinatorics & geometry, M.W. LIEBECK & J. SAXL (eds)
166 Surveys in combinatorics, 1991, A.D. KEEDWELL (ed)
167 Stochastic analysis, M.T. BARLOW & N.H. BINGHAM (eds)
168 Representations of algebras, H. TACHIKAWA & S. BRENNER (eds)
169 Boolean function complexity, M.S. PATERSON (ed)
170 Manifolds with singularities and the Adams-Novikov spectral sequence, B. BOTVINNIK
171 Squares, A.R. RAJWADE
172 Algebraic varieties, GEORGE R. KEMPF
173 Discrete groups and geometry, W.J. HARVEY & C. MACLACHLAN (eds)
174 Lectures on mechanics, J.E. MARSDEN
175 Adams memorial symposium on algebraic topology 1, N. RAY & G. WALKER (eds)
176 Adams memorial symposium on algebraic topology 2, N. RAY & G. WALKER (eds)
177 Applications of categories in computer science, M. FOURMAN, P. JOHNSTONE, & A. PITTS (eds)
178 Lower K- and L-theory, A. RANICKI
179 Complex projective geometry, G. ELLINGSRUD *et al*
180 Lectures on ergodic theory and Pesin theory on compact manifolds, M. POLLICOTT
181 Geometric group theory I, G.A. NIBLO & M.A. ROLLER (eds)
182 Geometric group theory II, G.A. NIBLO & M.A. ROLLER (eds)
183 Shintani zeta functions, A. YUKIE
184 Arithmetical functions, W. SCHWARZ & J. SPILKER
185 Representations of solvable groups, O. MANZ & T.R. WOLF
186 Complexity: knots, colourings and counting, D.J.A. WELSH
187 Surveys in combinatorics, 1993, K. WALKER (ed)
188 Local analysis for the odd order theorem, H. BENDER & G. GLAUBERMAN
189 Locally presentable and accessible categories, J. ADAMEK & J. ROSICKY
190 Polynomial invariants of finite groups, D.J. BENSON
191 Finite geometry and combinatorics, F. DE CLERCK *et al*
192 Symplectic geometry, D. SALAMON (ed)
193 Computer algebra and differential equations, E. TOURNIER (ed)
194 Independent random variables and rearrangement invariant spaces, M. BRAVERMAN
195 Arithmetic of blowup algebras, WOLMER VASCONCELOS
196 Microlocal analysis for differential operators, A. GRIGIS & J. SJÖSTRAND
197 Two-dimensional homotopy and combinatorial group theory, C. HOG-ANGELONI, W. METZLER & A.J. SIERADSKI (eds)
198 The algebraic characterization of geometric 4-manifolds, J.A. HILLMAN
199 Invariant potential theory in the unit ball of $\mathbf{C}^n$, MANFRED STOLL
200 The Grothendieck theory of dessins d'enfant, L. SCHNEPS (ed)
201 Singularities, JEAN-PAUL BRASSELET (ed)
202 The technique of pseudodifferential operators, H.O. CORDES
203 Hochschild cohomology of von Neumann algebras, A. SINCLAIR & R. SMITH
204 Combinatorial and geometric group theory, A.J. DUNCAN, N.D. GILBERT & J. HOWIE (eds)
207 Groups of Lie type and their geometries, W.M. KANTOR & L. DI MARTINO (eds)
208 Vector bundles in algebraic geometry, N.J. HITCHIN, P. NEWSTEAD & W.M. OXBURY (eds)

London Mathematical Society Lecture Note Series. 211

Groups '93 Galway / St Andrews

Galway 1993

Volume 1

Edited by

C.M. Campbell
University of St Andrews

T.C. Hurley
University College, Galway

E.F. Robertson
University of St Andrews

S.J. Tobin
University College, Galway

J.J. Ward
University College, Galway

CAMBRIDGE
UNIVERSITY PRESS

CAMBRIDGE UNIVERSITY PRESS
Cambridge, New York, Melbourne, Madrid, Cape Town, Singapore, São Paulo

Cambridge University Press
The Edinburgh Building, Cambridge CB2 2RU, UK

Published in the United States of America by Cambridge University Press, New York

www.cambridge.org
Information on this title: www.cambridge.org/9780521477499

First published 1995

A catalogue record for this publication is available from the British Library

ISBN-13 978-0-521-47749-9 paperback
ISBN-10 0-521-47749-2 paperback

Transferred to digital printing 2005

CONTENTS

Volume I

Contents of Volume II

PREFACE

This is the first of two volumes of the Proceedings of Groups 1993 Galway / St Andrews. There is a full contents of both volumes and papers written by authors with the name of the first author lying in the range A-K. Contained in this part are the papers of the main speakers J L Alperin, M Broué and P H Kropholler, and we would like especially to thank them for their contributions both to the conference and to this volume. There is some perturbation in strict alphabetical order at the join between the volumes and the second volume contains papers written by authors with the name of the first author lying in the range H-Z.

INTRODUCTION

An international conference 'Groups 1993 Galway / St Andrews' was held at University College, Galway, Ireland during the period 1 to 14 August 1993. This followed in the main the successful format developed in 1981, 1985 and 1989 by C M Campbell and E F Robertson. They invited T Hurley, S Tobin and J Ward to join them and continue the series in 1993 in Galway. Serious planning got under way when the five organisers met at a Warwick conference in March 1991, and decided to invite as principal speakers J L Alperin (Chicago), M Broué (Paris), P H Kropholler (London), A Lubotzky (Jerusalem) and E I Zel'manov (Madison). All of these agreed to give courses of about five lectures each, and articles based on these courses form a valuable part of these Proceedings - particularly so as in some cases they are strongly complementary in subject matter. Also, as it transpired, one speaker was awarded a Fields Medal exactly one year later at the 1994 ICM in Zurich; the organisers have great pleasure in congratulating Professor Zel'manov most heartily - and hope perhaps that this may augur well for future speakers in the series!

An invitation to J Neubüser (Aachen) to arrange a workshop on Computational Group Theory and the use of GAP was taken up so enthusiastically by him that the workshop became effectively a fully-fledged parallel meeting throughout the second week, with over thirty hours of lectures by experts and with practical sessions organised by M Schönert (Aachen). These Proceedings contain an article by Professor Neubüser based on a lecture he gave in the first week of the conference.

In addition there were sixteen invited one-hour lectures, and a very great number of research talks, which reflected the fact that the meeting was attended by 285 mathematicians (who were accompanied by about 150 family members) from 35 countries around the world. The articles in these Proceedings, other than those of the main speakers, are based on one-hour invited lectures and other research talks.

As in 1989 there are two volumes of the Proceedings, with the contributions arranged according to author's names in alphabetical order from A in Volume One to Z in Volume Two, except for a minor perturbation in alphabetical order at the join between the two volumes. They will form a stimulating record of recent work as well as being a valuable reference source, due to the wealth of material covered in the main courses. We thank all our contributors, and regret that we could not - because of space restrictions - accept several other worthy papers.

An unusual feature of this conference was the setting aside of one day for a special programme of lectures to honour the 65th birthday of Professor K W Gruenberg, in recognition of his many contributions to group theory. A second feature was the publication by the Galway Mathematics Department of a splendid memoir by Professor Lubotzky on Subgroup Growth, prepared by him as a background for his course of lectures. Another feature was the videoing of all the talks of the main speakers which form a lovely record of the events. These videos may be obtained from the Galway Mathematics Department.

There are many who helped to make the conference a memorable occasion; in particular we thank the staff of the Computing Services in University College Galway, our colleagues and students, members of the Secretariat in both Galway and St Andrews, and members of our own families who cheerfully helped out in various ways. We extend our thanks to the Administration of University College Galway for the Reception in the Aula Maxima; and we are grateful to the Mayor of Galway who gave a special reception in the Corporation's Council Chamber.

We gratefully acknowledge financial support for the academic program from the London Mathematical Society, the Royal Irish Academy, the Irish Mathematical Society, An Bord Gais, Irish Shell, Allied Irish Banks, Bank of Ireland, Bord Failte, the International Science Foundation, University College Galway, the University of St Andrews and the Deutsche Forschungsgemeinschaft.

As our final word, an focal scuir, we wish to thank Cambridge University Press for help with these Proceedings and Nik Ruškuc for undertaking so willingly the enormous task of reformatting files in many flavours of TEX. As in previous volumes, the present editors have endeavoured to produce a measure of uniformity - hopefully without distorting individual styles. For any inconsistencies in, or errors introduced by, our editing we of course accept responsibility.

Colin M Campbell
Thaddeus C Hurley
Edmund F Robertson
Sean J Tobin
James J Ward

GEOMETRY, STEINBERG REPRESENTATIONS AND COMPLEXITY

J.L. ALPERIN*[1] and G. MASON†[1]

*Mathematics Department, University of Chicago, IL 60637, U.S.A.
†Mathematics Department, University of California at Santa Cruz, CA 95064, U.S.A.

Group representation theory often relates quite different areas of mathematics and we shall give yet another example of this phenomenon. A construction from finite geometries will lead us to a new concept in representation theory which we shall then apply to the representation theory of Lie type groups. This, in turn, will involve ideas from the homological approach to modular representations. We shall, therefore, cover a spectrum of ideas.

One construction of finite projection planes involves the use of spreads. Suppose that V is a $2n$-dimensional vector space over a finite field k of characteristic p. A spread $\mathcal{S}$ is a collection of n-dimensional subspaces whose (set-theoretic) union is all of V but where the intersection of any two members of the collection is zero. A group of linear transformations of V preserves the spread $\mathcal{S}$ if its elements permute the members of $\mathcal{S}$.

Proposition 1. *If E is an elementary abelian 2-group of linear transformations of V which preserve $\mathcal{S}$ and $p = 2$ then there is a subgroup F of E with the following two properties:*

i) V is free as a kF-module;

ii) The space of fixed-points V^F, of V under F, equals V^E.

This is the key idea and we shall now formulate it in more generality. If E is an elementary abelian p-group and k is any field of characteristic p then the kE-module M is said to be subfree if there is a subgroup F of E with the two properties of the proposition. Before proceeding, let us note that this gives us some interesting parameters. The dimension of the subspace $V^F = V^E$ is called the breadth of M. Since M is a free kF-module we have that $\dim M = |F| \dim M^F$ so $|F|$ is independent of the choice of F. If $|F| = p^d$ then we say that d is the depth of M.

Returning to the geometry, we shall see that this property arises in a "layered" manner.

Proposition 2. *Under the same hypotheses as Proposition 1, let Q be a 2-group of linear transformations of V which preserve $\mathcal{S}$. Assume, moreover,*

[1]Supported by National Science Foundation grants.

that V^Q is not contained in any single member of $\mathcal{S}$. It follows that whenever N is a normal subgroup of the subgroup P of Q and P/N is elementary abelian then V^N is a subfree $k[P/N]$-module.

We now turn to the study of how these ideas arise in the representation theory of groups of Lie type. Assume that k is an algebraically closed field of characteristic p and let $G = SL(2,q)$, the special linear group over the q-element subfield k_q of k, where $q = p^e$. Let X be the subgroup of upper uni-triangular elements of G so X is a Sylow p-subgroup of order q. Let $X(\lambda)$ be the element of X which has λ as the first row second column entry. We can now state a surprising result [2].

Theorem 1. *If V is a kG-module, and $e > 1$, then the following are equivalent:*

i) V is simple of dimension a power of p;

ii) V is a subfree kX-module of breadth 1.

It is quite unexpected that simplicity can be described in these terms. The motivation is again geometric. If $p = 2$ then a non-simple module with property ii), written over a finite subfield of k, would be an interesting candidate for a spread left invariant by G. Geometric motivation again suggests the following question in the case $p = 2$: Which simple kG-modules are subfree on restriction to every 2-subgroup?

This theorem is based, in part, on a determination of the non-identity subgroups of X which act freely on particular simple modules. Such modules are necessarily of dimension a power of p so let us describe all the simple kG-modules of such a dimension. The basic Steinberg module St_1 for kG is the $(p-1)st$ symmetric power of the standard two-dimensional module so St_1 is p-dimensional and it is simple as well. Let $\sigma_1, \ldots, \sigma_d$ be d distinct automorphisms of k_q so $d \leq e$. Then Galois conjugate modules $\sigma_i(St_1), 1 \leq i \leq d$, are distinct and

$$S = \sigma_1(St_1) \otimes \ldots \otimes \sigma_d(St_1)$$

is of dimention p^d and simple. Such modules, called partial Steinberg modules, give all the simple kG-modules of dimension a power of p. The tensor product of all e conjugates is of dimension $p^e = q$ and is the Steinberg module.

This situation generalizes considerably so we shall give the adjunct to the theorem in this broader context. Now set $G = SL(n,q)$ and let X be the root subgroup which consists of matrices with ones on the main diagonal, zeros elsewhere, except perhaps in the first row second column entry. Let $X(\lambda)$ be the corresponding element of X. There is a basic Steinberg module St_1 for kG of dimension $p^{n(n-1)/2}$ and partial Steinberg modules

$$S = \sigma_1(St_1) \otimes \ldots \otimes \sigma_d(St_1)$$

(as well as the Steinberg module). The adjunct result is as follows [3].

Theorem 2. *The subgroup generated by $X(\lambda_1), \ldots, X(\lambda_d)$ is of order p^d and free on S if, and only if,* $\det(\sigma_i(\lambda_j)) \neq 0$.

This is only a special case of the general result: we can deal with p-subgroups of all orders, all simple kG-modules as well as some other groups of Lie type. It would be also nice to have a generalisation of Theorem 1 to the case of $SL(n,q)$ but there is a significant obstacle: a basic Steinberg module St_1 need *not* be subfree on restriction to the root subgroup X. However, the ideas in Proposition 2 are the way round this problem and we shall illustrate this first by looking at another Lie type group.

We let $p = 2$ and let $H = Sz(2^{2f+1})$ be the Suzuki group, $f \geq 1$. Here there is a basic Steinberg module St_1 of dimension four and partial Steinberg modules

$$T = \tau_1(St_1) \otimes \ldots \otimes \tau_c(St_1), \quad c \leq 2f+1,$$

which give all the simple kH-modules in fact (and the Steinberg module has dimension 4^{2f+1}). Let Y be a Sylow 2-subgroup of H, so $|Y| = 4^{2f+1}$, the center Z of Y is elementary abelian of order 2^{2f+1} as is Y/Z.

Theorem 3. *As a kZ-module, T is subfree of breadth 2^{2c+1} while T^Z is a subfree $k[Y/Z]$-module of breadth one.*

Presumably these conditions are also sufficient to characterize simple modules and give a result analogous to Theorem 1.

We have not indicated the ideas of the proofs. There are direct methods [2,3] but Carlson has shown that complexity theory and homological ideas are important [5]. In order to return to the special linear groups we require a first instalment of these techniques. Let E be an elementary abelian p-group and let $\operatorname{aug}(kE)$ be the augmentation ideal. If $g_1, \ldots, g_d$ is a minimal generating set of E then $V_E = \operatorname{aug}(kE)/\operatorname{aug}(kE)^2$ is a d-dimensional vector space with a basis consisting of the cosets of the elements $g_1 - 1, \ldots, g_d - 1$. If $x_1, \ldots, x_t$ are elements of $\operatorname{aug}(kE)$ whose cosets modulo $\operatorname{aug}(kE)^2$ are linearly independent elements of V_E then the units $1 + x_1, \ldots, 1 + x_t$ are of order p and generate an elementary abelian p-group of order p^t in the unit group of kE. Such a group is called a shifted subgroup (e.g. see [6]) of E. If M is a kE-module then M is called shifted subfree if there is a shifted subgroup F such that $M^E = M^F$ and M is a free kF-module. Again we can speak of the breadth and depth of M.

We now return to the group $G = SL(n,q)$. Let L be the subgroup stabilizing a fixed m-dimensional subspace, $0 < m < n$, of the standard n-dimensional space on which G acts. Let Q be the normal subgroup of L of all elements which induce the identity on the m-dimensional subspace and in its $(n-m)$-dimensional quotient so Q is an elementary abelian p-group of order $q^{m(n-m)} = p^{em(n-m)}$. We are going to state a result about the action of Q on the partial Steinberg module S (as above). Since L/Q contains a subgroup isomorphic with $SL(m,q) \times SL(n-m,q)$ and S^Q is a tensor product of partial Steinberg modules for this direct product, it becomes possible to use our result to give, inductively, a "layered" result about the action of a Sylow-p-subgroup of G on S analogous to the previous theorem. However, we shall restrict ourselves to Q.

Theorem 4. *The kQ-module S is shifted subfree of depth $dm(n-m)$.*

We also believe that when $q > p$ the partial Steinberg modules can be characterized in terms of subfree "layers" and the parameters involved and already have made considerable progress towards such a goal. The proof involves complexity theory, to which we turn for out last theorem.

Let G be an arbitary finite group and let M be a kG-module. The complexity $C_G(M)$ of M is a non-negative integer, a homological invariant of M. It is zero exactly when M is projective and it is one, when it is not zero, but there is a projective resolution

$$\to P_n \to \ldots \to P_1 \to P_0 \to M \to 0$$

of M such that $\dim P_n$ is bounded, independently of n, (i.e. $\dim P_n$ is bounded by a degree zero polynomial in n). Similarly, it is two, if it is not less than two and there is such a resolution where $\dim P_n$ is bounded bu a degree one polynomial in n, and so on for higher complexities.

Theorem 5. *The complexity of the basic Steinberg module St_1 for $SL(n,q)$ is $(e-1)[n^2/4]$.*

The way this result is connected with our previous results is as follows. The complexity $C_G(M)$ equals the maximum of the complexities $C_E(M_E)$ as E runs over all elementary abelian p-subgroups of G [1]. In turn, $C_E(M_E)$ can be calculated in two complementary fashions, using Carlson's rank variety [4] or Kroll's shifted subgroup approach [7]. The rank variety $V_E(M)$ is defined as follows, as a subset of V_E (notation as above). If X is in $\text{aug}(kE)$ and not in $\text{aug}(kE)^2$ then the coset $x + \text{aug}(kE)^2$ lies in $V_E(M)$ if, and only if, M is not a free module for $< 1 + x >$; this does not depend on the choice of coset representative. The variety $V_E(M)$ consists of these elements together with the zero vector, it is a homogeneous affine variety of dimension $C_E(M_E)$. If $|E| = p^s$ then there are subspaces of V_E of dimension $s - C_E(M_E)$ which intersect $V_E(M)$ in zero, by the homogeneity of $V_E(M)$, and no such subspaces of any larger dimension. This implies that there are shifted subgroups of E of order $p^{s-C_E(M_E)}$ for which M is a free module and none of any greater order with this property.

References

[1] J.L. Alperin and L. Evens, Representations, resolutions and Quillen's dimension theorem, *J. Pure Appl. Alg.* **22**(1981), 1-9.

[2] J.L. Alperin and G. Mason, On simple modules for $SL(2,q)$, *Bull. London Math. Soc.* **25**(1993), 17–22.

[3] J.L. Alperin and G. Mason, Partial Steinberg modules for finite groups of Lie type, *Bull. London Math. Soc.* **25**(1993), 553–557.

[4] J.F. Carlson, The varieties and the cohomology ring of a module, *J. Algebra* **85**(1983), 104–143.

[5] J.F. Carlson, Varieties and modules of small dimension, *Arch. Math.* **60**(1990), 425–430.

[6] L. Evens, *The Cohomology of Groups* (Oxford University Press, 1991).

[7] O. Kroll, Complexity and elementary abelian p-groups, *J. Algebra* **88**(1984), 155–172.

[8] G. Mason, Varieties attached to an $SL_2(2^k)$-module, *Proc. Amer. Math. Soc.* **116**(1992), 343–350.

THE STRUCTURE OF METABELIAN FINITE GROUPS

ZVI ARAD, ELSA FISMAN and MIKHAIL MUZYCHUK

Department of Mathematics and Computer Science, Bar-Ilan University, Ramat-Gan, 52900 Israel

1. Introduction

Let G be a group and $H \subset G$ a proper subgroup of G, then $\mathrm{Core}(H)$ is the maximum normal subgroup of G contained in H.

In 1955 Itô [I], using a surprisingly short commutator calculation, obtained the following classic result:

Theorem. (Itô 1955) *Let the group $G = AB$ be the product of two abelian subgroups A and B. Then G is metabelian. Furthermore, if G is finite, then either* $\mathrm{Core}(A)$ *or* $\mathrm{Core}(B)$ *is not trivial.*

The following natural question ([AFG, p.18]) arises: Describe all metabelian groups which are factorized by two abelian subgroups.

Our interest in factorizable groups by two abelian subgroups was inspired by the following conjecture stated at the Groups 1993 Galway / St Andrews Conference:

Conjecture. *Let $G = AB$ be a finite factorizable group where A is abelian and B is cyclic. Assume that $Z(G) = \mathrm{Core}(B) = 1$. Then $A \triangleleft G$.*

Counterexamples to this conjecture are constructed in Section 2, Theorem A.

The following Theorem B gives a description of finite factorizable groups $G = AB$ by two abelian subgroups such that $Z(G) = \mathrm{Core}(B) = 1$ and $A \ntriangleleft G$.

Theorem B. *Let $G = AB$ be a finite factorizable group by two abelian subgroups A and B. Assume $\mathrm{Core}(B) = Z(G) = 1$ and $A \ntriangleleft G$. Then the following hold:*

(i) $G = BG'$, $B \cap G' = 1$, $|G'| = |A|$ *and* $G' = G^{\omega}$;

(ii) $G'A$ *is nilpotent;*

(iii) $Z(G'A) = G' \cap A = \mathrm{Core}(A)$;

(iv) B *is a Carter subgroup of* G.

The proof of Theorem B can be found in Section 3 of this paper.

The converse of Itô's theorem for finite metabelian groups with trivial centre is stated in Theorem C of Section 4.

Theorem C. *Let G be a finite metabelian group with trivial centre. Then $G = CG'$, where C is an abelian Carter subgroup of G, and G is the semidirect product of two abelian subgroups C and $G' = G^{\omega}$.*

Corollary 1. *If G is a finite metabelian group and $Z_n(G)$ is the last term of the upper central series of G then $\overline{G} = G/Z_n(G)$ satisfies the assumptions of Theorem B and consequently $\overline{G}$ is a semidirect product of two abelian subgroups as described in Theorem B.*

Our notation is standard and taken mainly from [G].

2. Main Results

Let us construct counterexamples to the conjecture.

Consider the group $GL_2(\mathbb{Z}_{p^m})$, p an odd prime and $m \geq 2$, where $\mathbb{Z}_{p^m}$ is a residue ring modulo p^m.

Define $G = \left\{ \begin{pmatrix} \beta & \alpha \\ 0 & 1 \end{pmatrix} \middle| \beta \in \mathbb{Z}^*_{p^m},\ \alpha \in \mathbb{Z}_{p^m} \right\}$, where $\mathbb{Z}^*_{p^m}$ is the set of all units of $\mathbb{Z}_{p^m}$. The cyclic subgroup $B = \left\{ \begin{pmatrix} \beta & 1 \\ 0 & 1 \end{pmatrix} \middle| \beta \in \mathbb{Z}^*_{p^m} \right\} \subseteq G$ is of order $\left|\mathbb{Z}^*_{p^m}\right|$.

The subgroup $1 + p\mathbb{Z}_{p^m} = \{1 + p\ell \ : \ \ell \in \mathbb{Z}_{p^m}\}$ of $\mathbb{Z}^*_{p^m}$ is a cyclic group of order $p^{m-1} > 1$. Denote by β the generator of this subgroup. Now take A to be the cyclic subgroup of G generated by the matrix $a = \begin{pmatrix} \beta & 1 \\ 0 & 1 \end{pmatrix}$.

Theorem A. *The group G is the product of the two cyclic subgroups A and B. The following properties hold for G :*

(1) $\mathrm{Core}(B) = Z(G) = 1$;

(2) $A \ntriangleleft G$.

In particular, G is a counterexample to the conjecture.

To prove Theorem A we need the following proposition.

Proposition 2.1. *The following properties hold:*

(i) $|A| = p^m$;

(ii) $B \cap A = 1$;

(iii) $A \ntrianglelefteq G$.

PROOF. (i) Clearly, $|A|$ coincides with the order of a as an element of G. The k-th power of a has a form

$$a^k = \begin{pmatrix} \beta^k & \sum_{i=0}^{k-1} \beta^i \\ 0 & 1 \end{pmatrix}. \tag{1}$$

Since β is an element of order p^{m-1}, it follows from (1) that $p^{m-1} \mid |A|$ and

$$a^{p^{m-1}} = \begin{pmatrix} 1 & \sum_{i=0}^{p^{m-1}-1} \beta^i \\ 0 & 1 \end{pmatrix}.$$

The sum $\sum_{i=0}^{p^{m-1}-1} \beta^i$ is a sum of all elements of $1 + p\mathbb{Z}_{p^m}$. Hence

$$\begin{aligned} \sum_{i=0}^{p^{m-1}-1} \beta^i &= \sum_{i=0}^{p^{m-1}-1} (1+pi) = \frac{1+1+p(p^{m-1}-1)}{2} \cdot p^{m-1} \\ &= \frac{2-p}{2} \cdot p^{m-1} = p^{m-1}. \end{aligned}$$

(We note that all calculations are done in $\mathbb{Z}_{p^m}$.) Thus, we have

$$a^{p^{m-1}} = \begin{pmatrix} 1 & p^{m-1} \\ 0 & 1 \end{pmatrix}$$

which immediately implies that $a^{p^m} = \begin{pmatrix} 1 & 0 \\ 0 & 1 \end{pmatrix}$ and $|A| = p^m$.

(ii) Let $a^k = \begin{pmatrix} \beta^k & \sum_{i=0}^{k-1} \beta^i \\ 0 & 1 \end{pmatrix} \in B$. This implies $1 + \beta + \ldots + \beta^{k-1} = 0$. Multiplying both sides of the latter equality by $\beta - 1$ gives us $\beta^k = 1$, where $a^k = \begin{pmatrix} 1 & 0 \\ 0 & 1 \end{pmatrix}$.

(iii) Assume the contrary, i.e. $A \trianglelefteq G$. Let $\gamma \in \mathbb{Z}_{p^m}^*$ be an element such that $\gamma \not\equiv 1 \pmod p$. Such a γ exists because $p \neq 2$. Then one has

$$\begin{pmatrix} \gamma & 0 \\ 0 & 1 \end{pmatrix} \begin{pmatrix} \beta & 1 \\ 0 & 1 \end{pmatrix} \begin{pmatrix} \gamma^{-1} & 0 \\ 0 & 1 \end{pmatrix} = \begin{pmatrix} \beta & 1 \\ 0 & 1 \end{pmatrix}^k \tag{2}$$

for an appropriate k.

After computations we obtain that (2) is equivalent to the following equalities:

$$\beta = \beta^k, \quad \gamma = 1 + \ldots + \beta^{k-1}$$

whence $\beta^{k-1} = 1$ and $(\beta-1)\gamma = \beta^k - 1 = \beta - 1$. This gives us $(\beta-1)(\gamma-1) = 0$. But $\gamma \not\equiv 1 \pmod p$ and therefore $\gamma - 1$ is invertible in $\mathbb{Z}_{p^m}$. This immediately implies $\beta = 1$, a contradiction. □

PROOF OF THEOREM A. Clearly, $|G| = p^m|\mathbb{Z}^*_{p^m}| = |A| \cdot |B|$. On the other hand, $A \cap B$ is trivial. Therefore, $G = AB = BA$. Since $A \ntriangleleft G$, then to show that G is a counterexample, we have to prove that $\mathrm{Core}(B) = Z(G) = 1$.

Let $\begin{pmatrix} \delta & \alpha \\ 0 & 1 \end{pmatrix} \in Z(G)$. Then $\begin{pmatrix} 1 & 1 \\ 0 & 1 \end{pmatrix}\begin{pmatrix} \delta & \alpha \\ 0 & 1 \end{pmatrix} = \begin{pmatrix} \delta & \alpha \\ 0 & 1 \end{pmatrix}\begin{pmatrix} 1 & 1 \\ 0 & 1 \end{pmatrix}$, which gives us $\delta = 1$. Furthermore, $\begin{pmatrix} \gamma & 0 \\ 0 & 1 \end{pmatrix}\begin{pmatrix} 1 & \alpha \\ 0 & 1 \end{pmatrix} = \begin{pmatrix} 1 & \alpha \\ 0 & 1 \end{pmatrix}\begin{pmatrix} \gamma & 0 \\ 0 & 1 \end{pmatrix}$ should hold for all $\gamma \in \mathbb{Z}^*_{p^m}$. Therefore, $\gamma\alpha = \alpha$ for all $\gamma \in \mathbb{Z}^*_{p^m}$, whence $\alpha = 0$ (we recall that $p \neq 2$). Thus, $Z(G)$ is trivial.

Let $\begin{pmatrix} \delta & 0 \\ 0 & 1 \end{pmatrix} \in \mathrm{Core}(B)$. Then $\begin{pmatrix} 1 & 1 \\ 0 & 1 \end{pmatrix}\begin{pmatrix} \delta & 0 \\ 0 & 1 \end{pmatrix}\begin{pmatrix} 1 & 1 \\ 0 & 1 \end{pmatrix}^{-1} \in B$, whence $\begin{pmatrix} 1 & 1 \\ 0 & 1 \end{pmatrix}\begin{pmatrix} \delta & 0 \\ 0 & 1 \end{pmatrix}\begin{pmatrix} 1 & -1 \\ 0 & 1 \end{pmatrix} = \begin{pmatrix} \delta & 1-\delta \\ 0 & 1 \end{pmatrix} \in B$. This immediately implies that $\delta = 1$, i.e. $\mathrm{Core}(B)$ is trivial. □

3. Products of finite abelian groups

Throughout this section let $G = AB$ be a finite factorizable group by two abelian subgroups A and B.

To study the structure of such factorizable groups we need the following propositions and lemmas.

Proposition 3.1. $[G,G] = [A,B]$.

PROOF. The inclusion $[G,G] \supseteq [A,B]$ is evident. To show the inverse, it is sufficient to prove that $[A,B] \trianglelefteq G$. Let $[a,b] \in [A,B]$. Then $[a,b]^{b_1} = [a^{b_1},b] = b_1ab_1^{-1}bb_1a^{-1}b_1^{-1}b^{-1} = b_1ab_1^{-1}a^{-1} \cdot abb_1a^{-1}b_1^{-1}b^{-1} = [b_1,a][a,bb_1] = [a,b_1]^{-1} \cdot [a,bb_1] \in [A,B]$. The analogous calculations show $[a,b]^{a_1} \in [A,B]$. □

As a corollary we obtain

Proposition 3.2. $\langle A^g \rangle_{g\in G} = AG'$.

PROOF. The inclusions $A \subseteq AG' \trianglelefteq G$ imply $\langle A^g\rangle_{g\in G} \subseteq AG'$. Now take a commutator $[g,a]$, $g \in G$, $a \in A$. Clearly, $[g,a] \in \langle A^g\rangle_{g\in G}$. Therefore, $[G,A] \subseteq \langle A^g\rangle_{g\in G}$. Now the sequence of inclusions

$$[G,G] \supseteq [G,A] \supseteq [B,A] = [G,G]$$

gives us $G' = [G,G] = [G,A] \subseteq \langle A^g\rangle_{g\in G}$. Therefore, $AG' \subseteq \langle A^g\rangle_{g\in G}$. □

Proposition 3.3.

(i) $\mathrm{Core}(A) \subseteq Z(AG')$.

(ii) $A \cap G' \subseteq Z(AG')$.

PROOF. (i) Take any $a \in \mathrm{Core}(A)$. Then a belongs to all A^g, $g \in G$, and, therefore, a lies in the centre of $\langle A^g \rangle_{g \in G} = AG'$.

(ii) Since A and G' are abelian, this inclusion is evident. □

Since A and B appear symmetrically in the above propositions, these propositions remain true after the substitution of B instead of A.

Lemma 3.4. *Assume* $Z(G) = 1$. *Then*

(i) $A \cap B = 1$, $C_G(B) = B$, $C_G(A) = A$;

(ii) $\mathrm{Core}(A) = Z(AG') \supseteq A \cap G'$; $\mathrm{Core}(B) = Z(BG') \supseteq B \cap G'$.

PROOF. (i) The equality $A \cap B = 1$ is evident. Furthermore, $B \subseteq C_G(B) = B[C_G(B) \cap A]$. Hence $C_G(C_G(B) \cap A) \supseteq \langle A, B \rangle = G$ and $C_G(B) \cap A \subseteq Z(G) = 1$. The same arguments applied to A yield $C_G(A) = A$.

(ii) By the previous statement, $\mathrm{Core}(A) \subseteq Z(AG')$ and $A \cap G' \subseteq Z(AG')$. But $C_G(A) = A$, hence $Z(AG') \subseteq A$. The centre $Z(AG')$ is a characteristic subgroup of $AG' \trianglelefteq G$. Hence $Z(AG') \subseteq \mathrm{Core}(A)$ and this gives us $Z(AG') = \mathrm{Core}(A)$. □

If $Z(G)$ is trivial, then G is not nilpotent. On the other hand, G is metabelian, and therefore it contains a Carter subgroup, say C. By [S, Theorem VII 4a, p.227] G admits the following decomposition $G = CG^\omega$ where G^ω is the intersection of all elements of the lower central series of G. Moreover, by [S, Proposition VII 4b, p.229] $C \cap G^\omega \subseteq (G^\omega)' \subseteq G'' = 1$. Thus, we have $G = CG^\omega$, $C \cap G^\omega = 1$. The claim below gives the structure of the Carter subgroup in our particular case.

Lemma 3.5. *Let* $G = AB$ *where* A *and* B *are abelian. Assume that* $Z(G) = \mathrm{Core}(B) = 1$. *Then*

(i) B *is a Carter subgroup of* G;

(ii) $G = G'B$, $G' \cap B = 1$, $G' = G^\omega$, $|G'| = |A|$.

PROOF. The subgroup B is abelian. Therefore, it is a Carter subgroup iff it is self-normalized. By Lemma 3.4, part (ii), $1 = \mathrm{Core}(B) \supseteq B \cap G'$, whence $B \cap G' = 1$. This immediately implies that $N_G(B) = C_G(B)$. Indeed, if $g \in N_G(B)$, then $b^g b^{-1} \in B \cap G' = 1$ holds for all $b \in B$. Therefore, $g \in C_G(B)$. But part (i) of Lemma 3.4 yields $C_G(B) = B$. Thus we have shown that B is a Carter subgroup of G.

As mentioned above, the property of being a Carter subgroup implies the factorization $G = BG^\omega$, $B \cap G^\omega = 1$. Therefore,

$$|G^\omega| = \frac{|G|}{|B|} = |A|.$$

On the other hand, $|G^\omega| \leq |G'| \leq \frac{|G|}{|B|} = |A|$ (the latter inequality follows from the fact that $G' \cap B = 1$). Therefore, $G' = G^\omega$ and $|G'| = |G^\omega| = |A|$. □

Parts (ii) and (i) of Lemma 3.5 imply parts (i) and (iv) of Theorem B, respectively. To prove part (ii) of Theorem B we need the following lemma.

Lemma 3.6. ([AFG, Corollary 1.3.5, p.10]) *Let the finite group $G = AB = AK = BK$ be the product of three nilpotent subgroups A, B and K, where K is normal in G. Then G is nilpotent.*

Now we are able to prove parts (ii) and (iii) of Theorem B.

PROOF OF THEOREM B. (parts (ii) and (iii)) (ii) Consider the group AG'. Denote $B^* = AG' \cap B$. Then $AG' = AB^*$. Furthermore, $B^* \cap G' \subseteq B \cap G' = 1$ (see part (i) of the Theorem). By Lemma 3.5, $|G'| = |A|$, whence $AG' = B^*G'$. Thus, $AG' = AB^* = G'B^*$ where A, B^*, G' are abelian and, moreover, $G' \supseteq (AG')'$. This shows that the group AG' satisfies the conditions of Lemma 3.6, and, therefore, it is nilpotent.

(iii) By Lemma 3.4, one has

$$Z(G'A) = \text{Core}(A) \supseteq A \cap G'.$$

So it is necessary to show that $A \cap G' \supseteq Z(G'A)$. Let $B^* \subseteq B$ be as in part (ii), i.e.,

$$B^*A = B^*G' = AG', \quad B^* \cap A = B^* \cap G' = 1.$$

Take any $z \in Z(AG')$ and write it as a product of bg, $b \in B^*$, $g \in G'$. It is easy to check that $b \in Z(AG') = \text{Core}(A) \subseteq A$. Therefore, $b = 1$ and $z = g \in G'$. This shows that $Z(AG') \subseteq G'$. Combining this with $Z(AG') \subseteq A$ we get $Z(AG') \subseteq A \cap G'$. □

We conclude this section by summarizing the properties of the group G factorizing into the product of two abelian subgroups A and B.

Theorem 3.2. *Let $G = AB$ where A and B are abelian finite groups. Then at least one of the following holds:*

(i) Both $\text{Core}(A)$ *and* $\text{Core}(B)$ *are non-trivial;*

(ii) $\text{Core}(B) = 1$ *and* $Z(G) \neq 1$;

(iii) $\text{Core}(B) = Z(G) = 1$.

If (iii) *holds, then by Theorem B and Lemma 3.5 we have:*

(1) B *is a Carter subgroup of* G;
(2) $G'B = G,\ G' \cap B = 1,\ G' = G^{\omega}$;
(3) AG' *is a nilpotent group;*
(4) $Z(AG') = G' \cap A = \mathrm{Core}(A)$;
(5) $|A| = |G'|$.

4. The structure of metabelian finite groups

PROOF OF THEOREM C. By [S, Theorem VII 4a, p.227], $G = CG^{\omega}$ and $C \cap G^{\omega} \subseteq (G^{\omega})'$, where C is a Carter subgroup of G. Since $G^{\omega} \subseteq G'$ and G' is abelian it follows that $C \cap G^{\omega} = 1$. Clearly, $C_C(G^{\omega})$ is a normal subgroup of C.

Take $z \in C_C(G^{\omega}) \cap Z(C)$. Clearly, $z \in Z(G)$. Hence $C_G(G^{\omega}) \cap Z(C) = 1$. But C is nilpotent and $C_C(G^{\omega}) \triangleleft G$, hence $C_C(G^{\omega}) = 1$.

This immediately implies $C_G(G^{\omega}) = G^{\omega}$. But $G^{\omega} \subseteq G'$ and G' is abelian. Thus, $C_G(G^{\omega}) \supseteq G'$ implies $G^{\omega} \supseteq G'$ and $G^{\omega} = G'$.

Therefore, we have proved $G = CG'$, $C \cap G' = 1$. But $C' \subseteq G' \cap C = 1$ implies C is abelian, as desired. □

Using Theorem C we can strengthen part (ii) of Lemma 3.4.

Corollary 2. *Keep the assumptions of Lemma 3.4. Then* $Z(AG') = A \cap G'$; $Z(BG') = B \cap G'$.

PROOF. We prove the first equality only, because the second one has an analogous proof. Due to Lemma 3.4, the only inclusion we have to check is $Z(AG') \subseteq A \cap G'$. By Theorem C, $G = CG'$, and consequently, $C_G(G') = G'$ since $Z(G) = 1$. Therefore, $Z(AG') \subseteq C_G(G') \subseteq G'$. But, due to Lemma 3.4, $Z(AG') = \mathrm{Core}(A) \subseteq A$. Hence $Z(AG') \subseteq A \cap G'$, as desired. □

References

[AFG] Amberg, B., Franciosi, S., and Giovanni, F., *Products of Groups* (Oxford Mathematical Monographs, Oxford Science Publications, 1992).

[G] Gorenstein, D., *Finite Groups* (Harper and Row, New York, 1968).

[I] Itô, N., Über das Produkt von zwei abelschen Gruppen, *Math. Z.* **62**(1955), 400–401.

[S] Schenkman, E., *Group Theory* (D. Van Nostrand Princeton, N.J.-Toronto, Ontario-London, 1965).

TABLE ALGEBRAS OF EXTENDED GAGOLA TYPE AND APPLICATIONS TO FINITE GROUP THEORY

Z. ARAD*, E. FISMAN* and C.M. SCOPPOLA[†1]

*Department of Mathematics and Computer Science, Bar-Ilan University, Ramat-Gan, 52900 Israel
†Department of Mathematical Science, University of Rome, Rome, Italy

1. Introduction

In this paper every group G is a finite group.

S.M. Gagola, in [Ga], determined the structure of groups with an irreducible character which vanishes on all but two conjugacy classes. This paper is devoted to some generalisations of Gagola's theorem to semi-table algebras and table algebras (for the definitions see Section 2), which imply new results for finite groups. For brevity, we state in this introduction only some of our results. More details can be found in 2.11 and 3.3.

In this paper we introduce the concept of a semi-table algebra, a special case of a C-algebra. The concept of a C-algebra was first articulated by Kawada [Ka] following Hoheisel [Ho]. This concept explained and abstracted the duality between the character ring and the centre of a group algebra. It enables us to state the general Theorem 2.11 for semi-table algebras, which directly implies the following two corollaries, Theorem 1.1 and Theorem 1.3.

Theorem 1.1. *Let G be a group. Assume there exist $\chi \in Irr(G)$ and $g \in G$ such that $\xi(g) = 0$ for every irreducible character $\xi \notin \{1, \chi, \overline{\chi}\} \neq Irr(G)$. Let $N = \ker \chi$. Then the following hold:*

(i) G is a Frobenius group with Frobenius kernel N. In particular, N and G/N are of coprime order.

(ii) N is the unique maximal normal subgroup of G.

(iii) The nilpotent class of N is at most 2.

Moreover, $|G/N| = 3$ if $\chi \neq \overline{\chi}$ and $|G/N| = 2$ if χ is real. In the last case N is abelian.

In order to state our next result we need the following definition.

[1]This work was done during the third author's visit to Israel as an INDAM Sabbatical Fellow. He wishes to thank the Gelbart Research Institute for Mathematical Sciences at Bar-Ilan University for their hospitality and support.

Definition 1.2. (A. Mann [M]) Let G be a group and N a proper normal subgroup of G. Then (G, N) is a CF-pair if $xy \in cl_G(y)$ for every $x \in N$ and $y \in G - N$.

CF-pairs were first studied by Camina in [C], as generalisations of the Frobenius group (for more details see [CM], [CMS], [Ku], [Mc1] and [Mc2]).

Theorem 1.3. *Let G be a group. Assume there exist $\chi \in Irr(G)$, and $g \in G$, such that $\chi(y) = 0$ for every $y \in G - \{1_G, cl_G(g), cl_G(g^{-l})\} \neq 0$. Set $\mathcal{M} = Irr(G) - \{\chi, \overline{\chi}\}$ and $N = \bigcap_{\xi \in \mathcal{M}} \ker \xi$. Then the following properties hold:*

(i) *$N = \{1_G, cl_G(g), cl_G(g^{-l})\}$ is the unique minimal normal subgroup of G. In particular, N is an elementary abelian p-subgroup and p is odd if $cl_G(g) \neq cl_G(g^{-1})$.*

(ii) *(G, N) is a CF-pair.*

(iii) *$\chi \neq 1_{Irr(G)}, g \neq 1_G$ and $cl_G(g) = cl_G(g^{-1})$ if and only if $\chi = \overline{\chi}$.*

(iv) *Let $\epsilon = (1 + \delta_{\chi\overline{\chi}})/2$ and $|N| = p^n$, where $\delta_{\chi\overline{\chi}} = 1$ if $\chi = \overline{\chi}$ and $\delta_{\chi\overline{\chi}} = 0$ if $\chi \neq \overline{\chi}$. Then $|cl_G(g)| = (p^n - 1)\epsilon, \chi(1_G) = [G/N|(p^n - 1)\epsilon]^{1/2}$ and $\chi(g) = |G/N|^{1/2}[-1 \pm i(1 - \delta_{\chi\overline{\chi}})p^{n/2}]\epsilon$. In particular, $C_G(g)$ contains a Sylow p-subgroup of G and $g \in Z(G)$ if and only if either $|N| = 2$ and $\chi = \overline{\chi}$ or $|N| = 3$ and $\chi \neq \overline{\chi}$.*

As we mentioned before, the structure of G in the case that $\chi = \overline{\chi}$ was determined by Gagola. Theorem 1.4 contains further information for the case $\chi \neq \overline{\chi}$.

Theorem 1.4. *Let G, χ, g, N, p and n be as in 1.3, with $cl_G(g) \neq cl_G(g^{-1})$ and let P be a Sylow p-subgroup of G. Then G is a solvable group of odd order, $Z(P) \subseteq N, p \equiv -1(\bmod 4)$ and n is odd. Moreover, let H be a Hall p'-subgroup of G. Then the following properties hold:*

(i) *HN is a Frobenius group with Frobenius kernel N.*

(ii) *$|H| = (|N| - 1)/2 = |cl_G(g)|$.*

(iii) *$G = O_{p,p',p}(G), H \simeq O_{p,p'}(G)/O_p(G)$ is metacyclic and $O_p(G) = C_G(N)$.*

(iv) *If G is not Frobenius then $O_{p'}(G/N) = 1$ and $|P/N| = p^\alpha$ for some α even.*

(v) *One of the following holds:*

(a) *$|N| = 3$, $Z(G) = N$ and $G = P$.*

(b) *$G = O_p(G)H$.*

(c) *Let $\overline{G} = G/O_p(G), \overline{F} = F(\overline{G}), \overline{P} = P/O_p(G), \pi = \pi(\overline{G}/\overline{F})$ and L a Hall π-subgroup of G. Then L is cyclic, $L \cap \overline{F} \subseteq Z(\overline{G})$ and $\overline{P}$ acts fixed point- freely on $\overline{F}/Z(\overline{G})$. Moreover, $P^G = G$ if and only if $Z(\overline{G}) = 1$ if and only if $|\overline{P}| = 3$.*

Corollary 1.5. *Let χ be a nonreal faithful irreducible character of a group G. If $\chi(y) = 0$ for every $y \in G - \{1_G, cl_G(g), cl_G(g^{-1})\}$ for some $g \in G$ then χ^2 is a linear combination of χ and $\overline{\chi}$.*

Examples of groups of type (b) in Theorem 1.4 can be constructed using Gagola's method by choosing n odd, by letting p be a prime with $p \equiv -1 \pmod{4}$ and by requiring in [Ga, Theorem 6.3] that $G = O_p(G)H$ have an irreducible character which vanishes on all but 3 conjugacy classes $\{1_G, cl_G(g), cl_G(g^{-l})\}$, and that $|H| = (p^n - 1)/2$. MacDonald in [Mc2, Theorem 5.1] builds examples of type (a) in Theorem 1.4.

In the proof, results of [CM] and [Ku] are used.

Recently, T.G. Sorenson [S], a student of Gagola, investigated the structure of groups with an irreducible character which vanishes on all but three distinct and real conjugacy classes.

2. Table algebras and semi-table algebras

In this section every algebra A is a k-dimensional associative and commutative algebra over the complex field $\mathbf{C}$, with identity 1_A, for $k \in \mathbf{N}$, a natural number. Semi-table algebras and table algebras are defined as follows:

Definition 2.1. Let $\mathcal{B} = \{b_1 = 1_A, b_2, \ldots, b_k\}$ be a basis of the algebra A. Then $(A, \mathcal{B})$ is a *semi-table algebra* if the following hold:

(I) For all i, j, m, $b_i b_j = \sum_{m=1}^{k} b_{ijm} b_m$,with $b_{ijm} \in \mathbf{R}$, ($\mathbf{R}$ the real field).

(II) There is an algebra automorphism (denoted by $^-$) of A whose order divides 2, such that $b_i \in \mathcal{B}$ implies that $\overline{b_i} \in \mathcal{B}$ (then $\bar{i}$ is defined by $b_{\bar{i}} = \overline{b_i}$ and $b_i \in \mathbf{R}$ is real if $i = \bar{i}$).

(III) $b_{ij1} = \delta_{i\bar{j}} b_{i\bar{i}1}$ with $b_{i\bar{i}1} > 0$.

(IV) Let $e_1, \ldots, e_k$ be the set of primitive idempotents of A and let $b_j = \sum_{m=1}^{k} b_{jm} e_m$ for every $b_j \in \mathcal{B}$. There exists $i \in \{1, \ldots, k\}$ such that $b_{ji} - b_{j\bar{j}1}$ for every j. Without loss of generality we assume that $i = 1$.

(V) $b_{j\bar{j}1} > |b_{jm}|$ for every $m \in \{1, \ldots, k\}$.

If $\mathcal{B}$ satisfies (I) with b_{ijm} non negative, (II) and (III) then $(A, \mathcal{B})$ is a *table-algebra.* If $\mathcal{B}$ satisfies (I) - (IV) then $(A, \mathcal{B})$ is a *C-algebra.* G. Hoheisel proved, in [Ho], that if $(A, \mathcal{B})$ satisfies (I), (II) and (III) then A is semisimple.

The concepts C-algebra and table algebra were introduced by Kawada in [Ka] and by Z. Arad and H. Blau in [AB], respectively, as generalisations of the algebras $Z[\mathbf{C}(G)]$ and $Ch(G)$ (the algebra generated by the set of all the irreducible characters $\mathrm{Irr}(G)$). A related theory about commutative association schemes was developed in [BI].

We have proved (see 2.5) that if $(A, \mathcal{B})$ is a table algebra then there exists a set of real positive numbers $\{r_1, \ldots, r_k\}$ such that $(A, \mathcal{C})$ is a semi-table algebra for $\mathcal{C} = \{c_i = r_i b_i | b_i \in \mathcal{B}\}$ (called in [AB] a rescaling of $\mathcal{B}$). There are examples of semi-table algebras which are not table algebras (see [AF, p. 2984]) and C-algebras which are not semi-table algebras. For example: let $0 < r < \frac{1}{2}$ and $\mathcal{B} = \{b_1 = (1,1), b_2 = (r,-1)\} \subset \mathbf{C}^2$. It is easy to see that $(\mathbf{C}^2, \mathcal{B})$ is a C- algebra and not a semi-table algebra (note that $b_2^2 = rb_1 + (r-1)b_2$).

In this paper, we characterise semi-table algebras and table algebras which satisfy one of the following symmetric assumptions (see 2.10 and 2.11).

Hypothesis. $(\mathcal{A}, \mathcal{B})$ is a semi-table algebra and there exist $h, j \in \{1, \ldots, k\}$ such that $\mathcal{B} \neq \{b_1, b_j, \overline{b_j}\}$ and one of the following holds:

(A) $b_{h\ell} = 0$ for every $\ell \notin \{1, j, \overline{j}\}$.

(B) $b_{\ell j} = 0$ for every $\ell \notin \{1, h, \overline{h}\}$.

Clearly, in the case that $\mathcal{B} = \{\chi(1)\chi | \chi \in \mathrm{Irr}(G)\}$ and (A) holds with $j = \overline{j}$ we get the assumptions of Gagola's problem.

We first quote some related basic results.

Theorem 2.2. (Hoheisel [Ho]) *Let $(A, \mathcal{B})$ be a C-algebra. Set $|\mathcal{B}| = \sum_{\ell=1}^{k} b_{l1}, \beta_{i1} = |\mathcal{B}| / \left(\sum_{\ell=1}^{k} |b_{\ell i}|^2 / b_{\ell 1}\right), \beta_{ij} = \beta_{i1} b_{ji} / b_{j1} (b_{j1} = b_{\overline{j}\overline{1}}$ for every j as in 1.1.(iv)), $\beta_i = (\beta_{i1}, \ldots, \beta_{ik})$ and $\widehat{\mathcal{B}} = \{\beta_i \in \mathbf{C}^k | 1 \leq i \leq k\}$. Then the following properties hold for every $i, j, m \in \{1, \ldots, k\}$.*

(i) $b_{ijm} = b_{jim} = b_{\overline{jim}}$ and $b_{ijm} b_{m1} = b_{\overline{im}\overline{j}} b_{j1}$.

(ii) $\sum_{\ell=1}^{k} b_{ij\ell} b_{\ell tm} = \sum_{l=1}^{k} b_{it\ell} b_{\ell jm}$ for every t.

(iii) $b_{\widetilde{ij}} = \overline{b_{ij}}$ and the complex map $b_{ij} \to \overline{b_{ij}}$ defines a permutation $^\sim$ on the set $\{1, \ldots, k\}$ such that $b_{i\widetilde{j}} = \overline{b_{ij}}$ and $\beta_{\widetilde{j}1} = \beta_{j1}$. In particular, $\beta_{\widetilde{j}} \in \widehat{\mathcal{B}}$. (In general, we will use $^-$ instead of $^\sim$, unless otherwise specified).

(iv) Let $E = E(\mathcal{B})$, $M_i = M_i(\mathcal{B})$, $D_i = D_i(\mathcal{B})$, $F = F(\mathcal{B})$ and $\Phi = \Phi(\mathcal{B})$ be the $k \times k$ matrices defined by $(E)_{jm} = b_{jm}, (M_i)_{jm} = b_{ijm}, (D_i)_{jm} = \delta_{jm} b_{ij}, (F)_{jm} = \delta_{jm} b_{j1}$ and $(\Phi)_{jm} = \delta_{jm} \beta_{j1}$. Then (a)$M_i E = E D_i$, (b)$M_i M_j = M_j M_i$, (c)$F M_i^* = M_{\overline{i}} F$ and (d)$F^{-1} E \Phi E^* = |\mathcal{B}| I_k = \Phi E^* F^{-1} E$, where A^* is the transpose of the conjugate matrix of A, for $a \in M_k(\mathbf{C})$, and I_k is the identity matrix in $M_k(\mathbf{C})$.

(v) $\beta_i \beta_j = \sum_{\ell=1}^{k} \beta_{ij\ell} \beta_\ell$, where $|\mathcal{B}| \beta_{m1} \beta_{ijm} = \sum_{\ell=1}^{k} b_{\ell 1} \beta_{i\ell} \overline{\beta_{m\ell}} \in \mathbf{R}$.

Definition 2.3. ([ABE]) Let $(A, \mathcal{B})$ and $(A', \mathcal{B}')$ be C-algebras. A *C-algebra isomorphism* is an algebra isomorphism $\tau : A \to A'$ such that $\tau(\mathcal{B}) = B'$. When such τ exists $(A, \mathcal{B})$ and $(A', \mathcal{B}')$ are called *isomorphic*.

Theorem 2.4 (Kawada, [Ka]) *Let $(A, \mathcal{B})$ be a C-algebra. Set $\widehat{\mathcal{B}}$ as in 2.2. Then $(<\widehat{\mathcal{B}}>, \widehat{\mathcal{B}})$ is a C-algebra, called the* dual *C-algebra of $(A, \mathcal{B})$. Moreover, $(A, \mathcal{B})$ and $(<\hat{\hat{\mathcal{B}}}>, \hat{\hat{\mathcal{B}}})$ are isomorphic.*

Theorem 2.5. *Let A be an algebra and $\mathcal{B}$ a basis of A which satisfies 2.1.(I), (II) and (III). Then the following hold:*

(i) If $(A, \mathcal{B})$ is a C-algebra and $\widehat{\mathcal{B}}$ the dual basis of $\mathcal{B}$ then $\mathcal{B}$ and $\widehat{\mathcal{B}}$ have the same number of real elements and $|\mathcal{B}| = |\widehat{\mathcal{B}}|$, for $|\widehat{\mathcal{B}}|$ as defined in 2.2.

(ii) $(A, \mathcal{B})$ is a semi-table algebra if and only if the dual $(<\widehat{\mathcal{B}}>, \mathcal{B})$ is a semi-table algebra.

(iii) If $(A, \mathcal{B})$ is a table algebra then there exists a C-algebra homomorphism f of A to $\mathbf{C}$ such that $f(b_i) \in \mathbf{R}^+$ for every $1 \leq i \leq k$. Moreover, there exists a unique $j \in \{1, \ldots, k\}$ such that $b_{ij} = f(b_i)$ for every i and $b_{ij} \geq |b_{i\ell}|$ for every ℓ. In particular, $(A, \mathcal{C})$ is a semi- table algebra for $\mathcal{C} = \{c_i = f(b_i) b_i / b_{i\bar{i}1} | b_i \in \mathcal{B}\}$.

PROOF. (i) follows by [AF,2.6].

(ii) By 2.4, $(<\widehat{\mathcal{B}}>, \widehat{\mathcal{B}})$ is a C-algebra and by 2.2, $|\beta_{ij}| = \beta_{i1} |b_{ji}| / b_{j1} > \beta_{i1}$ if and only if $|b_{ji}| > b_{j1}$.

(iii) The first assertions follow by [AF,2.6]. Let $b_{ij} = f(b_i)$ and $c_i c_m = \sum_{\ell=1}^{k} c_{im\ell} c_\ell$ for every i, m, ℓ. It is easy to see that

$$c_{im\ell} = b_{ij} b_{mj} b_{\ell\bar{\ell}1} b_{im\ell} / b_{i\bar{i}1} b_{m\bar{m}1} b_{\ell j}$$

for every i, m, ℓ. In particular, $c_{i\bar{i}1} = b_{ij}^2 / b_{i\bar{i}1} = c_{ij} \geq |c_{i\ell}|$ and (iii) holds. □

Definition 2.6. Let $(A, \mathcal{B})$ be a C-algebra, $\mathcal{N} = \{b_i \in \mathcal{B} | i \in I\}$ for some $\phi \neq I \subset \{1, \ldots, k\}$ with $\bar{i} \in I$ if $i \in I$. Then

(i) $(\mathcal{B}, \mathcal{N})$ is a *generalised CF-pair* if $b_i b_j = b_{ijj} b_j$ for every $j \notin I$ and $i \in I$.

(ii) $\mathcal{N}$ will be called a *table C-subset* of $\mathcal{B}$ if the multiplication co-efficients $b_{ij\ell}$ are zero for every $i, j \in I$ and $\ell \notin I$.

(iii) Let ψ be an algebra homomorphism of A. If $(\psi(A), \mathcal{C})$ is a C-algebra, for $\mathcal{C} \cap \psi(b) \neq \emptyset$ for every $b \in \mathcal{B}$, and $\mathcal{M} = \mathcal{B} \cap \psi^{-1} < 1_{\psi(A)} >$ is a table $\mathcal{C}$-subset of $\mathcal{B}$, then we say that the *quotient C-algebra of $\mathcal{B}$ by $\mathcal{M}$* is defined by $\mathcal{C}$.

(iv) For $b_i \in \mathcal{B}$, we denote by $\ker b_i = \{\beta_j \in \widehat{\mathcal{B}} | b_{ij} = b_{i1}\}$ and by $\ker \mathcal{N} = \bigcap_{i \in I} \ker b_i$.

Note. The definition of a generalised CF-pair is a generalisation to C-algebras of 1.2, first introduced by two of the authors in [AF, 2.10].

Theorem 2.7 *Let $(A, \mathcal{B})$ be a C-algebra, $\mathcal{N} = \{b_i \in \mathcal{B} | i \in I\}$ and $\ker \mathcal{N} = \{\beta_j \in \widehat{\mathcal{B}} | j \in J\}$. Then the following hold:*

(i) *If $(\mathcal{B}, \mathcal{N})$ is a generalised CF-pair then $\mathcal{N}$ is a table C-subset of $\mathcal{B}$ and $(\widehat{\mathcal{B}}, \ker \mathcal{N})$ is a generalised CF-pair. In particular, $|\mathcal{B}| = |\mathcal{N}||\ker \mathcal{N}|$ and $\ker \mathcal{N}$ is a table C-subset of $\widehat{\mathcal{B}}$, i.e. the quotient C-algebra of $\mathcal{B}$ by $\mathcal{N}$ is defined by $\widehat{\ker \mathcal{N}}$.*

(ii) *$(\mathcal{B}, \mathcal{N})$ is a generalised CF-pair if and only if $b_{hm} = 0$ for every $h \notin I$ and $m \notin J$.*

PROOF. By [AF, 2.9 and 2.11] (i) holds and if $(\mathcal{B}, \mathcal{N})$ is a generalised CF-pair then $b_{hm} = 0$ for every $h \notin I$ and $m \notin J$.

Now assume that $b_{hm} = 0$ for every $h \notin I$ and $m \notin J$. Let $i \in I$ and $h \notin I$. Then $|\mathcal{B}| b_{ih\ell} b_{l1} \sum_{t=1}^{k} \beta_{t1} b_{it} b_{ht} \overline{b_{\ell t}} = b_{i1} \sum_{t \in J} \beta_{t1} b_{ht} \overline{b_{\ell t}} = b_{i1} \sum_{t=1}^{k} \beta_{t1} b_{ht} \overline{b_{\ell t}} = b_{i1} |\mathcal{B}| \delta_{h\ell} b_{\ell 1}$. Thus (ii) holds. □

Theorem 2.8. ([ABE, 4.1 and 4.2]) *Let $(A, \mathcal{B})$ be a C- algebra. Assume that $\mathcal{B}$ has r real elements. Set $b_i^0 = b_i$ if $b_i = \overline{b_i}, b_i^0 = b_i + b_{\bar{i}}$, if $b_i \neq b_{\bar{i}}$, $\mathcal{B}^0 = \{b_i^0 | b_i \in \mathcal{B}\}$ and $A^- =< \mathcal{B}^0 >$. Then the following properties hold:*

(i) *$(A^-, \mathcal{B}^0)$ is a C-algebra of dimension $r + (k - r)/2$, called the* real C-algebra induced by $(A, \mathcal{B})$. *Moreover, if $(A, \mathcal{B})$ is a table algebra so is $(A^-, \mathcal{B}^0)$.*

(ii) *$\widehat{\mathcal{B}}^0 = \widehat{\mathcal{B}^0}$ i.e. the dual of $(A^-, \mathcal{B}^0)$ equals the real C-algebra induced by $(< \widehat{\mathcal{B}} >, \widehat{\mathcal{B}})$.*

Corollary 2.9. *Let $(A, \mathcal{B})$ be a C-algebra. Then*

(i) *$|\mathcal{B}| = |\mathcal{B}^0|$;*

(ii) *If $(A, \mathcal{B})$ is a semi-table-algebra so is $(A^-, \mathcal{B}^0)$ and $\beta_\ell \in \ker b_j$ if and only if $\beta_\ell^0 \in \ker b_j^0$.*

PROOF. (i) Let $\{1, \ldots, k\} = I \cup J \cup \overline{J}$ where $I = \{i | i = \bar{i}\}$, $J \cap I = \emptyset = J \cap \overline{J}, \overline{J} = \{j | \bar{j} \in J\}$. i.e. if $j \neq \bar{j}$ then either $j \in J$ or $\bar{j} \in J$. Then $|\mathcal{B}^0| = \sum_{i \in I \cup J} b_{i1}^0 = \sum_{i \in I} b_{i1} + \sum_{i \in J} b_{i1} + b_{\bar{i}1} = |\mathcal{B}|$.

(ii) Clearly if $j = \bar{j}$ then $b_{j1}^0 = b_{j1} \geq |b_{j\ell}| = |b_{j\ell}^0|$ for every ℓ. If $j \neq \bar{j}$ then $b_{j1}^0 = b_{j1} + b_{\bar{j}1} \geq |b_{j\ell}| + |b_{\bar{j}\ell}| \geq |b_{j\ell} + b_{\bar{j}\ell}| = |b_{j\ell}^0|$. In particular, $b_{j1}^0 = b_{j\ell}^0$ if and only if $b_{j1} = b_{j\ell}$. □

From now on $(A, \mathcal{B})$ is a *semi-table* algebra.

Proposition 2.10. *Assume that $(A, \mathcal{B})$ satisfies (A) with $j = \bar{j}$. Then the following properties hold:*

(i) *$b_h = \overline{b_h}$.*

(ii) *$b_{hj} = -b_{h1}/\beta_{j1}$.*

(iii) $\beta_{j1} = b_{h1}/(|\mathcal{B}| - b_{h1})$. *In particular,* $b_{hj} = b_{h1} - |\mathcal{B}|$.

(iv) $b_{\ell j} = b_{\ell 1}$ *for every* $\ell \neq h$ *i.e.* $\beta_j \in \bigcap_{\ell \neq h} \ker b_\ell$.

PROOF. (i) By our assumptions and 2.2. (iii), $b_{\overline{h}\ell} = b_{h\overline{\ell}} = 0 = b_{h\ell} = b_{\overline{h}\overline{\ell}}$ for every $\ell \notin \{1, j\}$ and $b_{\overline{h}j} = b_{\overline{h}\overline{j}} = b_{hj}$. Hence $b_h = b_{\overline{h}}$.

(ii) Clearly $b_h \neq b_1$. So, by 2.2.(iv) $0 = \delta_{1h} = \sum_\ell \beta_{\ell 1} b_{h\ell} = \beta_{11} b_{h1} + \beta_{j1} b_{hj} = b_{h1} + \beta_{j1} b_{hj}$ as desired.

(iii) Again, by 2.2.(iv) and (ii), $|\mathcal{B}| b_{h1} = \sum_\ell \beta_{\ell 1} |b_{h\ell}|^2 = b_{h1}^2 + \beta_{j1} |b_{hj}|^2 = b_{h1}^2 + b_{h1}^2/\beta_{j1}$. So (iii) holds.

(iv) By 2.2.(iv), $0 = \sum_{\ell=1}^k b_{\ell j} = b_{hj} + \sum_{\ell \neq h} b_{\ell j}$. Therefore, by (iii), $\sum_{\ell \neq h} b_{\ell 1} = |\mathcal{B}| - b_{h1} = -b_{hj} = \sum_{\ell \neq h} b_{\ell j} \leq \sum_{\ell \neq h} |b_{\ell j}| \leq \sum_{\ell \neq h} b_{\ell 1}$. So $b_{\ell j} = b_{\ell 1}$ for every $\ell \neq h$, as desired. □

Theorem 2.11. *(i)* $(A, \mathcal{B})$ *satisfies (A) if and only if the dual* $(< \widehat{\mathcal{B}} >, \widehat{\mathcal{B}})$ *satisfies* $(\mathcal{B})$.

(ii) Assume that $(A, \mathcal{B})$ *satisfies the hypothesis (A). Set* $\mathcal{C} = \mathcal{B} - \{b_h, b_{\overline{h}}\}$ *and* $D = \{\beta_1, \beta_j, \beta_{\overline{j}}\}$. *Then* $(\mathcal{B}, \mathcal{C})$ *and* $(\widehat{\mathcal{B}}, \mathcal{D})$ *are generalized CF-pairs,* $\mathcal{D}$ *is the unique minimal table C-subset of* $\widehat{\mathcal{B}}$ *and* $\mathcal{C}$ *is the unique maximal table C-subset of* $\mathcal{B}$. *Moreover, the following hold:*

(a) $h \neq 1 \neq j$.

(b) $b_h = b_{\overline{h}}$ *if and only if* $\beta_j = \beta_{\overline{j}}$.

(c) $|\mathcal{C}| = |\mathcal{B}|/|\mathcal{D}|$

(d) Let $\epsilon = (1 + \delta_{j\overline{j}})/2$, *then:*

$$\beta_{j1} = (|\mathcal{D}| - 1)\epsilon, \quad b_{h1} = |\mathcal{C}|\beta_{j1} \ \textit{and}$$
$$\beta_{jh} = (-1 \pm i(1 - \delta_{j\overline{j}})|\mathcal{D}|^{1/2})\epsilon, \quad \textit{i.e. } b_{hj} = |\mathcal{C}|\beta_{jh}.$$

PROOF. Clearly (i) and (a) hold.

If $j = \overline{j}$ then 2.11 follows by 2.10 and 2.7. So we assume that $j \neq \overline{j}$. Let $(A^-, \mathcal{B}^0)$ be the real C-algebra induced by $(A, \mathcal{B})$. Then by 2.9, $(A^-, \mathcal{B}^0)$ satisfies the hypothesis (A) with $j = \overline{j}$. So by the above $(\mathcal{B}^0, \mathcal{C}^0)$ is a generalized CF-pair with $\ker \mathcal{C}^0 = \mathcal{D}^0$, In particular, $b_{\ell j}^0 = b_{\ell 1}^0$ for every $\ell \not\in \{1, h, \overline{h}\}$. So, by 2.9, $b_{\ell j} = b_{\ell 1}$ for every $\ell \notin \{1, h, \overline{h}\}$, in particular, $b_h \neq b_{\overline{h}}$, and by 2.7, $(\mathcal{B}, \mathcal{N})$ is a generalized CF-pair. By 2.10.(iii), $2b_{h1} = 2|\mathcal{C}|\beta_{j1}, (b_{hj} + b_{h\overline{j}}) = |\mathcal{C}|(\beta_{jh} + \beta_{\overline{j}h})$, $2\beta_{j1} = |\mathcal{D}| - 1$ and $\beta_{jh} + \beta_{\overline{j}h} = -1 = \beta_{j1}(b_{hj} + b_{h\overline{j}})/b_{h1}$ then $0 = \sum_\ell \beta_{\ell h}^2/\beta_{\ell 1} = 1 + 2[1 + (\beta_{jh} + \beta_{\overline{j}h})^2]/4\beta_{j1}$. So $-(\beta_{jh} - \beta_{\overline{j}h})^2 = 2\beta_{j1} + 1 = |\mathcal{D}|$ and 2.11 holds. □

By 2.11, 2.6.(i) and 2.2.(i) we get

Corollary 2.12. (A, B) *satisfies the hypothesis (A) with* $h \neq \overline{h}$, *then* b_h *is not contained in any proper table C-subset of* $\mathcal{B}$ *and* b_h^2 *is a linear combination of* b_h *and* $b_{\overline{h}}$.

3. Application to finite groups

In this section, we denote by G a finite group and by $\mathcal{B}(G) = \{\xi(1_G)\xi | \xi \in \text{Irr}(G)\}$. So $(\text{Ch}(G), \mathcal{B}(G))$ is a table algebra which is also a semi-table algebra and the dual semi-table algebra is $Z[(\mathbf{C}(G)]$ with the usual bases $Cla(G)$, obtained from the conjugacy classes of G.

3.1. PROOF OF THEOREM 1.1. Since $(\text{Ch}(G), \mathcal{B}(G))$ satisfies the hypothesis (B), then, by 2.11, $(Z[\mathbf{C}(G)], Cla(G))$ satisfies the hypothesis (A), (G, N) is a CF-pair and $\{1_{\text{Irr}(G)}, \chi(1_G)\chi, \chi(1_G)\overline{\chi}\}$ (which is isomorphic to $\mathcal{B}(G/N)$) is the unique minimal table subset of $\mathcal{B}(G)$ i.e. (ii) holds. Clearly, $|G/N| = 2$ if $\chi = \overline{\chi}$ and $|G/N| = 3$ if $\chi \neq \overline{\chi}$. So, by [ABE, Theorem B], (i) holds and (iii) follows by [Hu, V.8.8]. □

3.2. PROOF OF THEOREM 1.3. Since $(\text{Ch}(G), \mathcal{B}(G))$ satisfies the hypothesis (A), then 1.3 follows by 2.11, as in 3.1. □

Theorem 3.3. *Assume that $(Ch(G), \mathcal{B}(G))$ satisfies the hypothesis (A) with $\chi \neq \overline{\chi}$. Let $|N| = p^n$ and let P be a Sylow p-subgroup of G. Then the following hold:*

(i) Let T be any p'-subgroup of G. Then TN is a Frobenius group with Frobenius kernel N and complement T.

(ii) $|G|$ is odd. In particular, G is solvable and the p-length of G is at most 2.

(iii) Let $\overline{G} = G/O_p(G)$ and $\overline{F} = F(\overline{G})$. Then $\overline{F} = C_{\overline{G}}(\overline{F})$ is cyclic and $\overline{G}/\overline{F} \subseteq Aut(\overline{F})$ is abelian. So $G = O_{p,p'p}(G) = PH$ with H a Hall p'-subgroup of G which is metacyclic of order $|cl_G(g)|$ and $H \simeq O_{p,p'}(G)/O_p(G)$. In particular, $G = P$ if and only if $|N| = 3$.

(iv) $C_G(g) = P^y$ for some $y \in G$ and $C_G(N) = O_p(G)$.

(v) $p \equiv 1 (\bmod 4)$ and n is odd.

PROOF. (i) Follows by [CM, Lemma 4.3] and 1.3.

(ii) Assume that G is of even order and let σ be an involution of G. By (i), σ acts fixed point-freely on N. Hence by [Go, Theorem 10.1.4] $g^{-1} \in cl_G(g)$, a contradiction. So $|G|$ is odd and then by Feit-Thompson's well-known result, G is solvable. So (ii) follows by [Ku, Theorem 6.1.A].

(iii) By (i) $\overline{F}$ is cyclic. Then $\overline{F} \subseteq C_{\overline{G}}(\overline{F}) \subseteq \overline{F}$. So $\overline{G}/\overline{F} \subseteq \text{Aut}(\overline{F})$ is abelian. Thus G has a normal p-complement. By (i) and (ii) G has a Hall p'-subgroup H such that HN is Frobenius and (iii) follows by 1.3.

(iv) By 1.3 and (iii) $C_G(g) = P^y$ for some $y \in G$. So $O_p(G) = \bigcap_{y \in G} P^y = C_G(N)$.

(v) By (ii) and (iv) $|Cl_G(g)| = (p^n - 1)/2$ is odd. □

PROOF OF THEOREM 1.4. By [CM, Lemma 4.4] if G is not Frobenius then $O_{p'}(G/N) = 1$. So by 1.3 and 3.3 we need only to prove that (v)(c) holds if $G \supset O_p(G)H$. So assume that $G \supset O_p(G)H$ and let $\overline{S} = SO_p(G)/O_p(G)$ for every $S \subseteq G$. By 3.3, $\overline{G} \subseteq \mathrm{Aut}(N) \simeq GL(n,p)$, every Sylow p-subgroup of $\overline{H}$ is cyclic and $\overline{F}$ is cyclic. Let r_t be a prime divisor of $|H|$ such that $r_t|(p^t-1)$ and $r_t \nmid (p^i - 1)$ for $i < t$. (x can be embedded in a field of order p^t for any $x \in \overline{H}$ of order r_t). Since n and p are odd, then r_t exists for every $t|n$ except $p = 3$ and $t = 1$. Since $n > 1$, then, by [Hu,II.7.3] $x \in \overline{F}$ for every $x \in H$ of order r_n; so also $\overline{G}/\overline{F}$ is cyclic with $|\overline{G}/\overline{F}||n$. Let $r_t||\overline{H}/\overline{F}|, |\overline{H}/\overline{F}| = r_t^{\alpha_t}\beta_t$ and $p^t - 1 = r_t^{a_t}b_t$ with $(\beta_t, r_t) = 1 = (b_t, r_t)$. Thus $t|(n, r_t - 1)$ and so $r_t^{a_t+\alpha_t}|(p^n-1)$. Hence $r_t||\overline{F}|$ and so $\pi(\overline{F}) = \pi(H)$, and since $\overline{G}/\overline{F}$ and $\overline{F}$ are cyclic then L is cyclic. By [Ku, 6.1.13] $\overline{P}$ acts fixed point freely on $\overline{F}/Z(\overline{G})$. Finally, $\overline{P} \subseteq C(x)$ for any $x \in \overline{H}$ of order r_1. So $P^G = G$ if and only if $Z(\overline{G}) = 1$ if and only if $(|\overline{P}| - 1)/2 = 1$ if and only if $|\overline{P}| = 3$. □

PROOF OF COROLLARY 1.5. The corollary follows by Theorem 1.3, 2.6.(i) and 2.2.(i) or directly by 2.12. □

Acknowledgment. The authors would like to thank Harvey Blau for his helpful comments and suggestions.

References

[AB] Z. Arad and H. Blau, On table algebras and applications to finite group theory, *J. Algebra* **138**(1991), 137–185.

[ABE] Z. Arad, H. Blau, Y. Erez and E. Fisman, About real table algebras and applications to finite groups of extended Camin-Frobenius type, submitted.

[AF] Z. Arad and E. Fisman, On table algebra, C-algebras and applications to finite group theory, *Comm. Algebra* **19**(1991), 2955–3009.

[BI] E. Bannai and T. Ito, *Algebraic Combinatorics I: Association Schemes* (The Benjamin/Cummings Pub. Co., Inc., Menic Park, 1984).

[C] A.R. Camina, Some conditions which almost characterize Frobenius groups, *Israel J. Math.* **31**(1978), 153–160.

[CM] D. Chillag and I.D. MacDonald, Generalized Frobenius groups, *Israel J. Math.* **47**(1984), 111–122.

[CMS] D. Chillag, A Mann, C.M. Scoppola, Generalized Frobenius groups II, *Israel J. Math.* **62**(1988), 269–282.

[Ga] S.M. Gagola, Character vanishing on all but two conjugacy classes, *Pacific J. Math.* **109**(1983), 363–385.

[Go] D. Gorenstein, *Finite Groups (second edition)* (Chelsea Publ. Co., New York, 1980).

[Ho] G. Hoheisel, Uber Charaktere, *Monatsch. F. Math. Phys.* **48**(1939), 448–456.

[Hu] G. Huppert, *Endliche Gruppen I* (Springer-Verlag, Berlin, 1967).

[Ka] Y. Kawada, Uber den Dualitatssatz der Charaktere nicht commutativer Gruppen, *Proc. Phys. Math. Soc. Japan* **4**(1942), 97–109.

[Ku] E.B. Kuisch, *Extension and Induction of Group Representations* (Cordon Art, Baarn, 1991).

[M] A. Mann, Products of classes and characters in finite groups, *J.Algebra*, to appear.

[Mc1] I.D. MacDonald, Some *p*-groups of Frobenius and extra-special type, *Israel J. Math.* **40**(1981), 350–364.

[Mc2] I.D. MacDonald, More on *p*-groups of Frobenius type, *Israel J. Math.* **56**(1986), 335–394.

[S] T.G. Sorenson, Characters which vanish on all but three conjugacy classess, submitted.

ON THE SATURATION OF FORMATIONS OF FINITE GROUPS

A. BALLESTER–BOLINCHES* and L.M. EZQUERRO†

*Departament d'Àlgebra, Universitat de València, C/ Dr. Moliner 50, 46100 Burjassot (València), Spain
†Departamento de Matemática e Informática, Universidad Pública de Navarra, Campus de Arrosadía, 31006 Pamplona, Spain

1. Introduction and preliminaries

All groups considered in this paper are finite.

The reader is assumed to be familiar with the theory of saturated formations of finite groups. We shall adhere to the notation used in [4]; this book is the main reference for the basic notation, terminology and results.

First let us introduce the question we aim to analyze in this paper together with the specific notation which is used throughout. Given a formation $\mathfrak{X}$, its saturation, $\overline{\mathfrak{X}} = < \mathrm{Q}, \mathrm{R}_0, \mathrm{E}_\Phi > \mathfrak{X}$, is the smallest saturated formation containing $\mathfrak{X}$. The canonical local definition of $\overline{\mathfrak{X}}$ is denoted by $\overline{X}$: $\overline{\mathfrak{X}} = LF(\overline{X})$. Our purpose is to study the saturation of two different formations.

In the first paragraph we deal with the following situation. Let $\mathfrak{Y}$ be a class of simple groups containing the class $\mathfrak{A}$ of the abelian simple groups. Let $\mathfrak{F}$ be a saturated formation canonically defined by $\mathfrak{F} = LF(F)$. Let us denote by $\mathfrak{X}$ the class $\mathfrak{X} = \mathcal{Z}_{\mathfrak{F}}^{\mathfrak{Y}} = (G$: If F is a $\mathfrak{Y}$-chief factor of G covered by an $\mathfrak{F}$-projector of G, then F is an $\mathfrak{F}$-central chief factor of $G)$. The class $\mathcal{Z}_{\mathfrak{F}}^{\mathfrak{Y}}$ is a formation but in general not saturated.

It is clear that if H/K is an $\mathfrak{F}$-central chief factor of a group G then H/K is covered by the $\mathfrak{F}$-projectors. In general, $\mathfrak{F}$-projectors do not have the cover and avoidance property. Doerk [2], working within the soluble universe, studied the formation $\mathfrak{X} = \mathcal{Z}_{\mathfrak{F}}$ of all soluble groups G such that if F is a chief factor of G covered by an $\mathfrak{F}$-projector of G, then F is $\mathfrak{F}$-central in G. He proved that the formation $\mathfrak{X}$ is saturated if and only if there exists a set of primes π such that either $\mathfrak{F} = \mathfrak{S}_{\pi'}$ or $\mathfrak{F} = \mathfrak{S}_{\pi'}\mathfrak{H}$ for some non-empty formation $\mathfrak{H}$ such that $\mathfrak{H} = \mathfrak{S}_\pi \mathfrak{H}$ (cf.[2] Satz 4.9).

In the general non-soluble universe, Doerk's methods cannot be extended. We elaborate a different way to solve the question. We prove that $\mathcal{Z}_{\mathfrak{F}}^{\mathfrak{Y}}$ is a saturated formation if and only if $\mathfrak{F} = (1)$ or $\mathfrak{F} = \mathfrak{E}$, the class of all finite groups. In particular, (1) and $\mathfrak{E}$ are the only saturated formations $\mathfrak{F}$ such that in every group G, F is a $\mathfrak{F}$-central chief factor of G if and only if F is covered by an $\mathfrak{F}$-projector of G, i.e. $\mathcal{Z}_{\mathfrak{F}} = \mathfrak{E}$.

It is worth remarking that using our arguments, Doerk's theorem can be proved with a remarkably shorter proof.

The second problem is the following. Let $\mathfrak{H}$ be a saturated formation, $\mathfrak{H} = LF(H)$, where H is the canonical local definition of $\mathfrak{H}$, and $\mathfrak{F}$ a formation. Let us denote by $\mathfrak{H} \downarrow \mathfrak{F}$ the class of all groups whose $\mathfrak{H}$-projectors belong to $\mathfrak{F}$. The class $\mathfrak{X} = \mathfrak{H} \downarrow \mathfrak{F}$ is a formation but in general $\mathfrak{X}$ is not saturated. Moreover $\mathfrak{H} \downarrow \mathfrak{F} = \mathfrak{H} \downarrow (\mathfrak{H} \cap \mathfrak{F})$. So it can be supposed in the sequel that $\mathfrak{F} \subseteq \mathfrak{H}$. If $\mathfrak{F}$ is also saturated, we will denote $\mathfrak{F} = LF(F)$, where F is the canonical local definition of $\mathfrak{F}$. This means that $F(p) \subseteq H(p)$ for all $p \in \mathrm{char}\mathfrak{F}$ (see [4], Prop. IV.3.11.(c)). We will study the saturation $\overline{\mathfrak{X}}$ of $\mathfrak{X}$.

The same question in the soluble universe, namely the saturation of the formation $\mathfrak{H} \downarrow \mathfrak{F} = (G \in \mathfrak{S}\colon \mathrm{Proj}_{\mathfrak{H}}(G) \subseteq \mathfrak{F})$, was fully analyzed by Doerk in [3], (see [4] pp. 503-508). There, he obtains the canonical local definition $\overline{X}$ of $\overline{\mathfrak{X}}$, gives necessary and sufficient conditions for the saturation of $\mathfrak{H} \downarrow \mathfrak{F}$ (see [4] Th. VII 4.8), and proves the following.

Theorem. ([4] Th. VII 4.10) *Let $\mathfrak{H}$ be a saturated formation of characteristic π. Then the following statements are equivalent:*

(i) $\mathfrak{H} \downarrow \mathfrak{F}$ *is a saturated formation for all saturated formations (of soluble groups)* $\mathfrak{F}$.

(ii) $\mathfrak{H} = \mathfrak{S}_\pi$.

A closely related question, also in the soluble universe, is studied in [1]. In this case the saturation of the formation $\mathfrak{H}_{\mathfrak{F}}$, composed of all (soluble) groups whose $\mathfrak{H}$-normalizers are in $\mathfrak{F}$, is analyzed. There it is proved that if $\mathfrak{H}$ is a saturated formation and $\mathfrak{F}$ is a proper saturated subformation of $\mathfrak{H}$, the formation $\mathfrak{H}_{\mathfrak{F}}$ is saturated if and only if $\mathfrak{N}\mathfrak{F} \subseteq \mathfrak{H}$. With this it is clear that if $\mathfrak{H}$ is a saturated formation, the formation $\mathfrak{H}_{\mathfrak{F}}$ is saturated for every saturated formation $\mathfrak{F}$ if and only if $\mathfrak{H}$ is either (1) or $\mathfrak{S}$.

Again it can be said that the arguments of the soluble case fail in the general case. The following example illustrates that the obvious generalization of Doerk's theoerem does not hold.

Example. Let $\pi = \{3, 7\}$ and consider the saturated formations $\mathfrak{H} = \mathfrak{E}_\pi$ of all the π-groups and $\mathfrak{F} = \mathfrak{U}_\pi$ of all the supersoluble π-groups. Denote by G the group $PSL(2,7)$; G is a simple group of order $2^3.3.7$ and $G \in \mathfrak{H} \downarrow \mathfrak{F}$: the $\mathfrak{E}_\pi$-projectors of G are isomorphic to the Frobenius group of order 21, and consequently they are supersoluble. Let A_3 be the Frattini module of G with respect to GF(3). Since G is not 3-supersoluble we know that $\dim A_3 > 1$ (see [6] Theorem 3) and A is faithful: $C_G(A) = 1$. Consider the maximal Frattini extension E; the group E possesses a normal subgroup A such that $A \cong A_3$, $A \leq \Phi(E)$ and $E/A \cong G$.

Denote by S an $\mathfrak{E}_\pi$-projector of G. Consider $T \in \mathrm{Proj}_{\mathfrak{E}_\pi}(E)$ such that $A \leq T$ and $T/A \cong S$. Let $Z \in \mathrm{Syl}_7(T)$. Then $Z \cong Z_7$. We claim that $T \notin \mathfrak{F}$. This will imply that $E \notin \mathfrak{H} \downarrow \mathfrak{F}$ and therefore that $\mathfrak{H} \downarrow \mathfrak{F}$ is not saturated.

First recall that there exist exactly two non-isomorphic irreducible Z-modules over GF(3), namely the trivial module, denoted by K ($\dim K = 1$) and a faithful module, V say, such that $\dim V = 6$. The restricted module A_Z is completely reducible. If A_Z is a trivial homogeneous module, then $1 = C_Z(A_Z) = Z$, a contradiction. Therefore there exists a component of A_Z isomorphic to V.

If T were supersoluble then all chief factors of T below A would be cyclic. But this cannot happen by the above argument. So, T is not supersoluble and then $E \notin \mathfrak{H} \downarrow \mathfrak{F}$.

2. The saturation of $\mathcal{Z}_{\mathfrak{F}}^{\mathfrak{Y}}$

With the notation of the preliminaries, denote by $\mathfrak{X}$ the formation $\mathcal{Z}_{\mathfrak{F}}^{\mathfrak{Y}}$.

Lemma 1. *The saturated formation $\overline{\mathfrak{X}}$ is of full characteristic and $\overline{X}(p) = \mathfrak{S}_p\mathfrak{X}$ for all primes p such that $F(p)$ is properly contained in $\mathfrak{F}$.*

PROOF. First notice that every simple group is in $\mathfrak{X}$ and then in $\overline{\mathfrak{X}}$; therefore $\overline{\mathfrak{X}}$ is of full characteristic.

Let p be a prime such that $F(p) \neq \mathfrak{F}$ and let H be a group of minimal order in $\mathfrak{F} \setminus F(p)$; then H is a monolithic group and $Soc(H)$ is not a p-group. Since $\overline{\mathfrak{X}} \subseteq \mathfrak{N}\mathfrak{X}$ it follows that for every prime number q we have $\overline{X}(q) \subseteq \mathfrak{S}_q\mathfrak{X}$. Suppose that $\overline{X}(p) \neq \mathfrak{S}_p\mathfrak{X}$ and let G be a group of minimal order in $\mathfrak{S}_p\mathfrak{X} \setminus \overline{X}(p)$. Then G is monolithic and $Soc(G)$ is not a p-group. Let $D = G \times H$. It is known that there exists an irreducible and faithful D-module V over GF(p). Construct the semidirect product $Y = [V]D$. Notice that $D \in \mathfrak{X}$ and V is an $\mathfrak{F}$-eccentric chief factor of Y.

Suppose there exists $K \in \mathrm{Proj}_{\mathfrak{F}}(Y)$ with $V \leq K$. Then $K = V(K \cap D)$. Set $T = K \cap D$. It is clear that $T \in \mathrm{Proj}_{\mathfrak{F}}(D)$. This implies that $T^{F(p)} \leq C_T(V) = 1$ and then $T \in F(p)$. Moreover $T/(G \cap T) \cong H$. Hence $H \in F(p)$, a contradiction. Therefore V is not covered by K.

This means that if an $\mathfrak{F}$-projector of Y covers a $\mathfrak{Y}$-chief factor of Y, this chief factor is necessarily $\mathfrak{F}$-central in Y. In other words, $Y \in \mathfrak{X} \subseteq \overline{\mathfrak{X}}$. Then $G \in \overline{X}(p)$, a contradiction. □

Remark. The above proof holds in the soluble universe and is a considerable simplification of Doerk's proof in [2].

Lemma 2. *Let $\pi = \{p \in \mathrm{char}\mathfrak{F} : F(p) = \mathfrak{F}\}$. If $\mathfrak{X}$ is saturated, i.e. if $\mathfrak{X} = \overline{\mathfrak{X}}$, then the following condition holds:*

(*) If G is a primitive group of type 2 in $b(\mathfrak{F})$, then $Soc(G)$ is a π'-group.

PROOF. Suppose there exists a primitive group G of type 2 in $b(\mathfrak{F})$ whose socle is not a π'-group, i.e. there exists a prime p dividing $|Soc(G)|$ such that $p \in \pi$. No $\mathfrak{F}$-projector of G covers $Soc(G)$. Then $G \in \mathfrak{X}$ and $G \in \overline{X}(p)$. Let V be an irreducible and faithful G-module over GF(p) and construct the semidirect product $Y = [V]G$. Then $Y \in \overline{X}(p) \subseteq \mathfrak{X}$. Let $E \in \mathrm{Proj}_{\mathfrak{F}}(Y)$. Then $VE \in \mathfrak{F} = F(p)$. Therefore $V \leq E$ and V is an $\mathfrak{F}$-central chief factor of Y, i.e. $G \in F(p) = \mathfrak{F}$, a contradiction. □

Theorem A. *If $\mathfrak{X}$ is saturated, i.e. $\mathfrak{X} = \overline{\mathfrak{X}}$, then either $\mathfrak{F} = (1)$ or $\mathfrak{F} = \mathfrak{E}$.*

PROOF. Assume that $\mathfrak{F} \neq (1)$. Denote $\pi = \{p \in \mathrm{char}\mathfrak{F}: F(p) = \mathfrak{F}\}$.

We have $\mathfrak{S}_p\mathfrak{X} = \mathfrak{X}$ for every prime $p \in \pi'$. We see now that for all $p, q \in \mathrm{char}\mathfrak{F} \setminus \pi$, $F(p) = F(q)$.

If not, there would exist a group G of minimal order in $F(p) \setminus F(q)$. Consider an irreducible and faithful G-module V over GF(q) and construct the semidirect product $H = [V]G$. Then $H \in \mathfrak{X} = \mathfrak{S}_q\mathfrak{X}$. If $E \in \mathrm{Proj}_{\mathfrak{F}}(H)$, then $E \cong G$. Consider an irreducible and faithful H-module W over GF(p) and construct the semidirect product $Y = [W]H$. So, $Y \in \mathfrak{S}_p\mathfrak{X} = \mathfrak{X}$. But every $\mathfrak{F}$-projector of Y covers W. Then W is $\mathfrak{F}$-central in Y and $H \in \mathfrak{F}$, a contradiction. Therefore $F(p) = F(q)$ for all prime numbers p and q in $\mathrm{char}\mathfrak{F} \setminus \pi$.

If $\pi = \emptyset$, we have $\mathfrak{F} = (1)$, a contradiction. Therefore $\pi \neq \emptyset$.

Let $p \in \pi$. Suppose there exists a prime $q \in \pi'$ and let H be a group of minimal order in $\mathfrak{F} \setminus F(q)$. Consider a non-abelian simple group S such that p and q divide $|S|$ and construct the wreath product $G = S \wr H$ with respect to the regular action. Clearly the group G is a primitive group of type 2 in $b(\mathfrak{F})$. Then $Soc(G) \notin \mathfrak{E}_{\pi'}$ contradicts the condition (*) in lemma 2. Therefore π is the set of all prime numbers and this implies that $\mathfrak{S}_p\mathfrak{F} = \mathfrak{F}$ for every prime number p. Hence, $b(\mathfrak{F})$ is a class of primitive groups of type 2. By Lemma 2, this implies that $b(\mathfrak{F}) = \emptyset$. Then $\mathfrak{F} = \mathfrak{E}$. □

Remark. Obviously in the soluble universe, such a simple group S with two different prime divisors does not exist. In this case, since $F(p) = \mathfrak{H}$ for all primes $p \in \mathrm{char}\mathfrak{F} \setminus \pi$, we deduce that $\mathfrak{F} = \mathfrak{S}_\pi$ if $\mathrm{char}\mathfrak{F} = \pi$ and $\mathfrak{F} = \mathfrak{S}_\pi \mathfrak{H}$ if $\pi \subset \mathrm{char}\mathfrak{F}$. So, Doerk's result is obtained.

3. The saturation of $\mathfrak{H} \downarrow \mathfrak{F}$

With the notation of the preliminaries, denote $\mathfrak{X} = \mathfrak{H} \downarrow \mathfrak{F}$.

There exists a trivial case of saturation of the class $\mathfrak{H} \downarrow \mathfrak{F}$. If $\mathfrak{F} = \mathfrak{H}$, then $\mathfrak{H} \downarrow \mathfrak{H} = \mathfrak{E}$, the class of all (finite) groups. So, we assume in the sequel that $\mathfrak{F}$ is a *proper saturated subformation* of $\mathfrak{H}$.

Lemma 3. $\mathfrak{H} \downarrow F(p) \subseteq \overline{X}(p) \subseteq \mathfrak{S}_p\mathfrak{X}$.

PROOF. Since $\mathfrak{X} \subseteq \overline{\mathfrak{X}} \subseteq \mathfrak{N}\mathfrak{X}$, we have that $\overline{X}(p) \subseteq \mathfrak{S}_p\mathfrak{X}$. Consider a group G of minimal order in $(\mathfrak{H} \downarrow F(p))\backslash\overline{X}(p)$. The group G is monolithic and $Soc(G)$ is not a p-group. Hence there exists an irreducible and faithful G-module V over GF(p) and we can form the semidirect product $Y = [V]G$.

Let $T \in \text{Proj}_{\mathfrak{H}}(Y)$. Then TV/V is isomorphic to an $\mathfrak{F}$-projector of G and hence $TV/V \in F(p)$. Therefore $T \in F(p) \subseteq \mathfrak{F}$. This means that $Y \in \mathfrak{X} \subseteq \overline{\mathfrak{X}}$ and therefore $Y/O_{p'p}(Y) \cong G \in \overline{X}(p)$, a contradiction. Thus $\mathfrak{H} \downarrow F(p) \subseteq \overline{X}(p)$. □

Lemma 4. *If* $\mathfrak{F} \not\subseteq H(p)$ *then* $\overline{X}(p) = \mathfrak{S}_p\mathfrak{X}$.

PROOF. Consider a group G of minimal order in $\mathfrak{S}_p\mathfrak{X}\backslash\overline{X}(p)$. The group G is monolithic and $Soc(G)$ is not a p-group. Hence we have indeed that $G \in \mathfrak{X}$.

Consider a group H of minimal order in $\mathfrak{F}\backslash H(p)$. The group H is monolithic and $Soc(H)$ is not a p-group.

Consider a prime q such that q divides $|Soc(H)|$ and $q \neq p$. Obviously q is in char$\mathfrak{F}$.

Denote by P the projective cover of the trivial irreducible GF$(q)[H]$-module. By a theorem of Brauer, see [4] B.4.23, we have that $C_H(P) = \text{Ker}(H \text{ on } P) = O_{q'}(H) = 1$ and P is a faithful H-module.

The semidirect product $A = [P]H$ is in $\mathfrak{F}$. Since $Rad(P) \leq \Phi(A)$, it is enough to see that $A/Rad(P) \cong H \times C_q \in \mathfrak{F}$. If moreover $A \in H(p)$, then $H \in H(p)$, a contradiction. If $A^{H(p)} \leq Z(A)$ then $A^{H(p)} \leq C_A(P) = PC_H(P) = P$. This implies that $H \cong A/P \in H(p)$, a contradiction. Thus A is a group in $\mathfrak{F} \setminus H(p)$ such that $A^{H(p)} \not\leq Z(A)$.

Notice that if N is a minimal normal subgroup of A and $N \cap P = 1$ then $N \leq C_A(P) = P$, a contradiction. Hence $N \leq P$. Then N is an irreducible H-submodule of P, i.e. $N \leq Soc(P)$. Therefore $Soc(A) = Soc(P) \cong C_q$. This implies the existence of an irreducible and faithful A-module W over GF(p).

Construct the direct product $D = G \times A$ and consider the induced module $V = W^D$. Construct the semidirect product $Y = [V]D$ and notice that $C_D(V) = \text{Ker}(D \text{ on } W^D) = core_D(\text{Ker}(A \text{ on } W)) = 1$, i.e. V is a faithful D-module.

If $Y \in \mathfrak{H}$, then $Y = Y/O_{p'}(Y) \in H(p)$ and $D \in H(p)$. In fact $A \in H(p)$, a contradiction. Therefore $Y \notin \mathfrak{H}$.

Consider $E \in \text{Proj}_{\mathfrak{H}}(G)$. Since $G \in \mathfrak{X}$, we have that $E \in \mathfrak{F}$. It is clear that $T = E \times A \in \text{Proj}_{\mathfrak{H}}(D)$ and $T \in \mathfrak{F}$. Moreover $TC_V(T^{H(p)}) \in \text{Proj}_{\mathfrak{H}}(Y)$.

Denote $R = T^{H(p)} \cap A$. Notice that $C_V(T^{H(p)}) \leq C_V(R) = (C_W(R))^D$. Since R is a normal subgroup of A, we have that $C_W(R)$ is an A-submodule of the irreducible A-module W.

If $C_W(R) = 0$, then $C_V(T^{H(p)}) = 0$ and $T \in \mathrm{Proj}_{\mathfrak{H}}(Y)$. In general, if we consider any $T^* \in \mathrm{Proj}_{\mathfrak{H}}(Y)$, it can be proved that T^* is conjugate in T^*V to an $\mathfrak{F}$-projector of the given form. So, $Y \in \mathfrak{X} \subseteq \overline{\mathfrak{X}}$. In fact $Y \in \overline{X}(p)$ which implies $G \in \overline{X}(p)$, a contradiction.

So $C_W(R) = W$. But then $R \leq C_A(W) = 1$. This implies that $T^{H(p)} \leq Z(A) \times E$. Then $A/Z(A) \cong (A \times E)/(Z(A) \times E) = T/(Z(A) \times E) \in H(p)$ and $A^{H(p)} \leq Z(A)$, a contradiction.

So our claim is true and $\overline{X}(p) = \mathfrak{S}_p\mathfrak{X}$. □

Lemma 5. *If $\mathfrak{X} = \overline{\mathfrak{X}}$, i.e. if $\mathfrak{X}$ is saturated, and if for a prime p, we have $\mathfrak{F} \subseteq H(p)$, then $\overline{X}(p) = \mathfrak{H} \downarrow F(p)$.*

PROOF. Let G be a group of minimal order in $\overline{X}(p) \setminus (\mathfrak{H} \downarrow F(p))$. Since $\mathfrak{S}_p(\mathfrak{H} \downarrow F(p)) = \mathfrak{H} \downarrow F(p)$, G is monolithic and $Soc(G)$ is not a p-group. There exists an irreducible and faithful G-module V over GF(p) and we construct the semidirect product $Y = [V]G$. Then $Y \in \mathfrak{S}_p\overline{X}(p) = \overline{X}(p) \subseteq \overline{\mathfrak{X}} = \mathfrak{X}$.

If $E \in \mathrm{Proj}_{\mathfrak{H}}(G)$, then $E \in \mathfrak{F} \subseteq H(p)$ since $G \in \mathfrak{X}$. Now $EV \in \mathfrak{S}_pH(p) = H(p) \subseteq \mathfrak{H}$ and so, $EV \in \mathrm{Proj}_{\mathfrak{H}}(Y)$. This implies that $C_V(E^{F(p)}) = V$. Then $E^{F(p)} \leq C_E(V) \leq C_G(V) = 1$. Therefore $E \in F(p)$. Hence $G \in \mathfrak{H} \downarrow F(p)$, a contradiction.

So, our claim is true and $\overline{X}(p) = \mathfrak{H} \downarrow F(p)$. □

Corollary. *If $\mathfrak{X} = \overline{\mathfrak{X}}$, i.e. if $\mathfrak{X}$ is saturated, then*

$$\overline{X}(p) = \begin{cases} \mathfrak{H} \downarrow F(p), & \text{if } \mathfrak{F} \subseteq H(p); \\ \mathfrak{S}_p\mathfrak{X}, & \text{if } \mathfrak{F} \not\subseteq H(p). \end{cases}$$

Theorem B. *The formation $\mathfrak{X} = \mathfrak{H} \downarrow \mathfrak{F}$ is saturated, i.e. $\mathfrak{X} = \overline{\mathfrak{X}}$, if and only if the following two conditions hold:*

i) if $\mathfrak{F} \subseteq H(p)$ then $\overline{X}(p) = \mathfrak{H} \downarrow F(p)$, and

ii) if $\mathfrak{F} \not\subseteq H(p)$ then $H(p) \cap \mathfrak{F} = F(p)$.

PROOF. First suppose that $\mathfrak{X}$ is saturated. If $\mathfrak{F} \subseteq H(p)$ then (i) holds by the above corollary.

If $\mathfrak{F} \not\subseteq H(p)$, then $\overline{X}(p) = \mathfrak{S}_p\mathfrak{X}$ by Lemma 2. Let G be a group of minimal order in $(H(p) \cap \mathfrak{F}) \setminus F(p)$. The group G is monolithic and $Soc(G)$ is not a p-group. Let V be an irreducible and faithful G-module over GF(p) and construct the semidirect product $Y = [V]G$. Then $Y \in \mathfrak{S}_pH(p) = H(p) \subseteq \mathfrak{H}$. Since $G \in \mathfrak{F} \subseteq \mathfrak{X}$, we have also $Y \in \mathfrak{S}_p\mathfrak{X} = \mathfrak{X}$. Therefore we have indeed $Y \in \mathfrak{F}$. Then $Y \in F(p)$ implies $G \in F(p)$, a contradiction. Thus (ii) holds.

Conversely, let G be a group of minimal order in $\overline{\mathfrak{X}} \setminus \mathfrak{X}$. The group G is monolithic. Let p be a prime dividing $|Soc(G)|$. Then $O_{p'}(G) = 1$ and $G \in$

$\overline{X}(p)$. If $\mathfrak{F} \subseteq H(p)$, then apply (i) to conclude $G \in \mathfrak{H} \downarrow F(p) \subseteq \mathfrak{H} \downarrow \mathfrak{F} = \mathfrak{X}$, a contradiction.

If $\mathfrak{F} \not\subseteq H(p)$, then $\overline{X}(p) = \mathfrak{S}_p\mathfrak{X}$ by Lemma 2 and $G \in \mathfrak{S}_p\mathfrak{X}$. Consider $E \in \mathrm{Proj}_{\mathfrak{H}}(G)$. Then $EO_p(G)/O_p(G) \in \mathrm{Proj}_{\mathfrak{H}}(G/O_p(G)) \subseteq \mathfrak{F}$ and $E/(E \cap O_p(G)) \in \mathfrak{F}$. On the other hand, since $E \in \mathfrak{H}$, we have $E/O_{p'p}(E) \in \mathfrak{F} \cap H(p) = F(p)$ by (ii). Then $E \in \mathfrak{F}$ and $G \in \mathfrak{X}$, a contradiction.

Therefore $\mathfrak{X} = \overline{\mathfrak{X}}$. □

As it is said in the introduction, Doerk solves the problem of characterizing the saturated formations (of soluble groups) $\mathfrak{H}$ such that $\mathfrak{H} \downarrow \mathfrak{F}$ is saturated for all saturated formations $\mathfrak{F}$. Our aim is to give an answer to the similar problem in the general non-soluble universe. In fact we consider also the "symmetric" problem of characterizing the saturated formations $\mathfrak{F}$ such that $\mathfrak{H} \downarrow \mathfrak{F}$ is saturated for all saturated formations $\mathfrak{H}$.

Proposition. *If* $\mathrm{char}\mathfrak{F} = \mathrm{char}\mathfrak{H} = \pi$ *and for all* $p \in \pi$, $\mathfrak{S}_p\mathfrak{F} = \mathfrak{F}$, *then* $\mathfrak{X}$ *is saturated.*

Conversely, if $\mathfrak{F}$ *is a saturated formation such that* $\mathfrak{H} \downarrow \mathfrak{F}$ *is saturated for all saturated formations* $\mathfrak{H}$ *with* $\mathrm{char}\mathfrak{F} = \mathrm{char}\mathfrak{H} = \pi$, *then* $\mathfrak{S}_p\mathfrak{F} = \mathfrak{F}$ *for all* $p \in \pi$.

PROOF. Let G be a group of minimal order in $\mathrm{E}_\Phi(\mathfrak{X}) \setminus \mathfrak{X}$. The group G is monolithic and if $N = Soc(G)$, $N \leq \Phi(G)$. Suppose that N is a p-group. If $E \in \mathrm{Proj}_{\mathfrak{H}}(G)$, then $EN/N \in \mathrm{Proj}_{\mathfrak{H}}(G/N) \subseteq \mathfrak{F}$ and therefore $E/(E \cap N) \in \mathfrak{F}$. If $p \in \pi$, then $E \in \mathfrak{S}_p\mathfrak{F} = \mathfrak{F}$, a contradiction. Hence $p \notin \pi$ and p does not divide $|E|$. This implies $E \cap N = 1$. Again we obtain a contradiction. Therefore $\mathrm{E}_\Phi(\mathfrak{X}) = \mathfrak{X}$ and $\mathfrak{X}$ is saturated.

Conversely, we can suppose $|\pi| \geq 2$ (the case $|\pi| = 1$ is obvious) and assume that there exists a prime p such that $\mathfrak{S}_p\mathfrak{F} \neq \mathfrak{F}$. Let G be a group of minimal order in $\mathfrak{S}_p\mathfrak{F} \setminus \mathfrak{F}$. Then G is a primitive group of type 1 and $Soc(G)$ is a p-group. Consider $\mathfrak{H} = \mathfrak{N}_\pi \mathfrak{Y}$ for $\mathfrak{Y} = \mathrm{QR}_0(G/Soc(G))$. Then $\mathfrak{H} = LF(H)$ such that $H(q) = \mathfrak{S}_q\mathfrak{Y}$ for all $q \in \pi$ and $H(r) = \emptyset$ if $r \notin \pi$. Now $\mathfrak{H} \downarrow \mathfrak{F} = \mathfrak{H} \downarrow (\mathfrak{F} \cap \mathfrak{H})$ is saturated. Denote by T the canonical definition of $\mathfrak{F} \cap \mathfrak{H}$. Clearly $T(q) \subseteq F(q)$. If $\mathfrak{H} \cap \mathfrak{F} \not\subseteq \mathfrak{S}_p\mathfrak{Y}$, then by Theorem B, $H(p) \cap \mathfrak{H} \cap \mathfrak{F} = H(p) \cap \mathfrak{F} = \mathfrak{S}_p\mathfrak{Y} \cap \mathfrak{F} = T(p) \subseteq F(p)$. Therefore $G \in \mathfrak{F}$, a contradiction. Then $\mathfrak{H} \cap \mathfrak{F} \subseteq \mathfrak{S}_p\mathfrak{Y}$, $\mathfrak{S}_q \subseteq \mathfrak{H} \cap \mathfrak{F} \subseteq \mathfrak{S}_p\mathfrak{Y}$ for all $q \in \pi$, and $\mathfrak{S}_q \subseteq \mathfrak{Y}$, a contradiction. □

The problem of characterizing the saturated formations $\mathfrak{H}$ such that $\mathfrak{H} \downarrow \mathfrak{F}$ is saturated for all saturated formations $\mathfrak{F}$ turns out to be much more complicated to solve. The authors have not been able to give a complete characterization. It is clear that if $\mathfrak{H}$ is either (1) or $\mathfrak{E}$ or $\mathfrak{S}_p$ for some prime p, then $\mathfrak{H} \downarrow \mathfrak{F}$ is saturated for all saturated formations $\mathfrak{F}$. We conjecture that these are the only saturated formations with this property. The most complete result we can present at the moment is the following.

Theorem C. *Let $\mathfrak{H}$ be a saturated formation with $|\mathrm{char}\mathfrak{H}| \geq 2$ such that for every saturated formation $\mathfrak{F}$, the formation $\mathfrak{H} \downarrow \mathfrak{F}$ is saturated. Then $\mathfrak{H}$ is of full characteristic and for every prime p, $H(p) = \mathfrak{H}$. Therefore $\mathfrak{S}\mathfrak{H} = \mathfrak{H}$.*

SKETCH OF PROOF. Set $\pi = \mathrm{char}\mathfrak{H}$.

1. $\mathrm{char}(H(p)) = \pi$ for all $p \in \pi$.

If $p \in \pi$ and $q \in \pi \setminus \mathrm{char}(H(p))$, then $\mathfrak{S}_q \not\subseteq H(p)$. Consider $\mathfrak{F} = \mathfrak{S}_q$. Since $\mathfrak{H} \downarrow \mathfrak{F}$ is saturated, we can apply theorem B to conclude that $H(p) \cap \mathfrak{F} = F(p) = \emptyset$, a contradiction. Therefore $\mathrm{char}(H(p)) = \pi$.

2. $\mathfrak{H} = H(p)$ for all $p \in \pi$. Therefore $\mathfrak{S}_\pi \mathfrak{H} = \mathfrak{H}$.

Let G be a group of minimal order in $\mathfrak{H} \setminus H(p)$ for some prime $p \in \pi$. The group G is monolithic and $Soc(G)$ is not a p-group. Consider the formation function: $f(p) = \mathfrak{S}_p(\mathrm{QR}_0(G/Soc(G)))$ and $f(q) = H(q)$ for all $q \in \pi \setminus \{p\}$. Let $\mathfrak{F} = LF(f)$. Since f is full, we have $\mathfrak{F} = LF(F)$ where $F(r) = f(r) \cap \mathfrak{F}$ for all $r \in \pi$. Notice that since $G/Soc(G) \in H(p)$, $F(p) \subseteq H(p)$. Then $\mathfrak{F} \subseteq \mathfrak{H}$. Suppose that $G \in \mathfrak{F}$. Since $\mathfrak{H} \downarrow \mathfrak{F}$ is saturated we deduce by theorem B, that $H(p) \cap \mathfrak{F} = F(p) = f(p) \cap \mathfrak{F}$. If $q \in \pi$ and $q \neq p$ then by step 1, $\mathfrak{S}_q \subseteq H(p) \cap \mathfrak{F}$ and therefore $\mathfrak{S}_q \subseteq \mathrm{QR}_0(G/Soc(G))$. But this is a contradiction. Hence $G \notin \mathfrak{F}$. This implies that G is a primitive group of type 2 and $p \in \pi\,(Soc(G))$. But then $G \in \mathfrak{H}$ and $G \in H(p)$ against our choice of G. In conclusion, $H(p) = \mathfrak{H}$ as required.

3. If $|\pi| \geq 2$ then $\{2,3\} \subseteq \pi$.

Let p be the least prime in π, q the least prime in $\pi \setminus \{p\}$, and suppose that $q \geq 5$.

Consider the simple group $G = Alt(q)$. Every π-subgroup of G is of order $p^a q^b$ with $0 \leq b \leq 1$ and $0 \leq a$, i.e. every π-subgroup of G is soluble and therefore is in $\mathfrak{H}$. $\mathrm{Proj}_{\mathfrak{H}}(G)$ is the set of all maximal π-subgroups of G. Consider $\mathfrak{F} = \mathfrak{S}_{q'}\mathfrak{S}_q\mathfrak{S}_{q'} \cap \mathfrak{S}_\pi$. Then $G \in \mathfrak{H} \downarrow \mathfrak{F}$. This implies that all $\mathfrak{H}$-projectors of G are in $F(q) = \mathfrak{S}_q\mathfrak{S}_{q'} \cap \mathfrak{S}_\pi$ and hence they are p-nilpotent. Therefore $G \in \mathfrak{H} \downarrow \mathfrak{T}$ for $\mathfrak{T} = \mathfrak{S}_{p'}\mathfrak{S}_p \cap \mathfrak{H}$, and all $\mathfrak{H}$-projectors of G are in $T(p) = \mathfrak{S}_p$, a contradiction.

Hence, $q < 5$ which means that $p = 2$ and $q = 3$.

4. If $|\pi| \geq 2$, then the group $Alt(5)$ is in $\mathfrak{H}$ and hence $5 \in \pi$.

Consider $S = Alt(5)$ and suppose that $S \notin \mathfrak{H}$. Then the set $\mathrm{Proj}_{\mathfrak{H}}(S)$ is composed of all maximal π-subgroups of S. Consider $\mathfrak{F} = \mathfrak{S}_{3'}\mathfrak{S}_3\mathfrak{S}_{3'} \cap \mathfrak{S}_\pi$. Then $G \in \mathfrak{H} \downarrow \mathfrak{F}$. This implies that all $\mathfrak{H}$-projectors of G are in $F(3) = \mathfrak{S}_3\mathfrak{S}_{3'} \cap \mathfrak{S}_\pi$ and $Alt(4) \notin F(3)$, a contradiction.

Therefore $Alt(5) \in \mathfrak{H}$ and $5 \in \pi$.

5. If $|\pi| \geq 2$, then π is the set of all primes. Therefore $\mathfrak{S}\mathfrak{H} = \mathfrak{H}$.

Let p be the least prime such that $p \notin \pi$, consider the simple group $S = PSL(2,p)$ and suppose that $S \notin \mathfrak{H}$. Notice that p is the biggest prime dividing $|S|$. By Dickson's theorem (see [7] Th.3.6.25), $\mathrm{Proj}_{\mathfrak{H}}(S)$ is the set of all maximal π-subgroups of S. With similar arguments to those used in step 4, we prove that the prime p cannot exist and therefore $\mathfrak{H}$ is of full characteristic. □

Conjecture. Let $\mathfrak{H}$ be a saturated formation with $(1) \neq \mathfrak{H} \neq \mathfrak{E}$. Then the following statements are equivalent:

i) For every saturated formation $\mathfrak{F}$, the formation $\mathfrak{H} \downarrow \mathfrak{F}$ is saturated.

ii) $\mathfrak{H} = \mathfrak{S}_p$ for some prime p.

Remarks. In the attempt to prove this conjecture the following partial approaches were considered:

1) Denote by $\mathfrak{M}_0$ the class of all soluble groups and define for every natural number $k > 0$ the class $\mathfrak{M}_k$ composed of all groups X such that $\mathrm{Max}(X) \subseteq \mathfrak{M}_{k-1}$. Clearly $\mathfrak{S} = \mathfrak{M}_0 \subset \mathfrak{M}_1 \subset \mathfrak{M}_2 \subset \ldots$. Then $\mathfrak{M}_2 \subseteq \mathfrak{H}$. (Notice that the simple groups in $\mathfrak{M}_1$ are the minimal simple groups.)

2) The simple N-groups are in $\mathfrak{H}$.

This research has been supported by *Proyecto PB 90-0414-C03-03* of DG-ICYT, Ministerio de Educación y Ciencia of Spain.

References

[1] A. Ballester-Bolinches, K. Doerk and L. M. Ezquerro, A question in the theory of saturated formations of finite soluble groups, *Israel J. Math.*, to appear.

[2] K. Doerk, Zur Theorie der Formationen endlicher auflösbarer Gruppen, *J. Algebra* **13**(1969), 345–373.

[3] K. Doerk, Zur Sättingung einer Formation endlicher auflösbarer Gruppen, *Arch. Math. (Basel)* **28**(1977), 561–571.

[4] K. Doerk and T. O. Hawkes, *Finite Soluble Groups* (Walter De Gruyter, Berlin–New York, 1992).

[5] D. Gorenstein, *Finite Groups* 2nd edition (Chelsea, New York 1980).

[6] R. L. Griess and P. Schmid, The Frattini module, *Arch. Math. (Basel)*, **30**(1978), 256–266.

[7] M. Suzuki, *Group Theory I* (Springer-Verlag, Berlin–Heidelberg–New York, 1982).

LOCALLY CONSTRUCTED FORMATIONS OF FINITE GROUPS

A. BALLESTER-BOLINCHES* and C. JUAN-MARTÍNEZ†

*Departament d'Algebra, Universitat de València, C/ Dr. Moliner 50, Burjassot 46100, València, Spain
†Departamento de Economía Financiera y Matemática, E. U. de Estudios Empresariales, C/ Artes Gráficas 11, Valéncia 46010, Spain

All groups considered in this paper are finite.

The origin and further development of Class Theory has as a corner-stone the concepts of saturated formation and covering subgroup. These were first introduced by Gaschütz in 1963 in his paper *Zur Theorie der endlichen auflösbaren Gruppen* with the aim of building a general context in which the properties of existence and conjugacy of Sylow subgroups, Hall subgroups and Carter subgroups could appear as particular cases. Thus, working in a universe $\mathcal{V}$ of groups which is closed with respect to the usual closure operators, a formation is a class of groups $\mathcal{F}$ contained in $\mathcal{V}$ with the following properties:

1. Every homomorphic image of an $\mathcal{F}$-group is an $\mathcal{F}$-group.
2. If G/M and G/N are $\mathcal{F}$-groups, then $G/(M \cap N)$ is also an $\mathcal{F}$-group.

The formation $\mathcal{F}$ is said to be saturated if the group G belongs to $\mathcal{F}$ whenever the Frattini factor group $G/\Phi(G)$ is in $\mathcal{F}$.

For a given group $G \in \mathcal{V}$ and a class of groups $\mathcal{F}$, an $\mathcal{F}$-covering subgroup of G is a subgroup C of G belonging to $\mathcal{F}$ and such that whenever $C \leq H \leq G$, $K \trianglelefteq H$ and $H/K \in \mathcal{F}$, then $H = CK$. So C covers each $\mathcal{F}$-quotient of every intermediate group of G. When $\mathcal{V}$ is the universe of finite soluble groups and $\mathcal{F}$ is a saturated formation of soluble groups, Gaschütz proved that every group $G \in \mathcal{V}$ has a unique non-empty conjugacy class of $\mathcal{F}$-covering subgroups. Also it was shown by him that when $\mathcal{F}$ is the class of $\mathcal{S}_p$, $\mathcal{S}_\pi$ or $\mathcal{N}$, then the $\mathcal{F}$-covering subgroups are the Sylow, Hall and Carter subgroups respectively.

From this beginning, one can see that further developments in Class Theory can be studied from two different, but complementary, points of view. One dealing with the behaviour of a class related to the existence of certain classes of subgroups, mainly covering subgroups, projectors or injectors in the groups of $\mathcal{V}$, and the other one coming from characterizing such a class through closure operators.

Keeping this general point of view, we have focused our attention on that part of the work of Doerk in his *Habilitationsschrift* ([Do]) intended to describe in the universe $\mathcal{S}$ of all soluble groups, the formations $\mathcal{F}$,such that $D(\mathcal{F}) = \mathcal{F}$, where

$$D(\mathcal{F}) = (G \in \mathcal{S} \mid \mathrm{Cov}_{\mathcal{F}}(G) \neq \emptyset)$$

and

$$\mathrm{Cov}_{\mathcal{F}}(G) = \{C \leq G \mid C \text{ is an } \mathcal{F}\text{-covering subgroup of } G\}.$$

If we notice that for a saturated formation $\mathcal{F}$, $D(\mathcal{F}) = \mathcal{S}$, the name of totally non-saturated formation given by Doerk to this type of formation matches perfectly with its behaviour. This kind of formation can be characterized in the soluble universe by means of the following properties:

1. $b(\mathcal{F}) \cap \wp = \emptyset$, where $b(\mathcal{F})$ is the boundary of $\mathcal{F}$ and $\wp$ is the set of all primitive groups.
2. The Schunk class $E_{\Phi}(\mathcal{F})$ is equal to $\mathcal{N}\mathcal{F}$. So $E_{\Phi}(\mathcal{F})$ is a formation.
3. $\mathcal{F}$ is an E_K-closed formation, where E_K is a closure operator on formations defined by:

$$E_K(\mathcal{F}) = (G \in \mathcal{S} \mid \text{all chief factors of } G \text{ below } G^{\mathcal{F}} \text{ are complemented in } G)$$

The first part of our paper is an approach to building a theory concerning totally non-saturated formations developed in the more general universe $\mathcal{E}$ of all finite groups. First of all, we would like to point out that the above results cannot be transferred directly from $\mathcal{S}$ to $\mathcal{E}$. In fact, the equivalence

$$D(\mathcal{F}) = \mathcal{F} \iff b(\mathcal{F}) \cap \wp = \emptyset$$

does not hold in the general case, even if we work with the class

$$P(\mathcal{F}) = (G \in \mathcal{E} \mid \mathrm{Proj}_{\mathcal{F}}(G) \neq \emptyset),$$

where $\mathrm{Proj}_{\mathcal{F}}(G) = \{V \leq G \mid V \text{ is an } \mathcal{F}\text{-projector of } G\}$. For instance, let $\mathcal{F}$ be the formation

$$\mathcal{F} = (G \in \mathcal{E} \mid \forall N \trianglelefteq G,\ N \text{ is complemented in } G).$$

Then $P(\mathcal{F}) = D(\mathcal{F}) = \mathcal{F}$ but $\mathrm{Aut}(A_6) \in b(\mathcal{F}) \cap \wp$. So $b(\mathcal{F}) \cap \wp$ is non-empty.

In order to establish a definition of what we would consider as a totally non-saturated formation in $\mathcal{E}$, we first make the following generalization of the closure operator E_K.

Definition 1. Let $\mathcal{F}$ be a formation. Define the class of groups

$$E_S(\mathcal{F}) = (G \in \mathcal{E} \mid \text{each chief factor of } G \text{ below } G^{\mathcal{F}} \text{ is supplemented in } G).$$

Then E_S is a closure operator on formations and a formation $\mathcal{F}$ is said to be *totally non-saturated* whenever $E_S(\mathcal{F}) = \mathcal{F}$.

Theorem 1. *Let $\mathcal{F}$ be a formation. The following statements are pairwise equivalent:*

1. *$\mathcal{F}$ is totally non-saturated.*
2. *$b(\mathcal{F}) \cap \wp = \emptyset$.*
3. *$G/\Phi(G) \in \mathcal{F}$ if and only if $G/F'(G) \in \mathcal{F}$, where*

$$F'(G) = \mathrm{Soc}(G \mod \Phi(G)).$$

In particular, $E_\Phi(\mathcal{F}) = \mathcal{N}\mathcal{F}$ is a saturated formation.

Next, if we consider the class of finite groups

$$D(\mathcal{F}) = (G \in \mathcal{E} \mid \mathrm{Cov}_{\mathcal{F}}(G) \neq \emptyset)$$

with $P(\mathcal{F})$ defined as before, we have the following:

Proposition 2. *Let $\mathcal{F}$ be a formation. If $\mathcal{F}$ is totally non-saturated, then $P(\mathcal{F}) = \mathcal{F}$ (and so $D(\mathcal{F}) = \mathcal{F}$).*

The converse is not true as we have shown in our example above.

The second part of our paper is devoted to a study of formations containing the family of the totally non-saturated ones. They are the locally constructed formations. This name is due to Hofmann who studied them from a local point of view in the soluble universe. However these formations were studied by Doerk in his Habilitationsschrift. In the general universe $\mathcal{E}$, a formation $\mathcal{F}$ is said to be a locally constructed formation if

$$\mathcal{F} = \overline{\mathcal{F}} \cap E_S(\mathcal{F}),$$

where $\overline{\mathcal{F}}$ is the saturation of $\mathcal{F}$, i. e. the smallest saturated formation containing $\mathcal{F}$. Our next result provides precise conditions to ensure that a formation $\mathcal{F}$ is locally constructed. First let us introduce the following notation:

- For a finite group G and a chief factor H/K of G,

$$[H/K] * G = \begin{cases} [H/K](G/C_G(H/K)) & \text{if } H/K \text{ is abelian} \\ G/C_G(H/K) & \text{otherwise} \end{cases}.$$

- $\mathrm{Psi}(G) = \{[H/K] * G \mid H/K \text{ is a chief factor of } G\}$.

- If $\mathcal{X}$ is a class of groups, we denote:

$$\mathrm{Psi}(\mathcal{X}) = \bigcup\{\mathrm{Psi}(G) \mid G \in \mathcal{X}\}.$$

- $S(G) = \mathrm{Psi}(G) \cup W$ where

$$W = \{[V](G/C_G(V)) \mid V \cong H_1/K_1 \times \ldots \times H_r/K_r,\ H_i/K_i \in J_p,\ p \in \pi\},$$
$$\pi = \pi(G) \text{ and}$$
$$J_p = \{H/K \mid H/K \text{ is an abelian } p\text{-chief factor of } G \text{ regarded as } \mathrm{GF}(p)(G)\text{-module by conjugation}\}.$$

- If $\mathcal{X}$ is a class of groups, $S(\mathcal{X}) = \bigcup\{S(G) \mid G \in \mathcal{X}\}$.

Theorem 3. *Let $\mathcal{F}$ be a formation. Then the following statements are pairwise equivalent:*

1. *$\mathcal{F}$ is locally constructed.*
2. *$A_\Phi(\mathcal{F})$ is a saturated formation, where*

$$A_\Phi(\mathcal{F}) = (G \in \mathcal{E} \mid G \in E_\Phi(\mathcal{F}) \text{ and if } H/K \text{ is a Frattini chief factor of } G, \text{ then } [H/K] * G \in \mathcal{F}).$$

3. *$E_\Phi(\mathcal{F}) = \overline{\mathcal{F}}$.*
4. *Given a group $G \in E_\Phi(\mathcal{F})$ and a prime p, the semidirect product $[A](G/C_G(A)) \in \mathcal{F}$ for all irreducible $GF(p)(G)$-modules A in the principal block of $GF(p)(G)$.*
5. *For each group $G \in E_\Phi(\mathcal{F})$ and for each chief factor H/K of G, then $[H/K] * G \in \mathcal{F}$.*
6. *$T(\mathcal{F}) = \mathcal{F}$, where*

$$T(\mathcal{F}) = (G \in \mathcal{E} \mid G \in E_S(\mathcal{F}) \text{ and } \mathrm{Psi}(G) \subseteq \mathrm{Psi}(E_\Phi(\mathcal{F}))).$$

As one can see, the locally constructed formations contain the family of totally non-saturated and saturated formations and some formations that fall in neither of these families such as $\mathcal{A}$, the class of the abelian groups. In the following we define a closure operator A on formations and prove that a formation $\mathcal{F}$ is locally constructed if and only if $A(\mathcal{F}) = \mathcal{F}$.

Definition 3. Let $\mathcal{F}$ be a formation. Define the class $A(\mathcal{F})$ as:

$$A(\mathcal{F}) = (G \in \mathcal{E} \mid G \in E_S(\mathcal{F}) \text{ and } \mathrm{Psi}(G) \subseteq \mathrm{Psi}(E_\Phi R_0 S(E_\Phi(\mathcal{F})))\ ).$$

Theorem 4. *The operator A is a closure operator on formations $\mathcal{F}$ with full characteristic.*

SKETCH OF PROOF. Let $G \in A(A(\mathcal{F})) - A(\mathcal{F})$ of minimal order. Then G is primitive of type 1 and so $G \in E_S(\mathcal{F})$. A contradiction arises if we prove that $G \in \mathrm{Psi}(E_\Phi R_0 S(E_\Phi(\mathcal{F})))$. For $X \in \mathcal{F}$ such that $X/T \cong G/\mathrm{Soc}(G)$ and $O_{p',p}(X) = 1$, let us consider its maximal Frattini extension E with p-elementary abelian kernel K. For $B = \mathrm{Soc}(K)$ there exists an integer n so that $\mathrm{Soc}(G)$ can be viewed as a chief factor of a group $C = [P](X/O_p(X))$, where P denotes the Hartley subgroup on n copies of B. As C belongs to $E_\Phi R_0 S(E_\Phi(\mathcal{F}))$, we reach the contradiction $G \in \mathrm{Psi}(C)$. □

Corollary 5. *Let $\mathcal{F}$ be a formation with full characteristic. Then $A(\mathcal{F})$ is the smallest locally constructed formation containing $\mathcal{F}$. In particular, $\mathcal{F}$ is locally constructed if and only if $A(\mathcal{F}) = \mathcal{F}$.*

The next result provides a sufficient condition for a formation $\mathcal{F}$ to be locally constructed.

Theorem 6. *Let $\mathcal{F}$ be a formation. If $P(\mathcal{F}) \subseteq A(\mathcal{F})$, then $\mathcal{F}$ is locally constructed.*

Finally we would like to mention, on the one hand, that the paper of Bechtell *Locally complemented formations* ([Be]) could be considered as a local study of the totally non-saturated feature. On the other hand, the results of Doerk in the soluble universe concernig the formations $\mathcal{F}$ such that $E_\Phi(\mathcal{F}) = \overline{\mathcal{F}}$ show that such a formation can be stated as the intersection of $\overline{\mathcal{F}}$ and $E_K(\mathcal{F})$. As both of them admit a local treatment, the formation $\mathcal{F}$ itself could be studied locally. This is done by Hofmann in his paper *Locally constructed formations* ([Ho]).

The results of Bechtell and Hofmann can be extended easily to the general universe $\mathcal{E}$.

This work is part of the Proyecto PB90-0414-C03-01 of CAICYT (Ministerio de Educación y Ciencia, Spain).

References

[Be] Bechtell, H., Locally complemented Formations, *J. Algebra* **106**(1987), 412–429.

[Do] Doerk, K., *Über Homomorphe und Formationen endlicher auflösbarer Gruppen*, Habilitationsschrift, Mainz, 1971.

[DH] Doerk, K., Hawkes, T.O., *Finite soluble Groups* (Walter de Gruyter, Berlin-New York, 1992).

[Ho] Hofmann, M.C., Locally constructed formations, *Arch. Math.* **53**(1989), 528–537.

[Sa] Salomon, E., *Über lokale und Baerlokale formationen endlicher Gruppen*, Diplomarbeit, Mainz, 1983.

[Sc] Schaller, K., Über die maximale Formation in einem gesättigten Homomorph, *J. Algebra* **45**(1977), 453–464.

REFLECTIONS ON VIRTUALLY ONE-RELATOR GROUPS

KATALIN BENCSÁTH* and BENJAMIN FINE†

*Department of Mathematics, Manhattan College, Riverdale, New York 10471, U.S.A.
†Department of Mathematics, Fairfield University, Fairfield, Connecticut 06430, U.S.A.

1. Introduction

Group theoretical investigation and classification efforts take advantage of the various group theoretical constructions: the attempt is to recapture information about groups of interest from their various subgroups, homomorphic images and extensions. The particularly successful theory of one-relator groups - which greatly benefitted from the pioneering work of W. Magnus - provided motivation for investigating one-relator quotients of free products as well as extensions and automorphisms of free groups. Questions of SQ-universality and recognizability, stemming from decision problems, also added interest in one-relator groups possessing free quotients, and in various groups with free subgroups.

Recall that if **P** is a group property, then a group G is *virtually* **P** if it has a subgroup of finite index satisfying **P**. Alternatively we also call G a virtual **P**-group. G is **P**-*by-finite* if G has a normal subgroup of finite index satisfying **P**. If **P** is a subgroup inherited property, such as torsion-freeness, freeness, or solvability then virtually **P** and **P**-by-finite are equivalent.

The structure of virtually-free groups {free-by-finite groups} is rather well understood {see Section 3} and generalizes the structure of free groups in expected ways. Virtually free groups also have connections with automatic groups and hyperbolic groups {see Section 3}.

The present paper represents the start of a general program to extend knowledge about virtually one-relator groups. The aim is to line up known results and give some extensions of relevance to the proposed project. Since a good number of properties of free groups appear in one-relator groups it is hoped that a similar parallelism can be detected for virtually one-relator groups. However, since subgroups of one-relator groups need not be one-relator groups {see Section 3}, the class of virtually one-relator groups properly contains the class of one-relator-by-finite groups. Our focus is mostly on the latter class.

The main thrust of the program is to address the following major questions.

(1) Is there a structure theorem for one-relator-by-finite groups, analogous to that of free-by-finite groups?

It is quite clear that the class of one-relator groups is too broad to expect the answer to (1) to be "Yes" in general. However the question may still be practical to consider if we impose restrictions on either the relator {cyclically pinched, positive etc.} or the nature of the extension.

(2) Is it possible to find a purely algebraic proof of the Nielsen Realization Theorem of Zieschang[Z1]?

Basically, the Nielsen realization theorem gives sufficient conditions for a virtually-surface group to be a planar discontinuous group.

(3) If every subgroup of finite index in the finitely generated one-relator group G is again a one-relator group, must G be a surface group? [Ko]
(4) Which, if any, of the classical decision problems have positive solution for "which" virtually one-relator groups?
(5) What can be said about the (virtual) residual properties these groups have? (virtually have?)

Not surprisingly, the process of surveying the literature, with the above problems in mind, led to the formulation of additional questions and problems; Section 7 of the paper contains a list of them. The outline of the rest of paper is as follows. In Section 2 we discuss virtual properties and extensions, and go over some results of Schneebeli [Sch] concerning virtual properties and extension closed classes of groups. In Section 3 we review the structure theorem and some consequences for free-by-finite groups. We also discuss the connections between virtually free groups and automatic groups. In Section 4 we begin the study of virtually one-relator groups. Section 5 deals with a linear integral representation of virtually F-groups and some consequences of this representation. In Section 6, we present the Nielsen realization theorem of Zieschang.

2. General virtual properties and extensions

If **C** is a class of groups, the *virtual class*, denoted v**C**, consists of those groups which are virtually **C** {contain a subgroup of finite index in **C**}. The fact that a group G is in v**C** is also expressed by saying that G is virtually **C** or that G is virtually in **C**. Clearly, a group has all the virtual properties possessed by any of its finite extensions and its subgroups of finite index.

For classes that contain the trivial group and are closed with respect to isomorphisms, extensions and taking subgroups of finite index, Schneebeli [Sch] studied the question as to when v**C** is extension closed. His results apply to certain one-relator groups, so we recount them, after first reviewing some general ideas about extensions.

A *group extension* G of K by Q consists of a group G, a normal subgroup $K \lhd G$ with $G/K \simeq Q$. We can picture this as a short exact sequence

$$\{1\} \to K \xrightarrow{i} G \xrightarrow{p} Q \to \{1\}$$

with i the inclusion map and p the projection. K is called the *kernel.* If K has a complement $Q' \simeq Q$ to G in G, that is, $G = KQ'$ with $K \cap Q' = \{1\}$ then the extension "splits" and G is a *semi-direct product.* G is a *central extension* if $K \subseteq Z(G)$. Suppose $\{g_\alpha\}$ is a complete set of a coset representatives {a transversal} for K in G. These elements g_α are then in one-to-one correspondence with the elements of Q. Since K is normal in G conjugation by these elements induces an automorphism of K and thus $\{g_\alpha\}$ defines a homomorphism $f : Q \to \mathrm{Aut}K/\mathrm{Inn}K$. The theory of extensions deals with determining the structure of the extension G, given the data K, Q and $f : Q \to \mathrm{Aut}K/\mathrm{Inn}K$. If K is abelian the problem is fairly tractable and leads to the analysis of the first two cohomology groups of G {see [R] and [J]}.

Given the above data one can write down a presentation for G. In particular, suppose $K =< X; S >$, $Q =< Y; R >$ are presentations for K and Q, respectively. We will think of X and Y as subsets of the groups K and Q. Suppose $\overline{Y} = \{\overline{y}; y \in Y\}$ are members of a transversal for K in G. For each $r \in R$, let $\overline{r}$ be the expression obtained from the word r when each y is replaced by the corresponding $\overline{y}$. Then since $r = 1$ in Q, $p(\overline{r}) = 1$ under the projection map p. Hence $\overline{r} \in K$ for each $r \in R$ so there exists an X-word v_r with $\overline{r}v_r^{-1} = 1$ in G. Put

$$\overline{R} = \{\overline{r}v_r^{-1}; r \in R\}.$$

Next, since $K \lhd G$, each $\overline{y}$ conjugates each x to a word $W_{\overline{y},x} \in K$. That is, $\overline{y}x\overline{y}^{-1} = W_{\overline{y},x}$ for $x \in X$, $\overline{y} \in \overline{Y}$. Put

$$\overline{T} = \{\overline{y}x\overline{y}^{-1}W_{\overline{y},x}^{-1}; \overline{y} \in \overline{Y}, x \in X\}.$$

Theorem 2.1. ([J]) *Let G, K, Q be as above. Then a presentation for G is given by $G = < X, \overline{Y}; S, \overline{R}, \overline{T} >$.*

Corollary 2.1.

(1) If K and Q are finitely presented, so is G.

(2) For finitely generated groups, "virtually finitely presented" and "finitely presented" are equivalent.

Part (1) is a straightforward consequence of the theorem, while part (2) follows in the following manner. That "finitely presented" implies "virtually finitely presented" is clear. On the other hand, if H is a finitely presented subgroup of finite index in the finitely generated group G then there exists a normal subgroup N of G with $|H : N| < \infty$. Since H is finitely presented and N is of finite index in H, N is finitely presented as shown by the Reidemeister-Schreier method. Thus G is an extension of a finitely presented group N by a finite group G/N and hence by part (1) G is finitely presented.

We now turn to Schneebeli's results. Consider a class $\mathbf{C}$ of groups containing the trivial group that is closed with respect to isomorphisms and subgroups of finite index. Recall that a group G is *poly-*$\mathbf{C}$ if there exists a finite normal series

$$G = G_0 \supset G_1 \ldots \supset G_n = 1$$

with $G_i/G_{i+1} \in \mathbf{C}$ for each i. Schneebeli proved the following:

Theorem 2.2. ([Sch]) *Let* $\mathbf{C}$ *be as above and suppose that in each group* G *in* $\mathbf{C}$*, for each positive integer* j *there exist at most finitely many subgroups of index* j *in* G*. Then the following three statements are equivalent:*

(1) The class of virtually poly-$\mathbf{C}$ *groups is extension closed.*

(2) All poly-v$\mathbf{C}$ *groups are virtual poly-*$\mathbf{C}$ *groups.*

(3) If H *is in* $\mathbf{C}$ *and* G *is a central extension of the finite group* K *by* H *then* G *is a virtually poly-*$\mathbf{C}$ *group.*

If, in addition, each group in $\mathbf{C}$ *is torsion-free then the following statement is equivalent to the above.*

(4) If H *is in* $\mathbf{C}$ *and* G *is a central extension of* $\mathbb{Z}_p$ *by* H *then* G *is virtually torsion-free.*

Schneebeli used these criteria to give specific examples of extension closed classes. We now recall that one-relator groups with non-trivial centers must be two-generator groups and, if non-abelian, they must have infinite cyclic centers [LS]. Moreover there exists an algorithm [BT] to determine the center of a group of a one-relator presentation.

The results below are relevant to our discussion of virtually one-relator groups.

Theorem 2.3. ([Sch]) *Let* $\mathbf{C}$ *be the class of all finitely generated groups* G *such that* G *has non-trivial center and* $cd(G) \leq 2$*. Then the class of virtually poly-*$\mathbf{C}$ *groups is extension closed and poly-v*$\mathbf{C}$ *groups are virtually torsion-free.*

Noting that the class $\mathbf{C}$ described in the theorem contains all one-relator groups with non-trivial center, we have the following.

Corollary 2.3. *Suppose the torsion-free group* G *has a subgroup of finite index which is a one-relator group with non-trivial center. Then* G *itself has non-trivial center and* $cd(G) \leq 2$.

A *surface group* is the fundamental group of a closed surface {orientable or non-orientable} of finite genus g. Such a group has either of the one-relator presentations

$$< a_1, b_1, \ldots, a_g, b_g; [a_1, b_1] \ldots [a_g, b_g] = 1 > \text{ (orientable case)} \quad (2.1)$$
$$< a_1, , \ldots, a_g; a_1^2 \ldots a_g^2 = 1 > \text{ (non-orientable case).} \quad (2.2)$$

We take again from Schneebeli.

Theorem 2.4. ([Sch]) *The class of virtually poly-{surface groups} is extension closed.*

This theorem has the following immediate consequences.

Corollary 2.4. *Any extension of a surface group is virtually a poly-surface group.*

Corollary 2.5. *Let G be a surface group. If H maps epimorphically onto G with a central finite kernel then H is virtually a poly-surface group.*

3. Free-by-finite groups

Properties of free groups manifest themselves in one-relator groups in various ways. The class of free-by-finite groups, which coincides with the class of virtually free groups, is well understood. It seems therefore reasonable to hope that the properties of virtually free groups will serve as satisfactory initial guidelines for what might (or might not) be expected for finite extensions of one-relator groups.

For free-by-finite groups there is a very detailed structure theorem from which most of the so far detected properties follow. The general form of the structure theorem we present is due to Karrass, Pietrowski and Solitar [KPS]; Stallings [St2], relying on his theory of ends, also proved its main portion {see [C1]}.

Theorem 3.1. ([KPS]) *A finitely generated group G is free-by-finite if and only if G is an HNN group of the form*

$$G =< t_1, \ldots, t_n, K; t_1 L_1 t_1^{-1} = M_1, \ldots, t_n L_n t_n^{-1} = M_n > \quad (3.1)$$

where K is a tree product of finitely many finite groups {the vertices of K} and each associated subgroup L_i, M_i is a subgroup of a vertex of K.

In particular, if there is no torsion in G then the base K is trivial and G is a free group.

Corollary 3.1. ([St]) *A torsion-free finite extension of a free group is free.*

This corollary was a key consequence of Stallings' proof that finitely generated torsion-free groups with $cd(G) = 1$ are free. For foundations he developed the *theory of ends* briefly sketched below.

Let Γ be a connected locally finite graph with a distinguished vertex v_0 selected as the origin of Γ. Let $\Gamma^{(n)}$ denote all points connected to v_0 by a path of length less than n. The *number of ends* of Γ is

$$e(\Gamma) = \lim_{n \to \infty} (\text{the number of infinite components of } \Gamma \setminus \Gamma^{(n)}).$$

If $G =< X; R >$ then the *number of ends of* G is the number of ends of the Cayley graph of G. It can be shown that this number is independent of the presentation. Intuitively, the number of ends of G is the number of ways that the Cayley graph can go to infinity. It is clear from this interpretation that an infinite cyclic group has 2 ends. Stallings [St] proved a structure theorem for groups with infinitely many ends. This structure theorem can then be applied almost directly to derive the structure theorem for free-by-finite groups. For our purposes now, Stallings' most applicable results are the following.

Lemma 3.1. ([C1]) *The number of ends of a group G is a virtual "invariant" - that is, if $|G : H| < \infty$ then G and H have the same number of ends.*

Lemma 3.2. ([C1]) *A non-abelian free group has infinitely many ends.*

Theorem 3.2. ([St1], Structure theorem for groups with infinitely many ends) *Suppose that G is a finitely generated group with infinitely many ends. Then either G is a free product with amalgamation with a finite amalgamated subgroup or G is an HNN extension with one finite associated subgroup. Conversely, if G is either a free product with amalgamation with a finite amalgamated subgroup or G is an HNN extension with one finite associated subgroup then G has infinitely many ends except in the cases where $G = G_1 \star_K G_2$ with K finite, $|G_1 : K| = |G_2 : K| = 2$ or where G is an HNN extension with finite base K and associated subgroup K. In these last two exceptional cases G has 2 ends. In particular, if G is torsion-free then G has infinitely many ends if and only if G is a free product.*

The proof of Theorem 3.1 on free-by-finite groups then follows from Theorem 3.2 by induction on the rank of the normal subgroup of finite index. A portion of the proof arrives at the following result on amalgams of free-by-finite groups which is of independent interest.

Theorem 3.3. *Let $\mathcal{F}$ be the class of free-by-finite groups. Then*

(1) $\mathcal{F}$ is closed under free products with finite amalgamated subgroups;

(2) $\mathcal{F}$ is closed under HNN extensions with bases from $\mathcal{F}$ and finite associated subgroups.

We note that part (1) was proved independently by Gregorac [Gr].

Using a formula of Stallings in [St2], Karrass, Pietrowski and Solitar [KPS] also obtained an index formula analogous to that of Schreier, for free-by-finite groups.

Theorem 3.4. ([KPS]) *Let G be a free-by-finite group of the form (3.1). Let $p+1$ be the number of vertices in the tree product K, let f_i be the order of L_i and let e_i, v_i range over the orders of the edge groups and vertex groups of K, respectively. Suppose H is a free subgroup of index j in G. Then the rank r of H is given by:*

$$r = j(\frac{1}{f_1} + \ldots + \frac{1}{f_n} + \frac{1}{e_1} + \ldots + \frac{1}{e_p} - \frac{1}{v_1} - \ldots - \frac{1}{v_{p+1}}) + 1.$$

The structure theorem has some consequences which are relevant to virtually one-relator groups.

Corollary 3.2. ([KPS]) *If G is free-by-finite and N is the normal subgroup of G generated by all elements of finite order then G/N is a free group.*

Corollary 3.3. ([KPS]) *Let $G =< a, b, \ldots; R^m >$ with R non-trivial and not a proper power in the free group $F =< a, b, \ldots; >$. Then G is free-by-finite if and only if R is a primitive in F (i.e. part of some free basis of F).*

Recall that a group G satisfies the *torsion-free subgroup property* [F] if each torsion-free subgroup turns out to be free. We then have the following:

Corollary 3.4. *A free-by-finite group satisfies the torsion-free subgroup property.*

PROOF. Suppose G is free-by-finite and H is a torsion-free subgroup of G. Then H is also free-by-finite since if F is a free normal subgroup of finite index in G then $H \cap F$ is a free subgroup of finite index in H. Therefore H is a torsion-free virtually free group and by Stallings' theorem {Corollary 3.1} H must be free. □

McCool [Mc] studied the automorphism groups of free-by-finite groups. In particular, he obtained:

Theorem 3.5. ([Mc]) *Let G be a finitely generated free-by-finite group. Then* Aut(G) *is finitely generated and is virtually of finite cohomological dimension.*

A countable free group is linear {see [LS]} and a group possessing a linear subgroup of finite index is linear itself. Therefore finitely generated virtually free groups are linear. In fact we know more: we are assured of the existence of integral representations.

Theorem 3.6. *Let G be a finitely generated virtually free group. Then G has a faithful representation in $GL_n(\mathbb{Z})$ for some n.*

PROOF. Suppose the finitely generated group G is virtually free and F is a free subgroup of finite index. F has a faithful representation in $GL_2(\mathbb{Z})$ {see [N]}, and then since $|G : F| < \infty$, G has a faithful representation in $GL_n(\mathbb{Z})$ for some n [W]. □

Hence virtually free, finitely generated groups carry the (isomorphism preserved) properties of integral matrix groups. The foregoing argument can be extended to any virtually-Fuchsian group and we will discuss this more fully in Section 5.

By virtue of being linear, finitely generated virtually-free groups are residually finite. Since virtually-free groups are linear they are residually finite. This also follows from combining the facts that free groups are residually finite and virtually residually finite implies residually finite [C2]. A specialized generalization of this was proved by G. Baumslag who showed that all cyclic extensions of free groups are residually finite [B].

From an entirely different viewpoint, virtually-free groups are related to automatic groups {see [BGSS] for terminology} and to hyperbolic groups {see [ABC]}, via some surprising connections found by Müller and Schupp [MS] {see also [HT]}. Suppose $< X; R >$ is a finite presentation for a group G and let $W_X(G)$ be the set of words in X which represent the identity in G. $W_X(G)$ is a *context-free language* if it is recognized by a push-down automaton. This is a non-deterministic finite state automaton with a push down stack storage device [HU]. The group G is *context-free* if for some presentation $W_X(G)$ is a context-free language. Müller and Schupp proved the remarkable theorem that being context-free is equivalent to being virtually free.

Theorem 3.7. ([MS],[D]) *A finitely generated group G is virtually free if and only if it is context-free.*

Müller and Schupp's original proof also required "accessibility" - a technical condition on the group G. However Dunwoody [D] proved that all finitely presented groups are accessible.

A *hyperbolic group* roughly is a group whose Cayley graph, relative to some presentation, satisfies a certain hyperbolic geometric condition{see [ABC]}. The relationship to virtually free groups is reflected by the following geometric characterization of context-free groups.

Theorem 3.8. ([MS]) *A finitely generated group G is context-free, and thus virtually-free, if and only if G has some presentation for which there exists a constant M such that every closed path in the Cayley graph can be M-triangulated.*

Autebert, Boasson and Senizergues [ABS] gave a refinement of Müller and Schupp's theorem in terms of generating sets. Gilman [G] attempted to translate the results of [ABS] into a computational procedure for possibly deciding if the group of a finite presentation is virtually free. In the terminology of Gilman [G]

Definition. A subset Q of a group G is *primary* if it satisfies the following conditions:

(1) Q is finite and $1 \in Q$

(2) Q is closed under taking inverses

(3) For every $n \geq 3$ and sequence $q_1, \ldots, q_n$ of elements of Q, if the product $q_1 \ldots q_n \in Q$ then $q_i q_{i+1} \in Q$ for some i with $1 \leq i < n$.

Theorem 3.9. ([ABS],[G]) *The following are equivalent:*

(1) G is finitely generated and virtually-free.

(2) G is generated by a primary set.

(3) Every set of generators of G lies in a primary set.

4. Virtually one-relator groups

As a first characterization step we list some necessary conditions for a group to belong to the class of virtually one-relator groups. Throughout this section we assume that our one-relator groups are finitely generated.

Theorem 4.1. *Let G be a virtually one-relator group and H a one-relator group of finite index in G. Then*

(1) G is virtually torsion-free.

(2) If rank $H \geq 3$ then G contains a subgroup of finite index which maps epimorphically onto a free group of rank 2. In particular, in this case G is SQ*-universal.*

(3) G has finite cohomological dimension.

(4) If H has non-trivial center then G has a non-trivial center.

The first three parts of Theorem 4.1 follow immediately from the observation that the statements concern virtual properties. The final part comes from the work of Schneebeli {see Section 2}.

As mentioned before, subgroups of one-relator groups are not necessarily one-relator groups. In fact Curran [Cu] has given an example of a one-relator group G with the property that for every integer $j > 1$ there exists a normal subgroup of index j in G which is not a one-relator group. Thus the concepts "virtually one-relator" and "one-relator-by-finite" differ. Focusing on one-relator-by-finite groups we now utilize the presentation results discussed in Section 2.

A straightforward application of Theorem 2.1 to a one-relator kernel yields:

Theorem 4.2. *Let G be a one-relator-by-finite group. Then G has a presentation of the form*

$$\begin{aligned} G =< x_1, \ldots, x_n, g_1, \ldots, g_m; S = R_1 V_1^{-1} = \ldots = R_k V_k^{-1} = 1, \\ g_i x_j g_i^{-1} W_{ij}^{-1} = 1, \; i = 1, \ldots, m, \; j = 1, \ldots, n > \end{aligned} \tag{4.1}$$

where $< x_1, \ldots, x_n; S >$ *is a presentation for a normal subgroup H of finite index in G and* $V_1, \ldots, V_k, W_{ij}$, $i = 1, \ldots, m$, $j = 1, \ldots, n$ *are words in* $\{x_1, \ldots, x_n\}$.

Combining this with the Reidemeister-Schreier rewriting process we arrive at a presentation of a pleasingly simple form when G/H is finite cyclic.

Theorem 4.3. *G is a finite cyclic extension of a one-relator group if and only if G has a presentation*

$$G =< x_1, \ldots, x_n, g; S = g^m V = 1, \; g x_i g^{-1} = \alpha(x_i), \; i = 1, \ldots, n > \tag{4.2}$$

where V is a word in $x_1, \ldots, x_n$ *and* α *is an automorphism of the one-relator group* $< x_1, \ldots, x_n; S >$. *In case G is a split extension then* $V = 1$ *in the one-relator group* $< x_1, \ldots, x_n; S >$.

PROOF. Suppose G is one-relator-by finite cyclic. That G has a presentation of form (4.2) follows directly from Theorem 4.2 by realizing that conjugation by the lift of a generator of the finite cyclic quotient induces an automorphism α of the normal one-relator subgroup. Therefore the words W_{ij} in Theorem 4.1 are given by $\alpha(x_i)$.

Conversely, suppose G has a presentation of form (4.2). Then the normal subgroup generated by $x_1, \ldots, x_n$ has finite cyclic quotient with presentation $< g; g^m = 1 >$. Using the Reidemeister-Schreier rewriting process we find that this normal subgroup has the presentation $< x_1, \ldots, x_n; S >$ and is thus a one-relator group. □

Knowledge of the structure of the automorphism group of the one-relator group can translate into useful information about the automorphism α. For example, every automorphism of a cyclically pinched one-relator group, in particular a surface group, is induced by a free group automorphism [LMR].

5. Virtual F-groups and linear representations

A *planar discontinuous group* is a discontinuous group of isometries of the hyperbolic plane. If a planar discontinuous group consists of only orientation preserving isometries it is a *Fuchsian group*; if it contains any non-orientation preserving isometries it is a *Non-Euclidean Crystallographic* or *NEC* Group. Alternatively a Fuchsian group can be defined as a discrete subgroup of $PSL_2(\mathbb{R})$ or a conjugate of such a subgroup in $PSL_2(\mathbb{C})$. A finitely generated Fuchsian group G has a *Poincaré presentation* of the form

$$\begin{aligned} G =< e_1, \ldots, e_p, h_1, \ldots, h_t, a_1, b_1, \ldots, a_g, b_g; \\ e_i^{m_i} = 1, i = 1, \ldots, p, R = 1 > \end{aligned} \tag{5.1}$$

where $R = e_1 \ldots e_p h_1 \ldots h_t [a_1, b_1] \ldots [a_g, b_g]$ and $p \geq 0, t \geq 0, g \geq 0, p+t+g > 0$, and $m_i \geq 2$ for $i = 1, \ldots, p$.

A group with a presentation of the form (5.1) is called an F-*group*. If $g = 0$ then the corresponding F-group is a free product of cyclics while if $p = t = 0$ then the group is a surface group. Therefore the class of F-groups includes the class of free products of cyclics and thus all the countable free groups - and the class of orientable surface groups. An F-group G is faithfully represented by a Fuchsian group if $\mu(G) > 0$ where $\mu(G) = 2g - 2 + t + \sum_i^p (1 - 1/m_i)$. $2\pi\mu(G)$ represents the hyperbolic area of a fundamental polygon for G {see [LS] or [K]}.

The class of F-groups falls into the wider class of *one-relator products of cyclics* which have been studied extensively by Fine and Rosenberger and others {see [FR1],[FR2] and the references therein}. These are groups with presentations of the form

$$< x_1, \ldots, x_n; x_1^{m_1} = \ldots = x_n^{m_n} = R = 1 > \tag{5.2}$$

where $m_i = 0$ or $m_i \geq 2$ and R is a cyclically reduced word in the free product on $\{x_1, \ldots, x_n\}$. These groups have many of the linearity properties of Fuchsian groups. In particular, very close to Fuchsian groups are those *groups of* F-*type* which are one-relator products of cyclics where the relator R has the form $R = UV$ with $U = U(x_1, \ldots, x_p)$, $V = V(x_{p+1}, \ldots, x_n), 1 \leq p < n$ non-trivial words of infinite order in the free products on the respective generators they involve [FR3]. This is the analog of cyclically pinched one-relator groups where the generators are allowed to possibly have finite order.

Due to the above mentioned relationship this section's integral linear representation considerations for virtually F-groups will be valid for virtual surface groups.

Theorem 5.1. *Let G be a virtual F-group. Then G has a faithful representation in $GL_n(\mathbb{Z})$ for some n.*

PROOF. The proof essentially follows the argument of Morris Newman [N] who showed that any Fuchsian group has such a representation.

An F-group H is virtually torsion-free [LS], so, unless finite, it contains a torsion-free subgroup K of finite index. If H is finite it can clearly be represented faithfully in $GL_n(\mathbb{Z})$ for some n. If H is infinite then by a result of Hoare, Karrass and Solitar [HKS1] K is either a free group or a surface group. If K is free then it admits a faithful representation in some $GL_n(\mathbb{Z})$ and therefore so does H. On the other hand, if K is a surface group (a copy of) K occurs as a subgroup of a genus 2 surface group. By a result of Magnus [M] a genus 2 surface group has a faithful representation in $GL_2(\mathbb{Q})$ hence also in $GL_n(\mathbb{Z})$ for some n. The existence of the desired kind of representation for any F-group H thus follows.

Finally if G is virtually an F-group, with F-group H of finite index in it, then in turn G also has a faithful integral representation of some higher degree [W]. This completes the proof. □

We can now apply some results of Beidleman and Robinson [BR] to virtually F-groups. A group G has *property ν* if, whenever N is a non-nilpotent normal subgroup of G, there is a normal subgroup L of G contained in N so that N/L is finite and non-nilpotent. That is, any non-nilpotent normal subgroup of G must possess a finite non-nilpotent G-quotient.

Denote by $\phi_f(G)$ the intersection of all maximal subgroups of finite index in G with the convention that $\phi_f(G) = G$ if G has no maximal subgroups of finite index. $\phi(G)$ stands for the Frattini subgroup and $\text{Fit}(G)$ for the Fitting subgroup {see [W] or [R] for terminology}.

Beidleman and Robinson prove the following characterization of groups with property ν.

Theorem 5.2. ([BR]) *A group G has property ν if and only if the following hold:*

(1) $\phi_f(G) \subset \text{Fit}(G)$ *and* $\text{Fit}(G)$ *is nilpotent;*

(2) $\text{Fit}(G/\phi_f(G)) = \text{Fit}(G)/\phi_f(G)$.

The tie-in to virtually free groups is via the following lemmas.

Lemma 5.1. ([W],[BR]) *Any linear group over a finitely generated integral domain satisfies property* ν.

Since virtual F-groups were shown to be integral linear groups we have arrived at:

Corollary 5.1. *Virtual F-groups satisfy property* ν.

This establishes the following properties of virtually F-groups.

Theorem 5.3. *Suppose G is virtually an F-group. Then the following hold:*

(1) $\phi_f(G)$ *is nilpotent.*

(2) If $T \triangleleft G$ *and* $T \subseteqq \phi_f(G)$ *then* G/T *has property* ν.

(3) For T as above, $\mathrm{Fit}(G/T) = \mathrm{Fit}(G)/T$.

(4) $\mathrm{Fit}(G)$ *is nilpotent and coincides with the Hirsch-Plotkin radical and with the intersection of the centralizers of the chief factors of G.*

(5) G is an FGH-group in the sense of Lennox [Le].

We also have:

Theorem 5.4.

(1) Any virtual F-group G is an extension of a nilpotent group by a group which is a residually finite group with trivial Frattini subgroup.

(2) Given a finite subset $\{x_1, \ldots, x_n\}$ *of G together with the information that the subgroup N generated by it is normal in G, there is an algorithm to decide whether N is nilpotent.*

Virtual F-groups possess other properties of integral linear groups as well.

Theorem 5.5. *Let G be a virtual F-group.*

(1) If $K \triangleleft G$ *such that* $\phi(G) \subseteqq K$ *and* $K/\phi(G)$ *is locally nilpotent then K is nilpotent. In particular,* $\phi(G)$ *is nilpotent.*

(2) G has finite central height and nilpotent Hirsch-Plotkin radical.

Part (1) is a corollary of Theorem 4.17 in [W] while part(2) is a corollary of Theorem 4.23 in [W] which is due to Gruenberg [Gru].

6. Virtual F-groups - the Nielsen realization theorem

It can be proven algebraically as well as geometrically {see [LS] and the references therein and [HKS1], [HKS2]} that planar discontinuous groups contain surface groups as subgroups of finite index. It is natural then to ask the question: under what conditions are finite extensions of surface groups isomorphic to planar discontinuous groups? A large amount of the work on virtual F-groups and thus virtual surface groups has gone into various solutions of this so-called *Nielsen realization problem.* The results we outline here are described in detail in [Z1]. Zieschang originally gave the following positive answer to a question of A. Karrass about torsion-free finite extensions of surface groups.

Theorem 6.1. ([Z2],[Z3]) *A torsion-free finite extension of a surface group is also a surface group.*

Note the analogy in the above statement with Stallings' theorem {Corollary 3.1} on free-by-finite groups.

Given any group H we say that H is *geometrically realized* if H is isomorphic to a planar discontinuous group. For a planar discontinuous group G, the centralizer $C_G(K)$ of any surface group K contained in G, must be trivial. Therefore, to properly pose the Nielsen realization problem, this fact must be included as a necessary condition. To this end, following [EM] we say that G is an *effective extension* of the group K if

$$\text{whenever } g^{-1}xg = x \text{ for all } x \in K \text{ then } g \in K. \qquad (6.1)$$

If K is a surface group this means $g = 1$ since K is centerless. {See also the terminology "allowable extension" in [KE] and "faithful extension" in [CZ].} We are now ready to pose:

Nielsen realization problem. Suppose G is a finite effective extension of a surface group K. Under what conditions is G isomorphic to a planar discontinuous group– in other words, under what conditions is G geometrically realized?

For our discussion here, by a surface group K, we mean the fundamental group of a closed surface of finite genus g, so that K has a one-relator presentation of the form (2.1) or (2.2). Zieschang's work [Z1] handled surfaces with holes also but group-theoretically the corresponding fundamental groups are then free groups and therefore already covered {at least in terms of abstract group theory} by our previous discussion on virtually-free groups.

The Nielsen realization problem has an equivalent topological interpretation. Nielsen [Ni] proved that any automorphism of the fundamental group of a closed surface S is induced by a homeomorphism of S. The *mapping class group* or *homeotopy group* of S is $\text{Meom}(S)/\text{Meot}(S)$– that is, the group of homeomorpisms of S modulo the isotopy group of S. Theorems of Baer and Nielsen {see [Z1]} say that for a closed surface S, the mapping class group is isomorphic to $\text{Aut}\Pi_1(S)/\text{Inn}\Pi_1(S)$– the outer automorphism group of $\Pi_1(S)$, where $\Pi_1(S)$ is the fundamental group of S.

In this context the Nielsen realization problem corresponds to determining the finite subgroups of the mapping class groups or, equivalently, determining which finite groups can be viewed as finite groups of mapping classes of S. We can now rephrase the

Nielsen realization problem. Suppose G is a finite effective extension of a surface group $\Pi_1(S)$ with $A_0 = G/\Pi_1(S)$. Can the group A_0 be realized as a group of homeomorphisms of S; which is to say that to each class $a \in A_0$, there exists a homeomorphism $f_a : S \to S$ such that $\{f_a; a \in A_0\}$ form a group of mappings of S.

In [Z3] Zieschang stated the following {a complete proof is in [Z1]}:

Theorem 6.2. ([Z3]) *Let G be a finite effective extension of a surface group K. Suppose that G satisfies the following condition.*

(i) Whenever $x^a = y^b = (xy)^c$, with $x, y \in G, a, b, c \geq 2$, then x, y generate a cyclic subgroup.

Then G is isomorphic to a finitely generated planar discontinuous group.

Using a somewhat different approach Fenchel proved:

Theorem 6.3. ([Fe1],[Fe2]) *Let G be a finite effective extension of a surface group K. Suppose that G/K is a solvable group. Then G is isomorphic to a finitely generated planar discontinuous group.*

From a different viewpoint Eckman and Müller [EM] considered Poincaré duality groups {see [Z1]}. They proved:

Theorem 6.4. ([EM]) *Let G be a finite extension of a surface group K such that the first Betti number of G is positive. Then G is isomorphic to a planar discontinuous group.*

In preparation for Zieschang's extension of Theorem 6.2 consider the triangle group $D^\star$ with presentation $< u, v; u^a = v^b = (uv)^c = 1 >$. A *semi-triangle group* is a two-generator non-cyclic group $D =< x, y >$ of isometries of the hyperbolic plane such that x, y, xy have finite orders $a, b, c \geq 2$, respectively, and there is an epimorphism $\phi : D^\star \to D$ with torsion-free kernel. Now suppose K is a Fuchsian group and G is a finite effective extension represented as isometries of the hyperbolic plane. Zieschang [Z1] then gives the following extension of Theorem 6.2.

Theorem 6.5. ([Z1]) *Suppose G and K are as above.*

(a) *If all elements of G preserve orientation then the following is a complete list of possibilities:*
   (a1) *G can be realized by a Fuchsian group.*
   (a2) *G is a semi-triangle group. Each normal subgroup of G is torsion-free. Moreover, either G contains a triangle group or else no proper subgroup of G is a semi-triangle group.*

(b) *If G contains orientation reversing elements then either G is a non-Euclidean crystallographic group or else the subgroup of index 2 of orientation preserving elements is of the type described in (a2).*

A complete positive solution to the Nielsen realization problem was finally given in the early 1980's by Kerchoff [Ke], Eckman and Müller [EM] and Eckman and Linnell [EL]. {See also Collins and Zieschang [CZ].}

Theorem 6.6. *Every finite subgroup of the homeotopy group of a surface can be realized by a finite group of homeomorphisms. Every finite effective extension of a planar group is a planar group.*

In all the above work on the Nielsen realization problem, only effective extensions have been considered since this was necessary for F-group extensions. Group theoretically however there are many non-effective finite extensions - any direct product of a surface group with a finite group will do. A question of interest is whether in some sense only the direct products with finite groups are the non-effective extensions.

Lemma 6.1. *Let K be a surface group and G a finite extension of K. Then $C_G(K)$ {the centralizer of K in G} is finite and $< K, C_G(K) > \cong K \times C_G(K)$.*

PROOF. Since K is centerless we must have $K \cap C_G(K) = \{1\}$ and since $|G : K| < \infty$ we have $|C_G(K)| < \infty$. Further since $K \cap C_G(K) = \{1\}$ we must have $< K, C_G(K) > \cong K \times C_G(K)$. □

Lemma 6.2. *Suppose G is a finite extension of a surface group K satisfying condition (i) in Theorem 6.2. If $C_G(K)$ has a complement G_1 in G then G_1 is a planar discontinuous group.*

PROOF. Since $G = G_1C_G(K)$ with $|C_G(K)| < \infty$, K has finite index in G_1. Since $G_1 \cap C_G(K) = \{1\}, G_1$ is a finite effective extension of K. Since G satisfies condition (i) of Theorem 6.2 so does G_1. Therefore Theorem 6.2 applies to show that G_1 is a planar discontinuous group. □

Question 1. Suppose that the centralizer of a surface group K in its finite extension G satisfying the conditions of Theorem 6.2 has a planar discontinuous group G_1 as a complement. Must G_1 necessarily be normal in G - that is, must G be a split extension of a planar discontinuous group by a finite group centralizing a surface group subgroup?

Question 2. Suppose G is a finite extension of $H \times K$ with H finite and K a surface group. Under what conditions is G a split extension of a planar discontinuous group K_1 by a subgroup H_1 of H?

7. Problems and questions

We now present a list of problems and questions on virtually-one-relator and one-relator-by-finite groups. It is quite likely that deeper involvement with some of the suggested questions will reveal the need for more precise or restrictive formulation or that we failed to notice readily available answers. Still, we hope that those which do not appear intractable can hold some interest and inspire further contributions. We have included also the major issues to consider in the proposed program for virtually one-relator groups mentioned in the introductory section.

(1) Is there a structure theorem for one-relator-by finite groups analogous to that for free-by-finite groups? More specifically, what restrictions or conditions must be placed on either the one-relator group or on the nature of the extension to develop such a structure theorem?

(2) Is it possible to give a purely algebraic proof of the Nielsen realization theorem? If a group is a finite extension of a surface group does it have to be isomorphic to either a Fuchsian group or an NEC group? If so, can a presentation demonstrating that fact be found?

(3) If G is a one-relator group with the property that every subgroup of finite index is again a one-relator group, must G be a surface group [Ko]?

(4) Which, if any, of the classical decision problems have positive solutions for virtually-one-relator groups?

(5) What can be said about the residual properties of (which) virtually-one-relator groups?

(6) It is known [Dy] that a non-trivial free group is not isomorphic to the (full) automorphism group of any group. Which virtually free groups have the same property? Which virtually one-relator groups have it?

(7) For which virtually one-relator groups can the center be determined in the sense of the Baumslag-Taylor result for one-relator groups?

(8) Can the intersection of all one-relator subgroups of a given virtually one-relator group be determined?

(9) When is the set of isomorphism classes of the one-relator subgroups of finite index in a (virtually) one-relator group finite?

(10) Does the isomorphism problem for virtually-one-relator groups have a solution in a sense similar to that of one-relator groups? The question seems more approachable for one-relator-by-finite groups.

(11) What further analogues of the Whitehead theorem - about recognizing from a one-relator presentation that the group is actually free - go over to virtually-free, virtually one-relator, virtually free-by-cyclic, etc. groups?

(12) The generalized free product of two finitely generated abelian groups with one subgroup amalgamated is virtually residually free. When is such a generalized free product of virtually one-relator groups (virtually) residually free?

(13) For which class **C** of virtually one relator groups is it true that vv**C** implies v**C**?

(14) Which virtually one-relator groups are residually one-relator groups?

(15) Certain cyclically pinched one-relator groups are subgroup separable. Which virtually one-relator groups are subgroup separable?

(16) Fuchsian groups are conjugacy separable [FR2]. Which virtually F-groups are conjugacy separable?

(17) Suppose G is a finite extension of a surface group K satisfying condition (i) in Theorem 6.2. If $C_K(G)$ has a complement G_1 in G must G be a split extension of a planar discontinuous group by a finite group centralizing a surface group subgroup {see Section 6}?

(18) Suppose G is a finite extension of $H \times K$ with H finite and K a surface group. Under what conditions is G a split extension of a planar discontinuous group K_1 by a subgroup H_1 of H {see Section 6}?

(19) Can the index be found from the information that a specific subgroup is a one-relator subgroup of a given virtually one-relator group?

(20) Which virtually one-relator groups have the torsion-free subgroup property?

Acknowledgements. The authors are pleased to acknowledge the support of the New York University Faculty Resource Network Scholars-in-Residence program which sponsored the collaboration during the summer of 1993. We also thank the Courant Institute of Mathematical Sciences for their kind hospitality, and especially, Sylvain Cappell for many enlightening conversations and Ed Friedman for technical assistance.

References

[ABC] J. Alonzo, T. Brady, D. Cooper,T. Delzant, V. Ferlini, M. Lustig, M. Mihalik, M. Shapiro and H. Short, *Notes on Negatively Curved Groups*, MSRI preprint, 1989.

[ABS] J.M. Autebert, L. Boasson and G. Senizergues, Groups and NTS languages *J. Comput. System Sci.* **35**(1987), 243–267.

[B1] G. Baumslag, Finitely generated cyclic extensions of free groups are residually finite, *Bull. Austral. Math. Soc.* **5**(1971), 87–94.

[B2] G. Baumslag, Groups with one defining relator, *Bull. Austral. Math. Soc.* **4**(1964), 385–392.

[BT] G. Baumslag and T. Taylor, The centre of groups with one defining relator, *Math. Ann.* **175**(1968), 315–319.

[BGSS] G. Baumslag, S. Gersten, H. Short and M. Shapiro, Automatic Groups and amalgams, in *Algorithms and classification in Combinatorial Group Theory* (Springer-Verlag, 1992), 175–195.

[BeR] J. Beidelman and D. Robinson, On the structure of the normal subgroups of a group: nilpotency, to appear.

[CM] B. Chandler and W. Magnus, *The History of Combinatorial Group Theory* (Springer-Verlag, 1982).

[C1] D. Cohen, *Groups of Cohomological Dimension One* (Lecture Notes in Math. **245**, Springer-Verlag, 1972).

[C2] D. Cohen, *Combinatorial Group Theory: a Topological Appraoch* (London Mathematical Society Student Texts **14**, 1989).

[CZ] D. Collins and H. Zieschang, Combinatorial Group Theory and Fundamental Groups, in *Algebra VII - Encyclopedia of Mathematical Sciences 58* (Springer-Verlag, 1993).

[Cu] P. M. Curran, Subgroups of finite index in certain classes of finitely presented groups, *J. Algebra* **122**(1989), 118–129.

[D] M.J. Dunwoody, The accessibility of finitely presented groups, *Invent. Math.* (1985), 449–457.

[Dy] J. Dyer, A remark on automorphism groups, *Contemp. Math.* **33**(1984), 208-211.

[EL] B. Eckman and P.A. Linnell, Poincaré duality groups of dimension 2, *Comment. Math. Helvetici* **58**(1983), 111–114.

[EM] B. Eckman and H. Müller, *Virtual surface groups*, ETH preprint, 1980.

[EM] B. Eckman and H. Müller, Plane motion groups and virtual Poincaré duality of dimension 2, *Invent. Math.* **69**(1982), 293–310.

[Fe1] W. Fenchel, Estensioni gruppi descontinui e transformazioni periodiche delle surficie, *Rend. Accad. Naz. Lincei* **5**(1948), 326–329.

[Fe2] W. Fenchel, Bemarkogen om endliche gruppen af abbildungsklasser, *Mat. Tidskrift* B (1950), 90–95.

[F] B. Fine, Groups whose torsion-free subgroups are free, *Bull. Acad. Sin.* **12**(1984), 31–36.

[FHR] B.Fine,J.Howie and G.Rosenberger, One-relator quotients and free products of cyclics, *Proc. Amer. Math. Soc.* **102**(1988) 1–5.

[FR1] B.Fine and G.Rosenberger, Complex representations and one-relator products of cyclics, *Contemp. Math.* **74**(1987), 131–149.

[FR2] B.Fine and G.Rosenberger, *Generalizing Algebraic Properties of Fuchsian Groups* (London Math. Soc. Lecture Notes Series **159**, 1, 1989), 124-128.

[FR3] B.Fine and G.Rosenberger, Conjugacy separability of Fuchsian groups and related questions, *Contemp. Math.* **109**(1990), 11–18.

[Ge] S.M. Gersten, *Essays in Group Theory* (MSRI Publications 8, Springer-Verlag, 1987).

[G] R.Gilman, Verifying that a group is virtually free, *Internat. J. Algebra Comput.* **1**(1991), 339–351.

[Gre] R. Gregorac, On generalized free products of finite extensions of free groups, *J. London Math. Soc.* **41**(1966), 662-666.

[Gru] K. Gruenberg, The hypercentre of linear groups, *J. Algebra* **8**(1968), 34–40.

[HT] T. Herbst and R.M. Thomas, *Group presentations,formal languages and characterizations of one-counter groups* (Technical Report CSD-47 U. Leicester Dept. of Computing Studies, 1991).

[HKS1] A. Hoare, A. Karrass and D. Solitar, Subgroups of Finite Index of Fuchsian Groups, *Math. Z.* **120**(1971), 289–298.

[HKS2] A. Hoare, A. Karrass and D. Solitar, Subgroups of infinite index in fuchsian groups, *Math. Z.* **125**(1972), 59–69.

[HU] J.D. Hopcroft and J.D. Ullman, *Introduction to Automata Theory* (Addison-Wesley, 1979).

[J] D.L. Johnson, *Presentations of Groups* (London Math. Society Student Texts **15**, 1990).

[KPS] A.Karrass, A. Pietrowski and D.Solitar, Finite and Infinite Cyclic Extensions of Free Groups, *J. Austral. Math. Soc.* **16**(1972), 458–466.

[Ke] S.P. Kerchoff, The Nielsen realization problem, *Ann. Math.* **117**(1983), 235–265.

[Kr] S. Krstic, Finitely generated virtually free groups, *Proc. London Math. Soc.* **64**(1992), 49–69.

[Ko] V. Mazurov (ed.), *The Kourovka Notebook: Unsolved Problems in Group Theory* (AMS Translations 121).

[Le] J.C. Lennox, The Fitting-Gaschutz-Hall relation in certain soluble by finite groups, *J. Algebra* **24**(1973), 219-225.

[LMR] M. Lustig, Y. Moriah and G. Rosenberger, Automorphisms of Fuchsian groups and their lifts to free groups, *Canad. J. Math* **41**(1989), 123–131.

[LS] R. Lyndon and P. Schupp, *Combinatorial Group Theory* (Springer-Verlag, 1977).

[M] W. Magnus, The uses of 2 by 2 matrices in combinatorial group theory: a survey, *Resultate Math.* **4**(1981), 171–192.

[Mc] J. McCool, The automorphism groups of finite extensions of free groups, *Bull. London Math. Soc.* **20**(1988), 131-135.

[Mu] T. Müller, Combinatorial aspects of finitely generated virtually free groups (London Math. Soc. Lecture Notes Series **160**, 1989), 386–395.

[MS] D.E. Müller and P.E. Schupp, Groups, the thoery of ends and context-free languages, *J. Comput. System Sci.*, **26**(1983), 295–310.

[N] M. Newman, A note on fuchsian groups, *Illinois J. Math.* **29**(1985), 682–686.

[Ni] J.Nielsen, Abbildungsklassen enlicher ordnung, *Acta Math.* **75**(1943), 23–115.

[R] D.J.S. Robinson, *A Course in the Theory of Groups* (Springer-Verlag, 1982).

[Sch] H. R. Schneebeli, On virtual properties and group extensions, *Math. Z.* **159**(1978), 159–167.

[St1] J. Stallings, On torsion-free groups with infinitely many ends, *Ann. of Math.* **88**(1968), 312–334.

[St2] J. Stallings, Groups of cohomological dimension one, in *Proceedings of Symposia in Pure Mathematics XVII* (AMS, 1970), 124–128.

[W] B.A.F. Wehrfritz, *Infinite Linear Groups* (Springer-Verlag, 1973).

[Z1] H. Zieschang, *Finite Groups of Mapping Classes of Surfaces* (Lecture Notes in Math. **875**, Springer-Verlag, 1980).

[Z2] H.Zieschang, On extensions of fundamental groups of surfaces and related groups, *Bull. Amer. Math. Soc.* **77**(1971), 1116–1119.

[Z2] H.Zieschang, Addendum to: On extensions of fundamental groups of surfaces and related groups, *Bull. Amer. Math. Soc.* **80**(1974), 366–367.

[ZVC] H. Zieschang, H. Vogt and E. Coldeway, *Surfaces and Planar Discontinuous Groups* (Springer-Verlag, 1980).

RICKARD EQUIVALENCES AND BLOCK THEORY

M. BROUÉ

École Normale Supérieure, L.M.E.N.S.–D.M.I. (C.N.R.S., U.A. 762), 45 rue d'Ulm, F–75005 Paris, France
E-mail: Michel.Broue@ens.fr

1991 Mathematics Subject Classification: 20, 20G.

1. Introduction

Control of fusion

Let G be a finite group, and let p be a prime number.

Definition 1.1. We say that a subgroup H of G controls the fusion of p-subgroups of G if the following two conditions are fulfilled:

(C1) H contains a Sylow p-subgroup S_p of G,

(C2) whenever P is a subgroup of S_p and g is an element of G such that $gPg^{-1} \subseteq S_p$, there exist z in the centralizer $C_G(P)$ of P in G, and h in H, such that $g = hz$.

Example 1.2. (The basic example) We denote by $O_{p'}(G)$ the largest normal subgroup of G with order prime to p. Then if H is a subgroup of G which "covers the quotient" $G/O_{p'}(G)$ (*i.e.*, if $G = HO_{p'}(G)$), then H controls the fusion of p-subgroups of G.

The following two results provide fundamental examples where the converse is true. The first one is due to Frobenius and was proved in 1905. The second one was proved by Glauberman for the case $p = 2$ (see [Gl]), and for odd p it is a consequence of the classification of non abelian finite simple groups (see also [Ro] for an approach not using the classification).

Theorem 1.3.

(Fr) Assume that a Sylow p-subgroup S_p of G controls the fusion of p-subgroups of G. Then $G = S_pO_{p'}(G)$.

(Gl) Assume that there exists a p-subgroup P of G whose centralizer $C_G(P)$ controls the fusion of p-subgroups of G. Then $G = C_G(P)O_{p'}(G)$.

Groups with abelian Sylow p-subgroups

A classical example where a subgroup controls the fusion of p-subgroups is given by an old result of Burnside:

> *Assume that the Sylow p-subgroups of G are abelian. Let H be the normalizer of one of them. Then H controls the fusion of p-subgroups of G.*

If G is p-solvable, this is once again a particular case of 1.2, since it is not difficult to prove the following result.

Proposition 1.4. *Let G be p-solvable and let H be the normalizer of a Sylow p-subgroup. If the Sylow p-subgroups of G are abelian, then $G = HO_{p'}(G)$.*

The situation may look quite different if G is a non abelian simple group with abelian Sylow p-subgroups. Indeed, in this case, we have $G \neq HO_{p'}(G)$ whenever H normalizes a non-trivial p-subgroup of G.

For example, there seems to be an enormous difference between the Monster group, a non abelian simple group of order

$$2^{46} \cdot 3^{20} \cdot 5^{9} \cdot 7^{6} \cdot 11^{2} \cdot 13^{3} \cdot 17 \cdot 19 \cdot 23 \cdot 29 \cdot 31 \cdot 41 \cdot 47 \cdot 59 \cdot 71 \simeq 8.10^{53},$$

and the normalizer of one of its Sylow 11-subgroups, a group of order 72600, isomorphic to $(C_{11} \times C_{11}) \rtimes (C_5 \times \mathrm{SL}_2(5))$ (here we denote by C_m the cyclic group of order m).

Nevertheless, there still is a strong connection between G and the normalizer of a Sylow p-subgroup, which is a kind of generalization of the "factorization situation" given by Theorem 1.3. In order to express this connection, we need to introduce the language of block theory.

The principal block

Let K be a finite extension of the field of p-adic numbers $\mathbb{Q}_p$ which contains the $|G|$-th roots of unity. Thus the group algebra KG is a split semi-simple K-algebra. Let $\mathcal{O}$ be the ring of integers of K over $\mathbb{Z}_p$. We denote by $\mathfrak{p}$ the maximal ideal of $\mathcal{O}$, and we set $k := \mathcal{O}/\mathfrak{p}$. If JkG denotes the Jacobson radical of the group algebra kG, the algebra kG/JkG is a split semi-simple k-algebra.

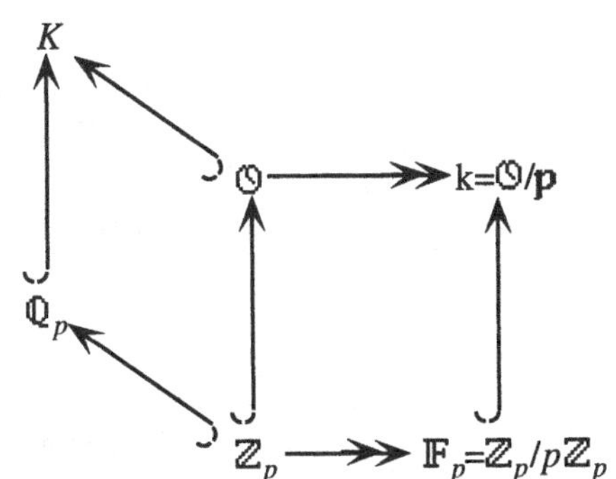

The decomposition of the unity element of $\mathcal{O}G$ into a sum of orthogonal primitive central idempotents $1 = \sum e$ corresponds to the decomposition of the algebra $\mathcal{O}G$ into a direct sum of indecomposable two-sided ideals $\mathcal{O}G = \bigotimes B$ ($B = \mathcal{O}Ge$), called the *blocks* of $\mathcal{O}G$. For B a block of $\mathcal{O}G$, we set $KB := K \otimes_{\mathcal{O}} B$ and $kB := k \otimes_{\mathcal{O}} B$.

By reduction modulo $\mathfrak{p}$, a primitive central idempotent remains primitive central, and consequently $kG = \bigotimes kB$ is still a decomposition into a direct sum of indecomposable two-sided ideals, called the blocks of kG.

$$\begin{array}{ccccc} \mathcal{O}G & = & \bigotimes & B \\ \downarrow & & & \downarrow \\ kG & = & \bigotimes & kB \end{array}$$

The augmentation map $\mathcal{O}G \to \mathcal{O}$ factorizes through a unique block of $\mathcal{O}G$ called *the principal block* and denoted by $B_p(G)$.

If B is a block of $\mathcal{O}G$, we denote by $\mathrm{Irr}(B)$ the set of all isomorphism classes of irreducible representations of the algebra KB. The set $\mathrm{Irr}(B)$ will be identified with a subset of the set $\mathrm{Irr}(G)$ of characters of irreducible representations of G over K.

For $\chi \in \mathrm{Irr}(G)$, we denote by $\ker(\chi)$ the kernel of the corresponding representation of G. In other words, we have

$$\ker(\chi) = \{z \in G\,;\ (\forall g \in G)(\chi(zg) = \chi(g))\}.$$

The factorization $G = HO_{p'}(G)$ can be interpreted in terms of principal blocks as follows.

Proposition 1.5.

(1) *We have* $\bigcap_{\chi \in \mathrm{Irr}(B_p(G))} \ker(\chi) = O_{p'}(G)$.

(2) *Let H be a subgroup of G. Assume that $G = HO_{p'}(G)$. Then the map* Res_H^G *induces a bijection*

$$\mathrm{Res}_H^G\ :\ \mathrm{Irr}(B_p(G)) \xrightarrow{\sim} \mathrm{Irr}(B_p(H)).$$

What happens in the general case (where the Sylow p-subgroups of G are abelian and H is the normalizer of one of them) will first be illustrated by the example of the group $G = \mathfrak{A}_5$. The bijection Res_H^G of the previous proposition is replaced by a "bijection with signs" between $\mathrm{Irr}(B_p(G))$ and $\mathrm{Irr}(B_p(H))$.

The case of $G = \mathfrak{A}_5$

Let G be the alternating group on five letters. Then $|G| = 2^2 \cdot 3 \cdot 5$, and for all prime number p which divides $|G|$, the Sylow p-subgroups of G are abelian. Let us examine the principal p-blocks of G and their connections with the corresponding Sylow normalizers.

	(1)	(2)	(3)	(5)	(5′)
1	1	1	1	1	1
χ_4	4	0	1	-1	-1
χ_5	5	1	-1	0	0
χ_3	3	-1	0	$(1+\sqrt{5})/2$	$(1-\sqrt{5})/2$
χ_3'	3	-1	0	$(1-\sqrt{5})/2$	$(1+\sqrt{5})/2$

Table 1.6. (Character table of $\mathfrak{A}_5$)

$$\begin{aligned}\operatorname{Irr}(B_2(G)) &= \{1\,,\,\chi_5\,,\,\chi_3\,,\,\chi_3'\}\,,\\ \operatorname{Irr}(B_3(G)) &= \{1\,,\,\chi_4\,,\,\chi_5\}\,,\\ \operatorname{Irr}(B_5(G)) &= \{1\,,\,\chi_4\,,\,\chi_3\,,\,\chi_3'\}\,.\end{aligned}$$

For each $p \in \{2,3,5\}$, let us denote by S_p a Sylow p-subgroup of G. We shall point out that there exists an isomorphism

$$I_p\ \mathbb{Z}\operatorname{Irr}(B_p(N_G(S_p))) \xrightarrow{\sim} \mathbb{Z}\operatorname{Irr}(B_p(G))$$

such that:

(I1) I_p is an isometry,

(I2) it preserves the character degrees modulo p,

(I3) it preserves the values on the p-elements.

The preceding properties express the fact that I_p is an *isotypy* between $B_p(G)$ and $B_p(N_G(S_p))$, as we shall explain below in §2.

The case $p = 2$

The normalizer $N_G(S_2)$ of a Sylow 2-subgroup S_2 of G is isomorphic to the alternating group $\mathfrak{A}_4$. The principal block $B_2(N_G(S_2))$ is its only 2-block. Let us change the sign of certain irreducible characters in its character table.

	(1)	(2)	(3)	(3′)
1	1	1	1	1
$-\alpha_3$	-3	1	0	0
$-\alpha_1$	-1	-1	$(1+\sqrt{-3})/2$	$(1-\sqrt{-3})/2$
$-\alpha_1'$	-1	-1	$(1-\sqrt{-3})/2$	$(1+\sqrt{-3})/2$

Table 1.7. (Character table of $\mathfrak{A}_4$)

We denote by I_2 the map

$$I_2 : \begin{pmatrix} 1 \\ -\alpha_3 \\ -\alpha_1 \\ -\alpha_1' \end{pmatrix} \mapsto \begin{pmatrix} 1 \\ \chi_5 \\ \chi_3 \\ \chi_3' \end{pmatrix}.$$

The case $p = 3$

The normalizer $N_G(S_3)$ of a Sylow 3-subgroup S_3 of G is isomorphic to the symmetric group $\mathfrak{S}_3$. The principal block $B_3(N_G(S_3))$ is its only 3-block.

	(1)	(2)	(3)
1	1	1	1
β_1	1	−1	1
β_2	2	0	−1

Table 1.8. (Character table of $\mathfrak{S}_3$)

We denote by I_3 the map

$$I_3 : \begin{pmatrix} 1 \\ \beta_1 \\ \beta_2 \end{pmatrix} \mapsto \begin{pmatrix} 1 \\ \chi_4 \\ \chi_5 \end{pmatrix}.$$

The case $p = 5$

The normalizer $N_G(S_5)$ of a Sylow 5-subgroup S_5 of G is isomorphic to the dihedral group D_5. The principal block $B_5(N_G(S_5))$ is its only 5-block. Let us change the sign of certain irreducible characters in its character table.

	(1)	(2)	(5)	(5′)
1	1	1	1	1
$-\gamma_1$	−1	1	−1	−1
$-\gamma_2$	−2	0	$(1+\sqrt{5})/2$	$(1-\sqrt{5})/2$
$-\gamma_2'$	−2	0	$(1-\sqrt{5})/2$	$(1+\sqrt{5})/2$

Table 1.9. (Character table of D_5)

We denote by I_5 the map

$$I_5 : \begin{pmatrix} 1 \\ -\gamma_1 \\ -\gamma_2 \\ -\gamma_2' \end{pmatrix} \mapsto \begin{pmatrix} 1 \\ \chi_4 \\ \chi_3 \\ \chi_3' \end{pmatrix}.$$

Remark. We can also notice that

- if one changes 5 to -3, then $I_2(\alpha)$ takes also on the 3-elements of G the same values as α takes on the 5-elements of $N_G(S_2)$,
- $I_3(\beta)$ takes also on the 2-elements of G the same values as β takes on the 5-elements of $N_G(S_3)$,
- $I_5(\gamma)$ takes also on the 2-elements of G the same values as γ takes on the 3-elements of $N_G(S_5)$,

i.e., in other words, if $\{p, q, r\} = \{2, 3, 5\}$, not only do ζ and $I_p(\zeta)$ take the same values on non trivial p-elements, but (with suitable trick for $p = 2$) they exchange values on non trivial q-elements and r-elements. This last property will not be explained in what follows.

2. Isotypies

The case of $\mathfrak{A}_5$ provides particular examples of what should replace the factorization type theorem (see above Theorem 1.3 and Proposition 1.3 in the case where H is the normalizer of an abelian Sylow p-subgroup.

Notation

Various class functions.

For $R = K$ or $\mathcal{O}$, we denote by $\mathrm{CF}(G, R)$ the R-module of all class functions from G into R. For B a block of G, we denote by $\mathrm{CF}(B, K)$ the subspace of $\mathrm{CF}(G, K)$ consisting of functions which are linear combination of characters of KB, and we set $\mathrm{CF}(B, \mathcal{O}) := \mathrm{CF}(G, \mathcal{O}) \cap \mathrm{CF}(B, K)$.

We denote by $\mathrm{CF}_{p'}(G, R)$ the submodule of $\mathrm{CF}(G, R)$ consisting of all class functions on G which vanish outside the set $G_{p'}$ of p'-elements of G. For B a block of G, we set $\mathrm{CF}_{p'}(B, R) := \mathrm{CF}_{p'}(G, R) \cap \mathrm{CF}(B, R)$.

Remark. The Brauer character of a kG-module is usually defined as a class function on $G_{p'}$. Extending it by 0 outside $G_{p'}$, *we shall make here the convention that the Brauer characters are elements of* $\mathrm{CF}_{p'}(G, \mathcal{O})$ (hence elements of $\mathrm{CF}_{p'}(B, \mathcal{O})$ for the characters of B-modules). It results from our convention that the set $\mathrm{BrIrr}(B)$ of Brauer irreducible characters of B-modules is an $\mathcal{O}$-basis of $\mathrm{CF}_{p'}(B, \mathcal{O})$.

Finally, we denote by $\mathrm{CF}_{p'}^{\mathrm{pr}}(G, \mathcal{O})$ the dual submodule of $\mathrm{CF}_{p'}(G, \mathcal{O})$ in $\mathrm{CF}_{p'}(G, K)$, consisting of all elements of $\mathrm{CF}_{p'}(G, K)$ whose scalar product with the elements of $\mathrm{CF}_{p'}(G, \mathcal{O})$ belongs to $\mathcal{O}$, and we set $\mathrm{CF}_{p'}^{\mathrm{pr}}(B, \mathcal{O}) := \mathrm{CF}_{p'}^{\mathrm{pr}}(G, \mathcal{O}) \cap \mathrm{CF}(B, K)$. From our previous convention it follows that the set $\mathrm{Prim}(B)$ of characters of indecomposable projective $\mathcal{O}G$-modules is an $\mathcal{O}$-basis of $\mathrm{CF}_{p'}^{\mathrm{pr}}(G, \mathcal{O})$.

Decomposition maps

For x a p-element of G, we denote by

$$d_G^x : \mathrm{CF}(G, \mathcal{O}) \to \mathrm{CF}_{p'}(C_G(x), \mathcal{O})$$

the linear map defined by

$$d_G^x(\chi)(y) := \begin{cases} \chi(xy) & \text{if } y \text{ is a } p\text{-regular element of } C_G(x) \\ 0 & \text{if } y \text{ is a } p\text{-singular element of } C_G(x). \end{cases}$$

It results from Brauer's second and third main theorems (see for example [Fe]) that the map d_G^x sends $\mathrm{CF}(B_p(G), \mathcal{O})$ into $\mathrm{CF}_{p'}(B_p(C_G(x)), \mathcal{O})$, and so induces by restriction a map still denoted by

$$d_G^x : \mathrm{CF}(B_p(G), \mathcal{O}) \to \mathrm{CF}_{p'}(B_p(C_G(x)), \mathcal{O}).$$

Isotypies

From now on, the following hypothesis and notation will be in force : We denote by G a finite group whose Sylow p-subgroups are *abelian*. We denote by S_p one of the Sylow p-subgroups, and we set $H := N_G(S_p)$.

The following definition is a slight modification of the analogous definition given in [Br1] (see Remarque 2 following definition 4.6 in *loc. cit.*).

Remark. A more general definition is available for non-principal blocks with *abelian defect groups*. Its statement requires the use of the "local structure" associated with a block (see [AlBr] or [Br4]). We refer the reader to [Br1] for details.

Definition 2.1. An isotypy I between $B_p(G)$ and $B_p(H)$ is the datum, for every p-subgroup P of S_p, of a bijective isometry

$$I(P) : \mathbb{Z}\mathrm{Irr}(B_p(C_H(P))) \xrightarrow{\sim} \mathbb{Z}\mathrm{Irr}(B_p(C_G(P)))$$

such that the following conditions are fullfilled:

(Equi) (Equivariance) For all $h \in H$, we have $I(P)^h = I(P^h)$.

(Com) (Compatibility condition) For every subgroup P of S_p and every $x \in S_p$, we still denote by

$$I(P) : \mathrm{CF}(B_p(C_H(P)), K) : \xrightarrow{\sim} \mathrm{CF}(B_p(C_G(P)), K)$$

the bijective isometry defined by linear extension of $I(P)$. The following diagram is commutative:

$$\begin{array}{ccc} CF(B_p(C_H(P)),K) & \xrightarrow{I(P)} & CF(B_p(C_G(P)),K) \\ \downarrow d^x_{C_H(P)} & & \downarrow d^x_{C_G(P)} \\ CF(B_p(C_H(P<x>))K) & \xrightarrow{I(P<x>)} & CF(B_p(C_G(P<x>)),K) \end{array}$$

(Triv) $I(S_p)$ is the identity map.
Moreover, we say that the isotypy I is normalized if $I(P)(1_{C_H(P)}) = 1_{C_G(P)}$ for all $P \subseteq S_p$.

Let us list some of the straighforward properties of an isotypy. In what follows, for every subgroup P of S_p we denote by $R(P)$ the inverse map of $I(P)$. We set $I(1) = I(\{1\})$ and $R(1) = R(\{1\})$.

Proposition 2.2.

(Loc) (Local isotypies) Let P be any subgroup of S_p. Set $I_P(Q) := I(PQ)$. The collection of maps $(I_P(Q))_{Q \subseteq S_p}$ defines an isotypy between $B_p(C_H(P))$ and $B_p(C_G(P))$.

(Int) (Integrality property) By restriction, the map $I(1)$ induces bijective isometries

$$\begin{cases} I(1) : \mathrm{CF}(B_p(H), \mathcal{O}) & \xrightarrow{\sim} \mathrm{CF}(B_p(G), \mathcal{O}) \\ I(1) : \mathrm{CF}_{p'}(B_p(H), \mathcal{O}) & \xrightarrow{\sim} \mathrm{CF}_{p'}(B_p(G), \mathcal{O}) \\ I(1) : \mathrm{CF}^{\mathrm{pr}}_{p'}(B_p(H), \mathcal{O}) & \xrightarrow{\sim} \mathrm{CF}^{\mathrm{pr}}_{p'}(B_p(G), \mathcal{O}). \end{cases}$$

PROOF. (Loc) is obvious. We give a proof of (Int). By 2.1, (Com), applied with $x = 1$, we see first that $I(1)$ sends $\mathrm{CF}_{p'}(B_p(H), K)$ into $\mathrm{CF}_{p'}(B_p(G), K)$. Moreover, since, for any finite group G, the ordinary decomposition map $d^1_G : \mathbb{Z}\mathrm{Irr}(B_p(G)) \to \mathbb{Z}\mathrm{BrIrr}(B_p(G))$ is onto, and since $\mathrm{BrIrr}(B_p(G))$ is an $\mathcal{O}$-basis of $\mathrm{CF}_{p'}(B_p(G), \mathcal{O})$, we see that $I(1)$ and $R(1)$ define inverse isometries between $\mathrm{CF}_{p'}(B_p(H)), \mathcal{O})$ and $\mathrm{CF}_{p'}(B_p(G), \mathcal{O})$. It then follows by adjunction that they induce inverse isometries between $\mathrm{CF}^{\mathrm{pr}}_{p'}(B_p(H), \mathcal{O})$ and $\mathrm{CF}^{\mathrm{pr}}_{p'}(B_p(G), \mathcal{O})$.

To check that $I(1)$ and $R(1)$ induce inverse isomorphisms between the modules $\mathrm{CF}(B_p(H), \mathcal{O})$ and $\mathrm{CF}(B_p(G), \mathcal{O})$, it suffices to check that the image of $\mathrm{CF}(B_p(H), \mathcal{O})$ under $I(1)$ is contained in $\mathrm{CF}(B_p(G), \mathcal{O})$, *i.e.* that, for $\zeta \in \mathrm{CF}(B_p(H), \mathcal{O})$ and $g \in G$, we have $I(1)(\zeta)(g) \in \mathcal{O}$. Let x be the p-component of g and let x' be its p'-component. We have $I(1)(\zeta)(g) = (I(\langle x \rangle)(d^x_H(\zeta)))\,(x')$ and the result follows from the fact that $I(\langle x \rangle)$ sends $\mathrm{CF}_{p'}(C_H(x), \mathcal{O})$ into $\mathrm{CF}_{p'}(C_G(x), \mathcal{O})$, by the "local isotypies" property (Loc) and what precedes. □

Remark. The integrality properties (Int) show that the map $I(1)$ is a "perfect isometry" as defined in [Br1].

It follows in particular (see [Br1] or [Br5]) that an isotypy induces an isomorphism between the associated "Cartan-decomposition" triangles between Grothendieck groups (see [Se] or [Br1] for the definitions of the triangles).

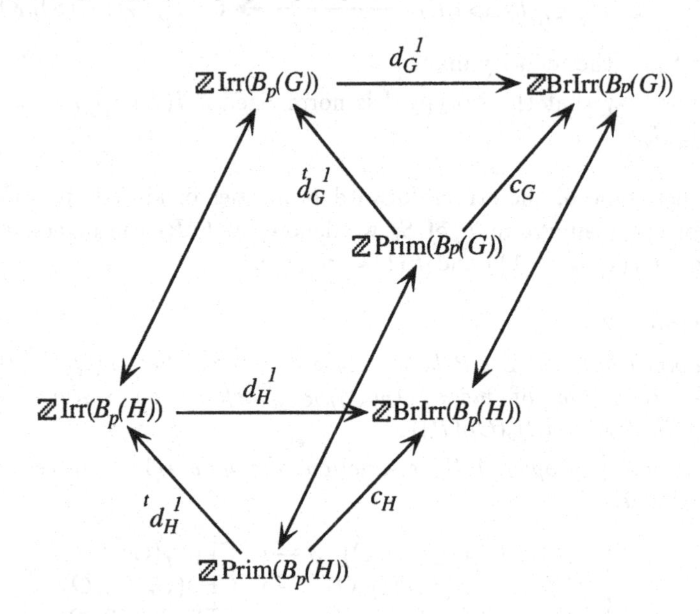

The character of an isotypy

Any linear map $F : \mathbb{Z}\mathrm{Irr}(H) \to \mathbb{Z}\mathrm{Irr}(G)$ between the character groups of H and G defines a character μ_F of $H \times G$ by the formula

$$\mu_F := \sum_{\zeta\in\mathrm{Irr}(H)} \zeta \otimes F(\zeta) .$$

Let $I = (I(P))_{P\subseteq S_p}$ be an isotypy between $B_p(G)$ and $B_p(H)$. We set $\mu_P := \mu_{I(P)}$.

Then the defining properties of an isotypy (*cf.* 2.1) translate as follows:

(Equi) For all $h \in H$, we have $\mu_P^h = \mu_{P^h}$.

(Com) $\mu_P(xx', yy') = \begin{cases} 0 & \text{if } x \text{ and } y \text{ are not conjugate in } G, \\ \mu_{P\langle x\rangle}(x', y') & \text{if } x = y . \end{cases}$

The following statement describes the case of a p-group. If G is an (abelian) p-group, then $H = G$, and $B_p(G) = \mathcal{O}G$.

Lemma 2.3. *Let G be an abelian p-group. Let $I = (I(P))_{P \subseteq G}$ be a "self-isotypy" of $\mathcal{O}G$. If $p = 2$, assume moreover that I is normalized. Then $I(P)$ is the identity for all p-subgroups P of G.*

PROOF. By definition of an isotypy (property (Triv)), we know that $I(G)$ is the identity. We prove by induction on $|G : P|$ that $I(P) = \mathrm{Id}$ for any subgroup P of G. Let P be a proper subgroup of G. By the induction hypothesis, we may assume that $I(P') = \mathrm{Id}$ whenever P' is a subgroup of G which strictly contains P, and we must prove that $I(P) = \mathrm{Id}$. For $x \in G$, the map d_G^x is identified with the map $\chi \mapsto \chi(x)$. For $\chi \in \mathrm{Irr}(G)$ we have $I(P)(\chi)(x) = d_G^x(I(P)(\chi)) = I(P\langle x\rangle)(d_G^x(\chi))$ and so $I(P)(\chi)(x) = \chi(x)$ for all $x \notin P$. Now let $y \in P$. For $x \notin P$, we have $xy \notin P$, so $I(P)(\chi)(xy) = \chi(xy)$. Set $I(P)(\chi) = \epsilon_\chi \chi'$ where $\epsilon_\chi = \pm 1$ and $\chi' \in \mathrm{Irr}(G)$. It follows that

$$\chi'(z) = \begin{cases} \epsilon_\chi \chi(z) & \text{for } z \notin P \\ \chi(z) & \text{for } z \in P\,. \end{cases}$$

For p odd, this implies that $\epsilon_\chi = 1$, and so $I(P) = \mathrm{Id}$.

Assume $p = 2$. Since I is normalized, $I(P)$ fixes the trivial character, and so for $x \in P$ we have $I(P)(\chi)(x) = I(P\langle x\rangle)(d_G^x(\chi)) = I(P)(d_G^x(\chi)) = d_G^x(\chi) = \chi(x)$ which also shows that $\epsilon_\chi = 1$ and that $I(P) = \mathrm{Id}$. □

Remark. If $G = \mathbb{Z}/2\mathbb{Z}$ and if $\mathrm{Irr}(G) = \{1, \sigma\}$, then $I := \{I(1), I(G)\}$ where $I(G) = \mathrm{Id}$ and

$$I(1) : \begin{cases} 1 \mapsto -\sigma \\ \sigma \mapsto -1 \end{cases}$$

is a non trivial self-isotypy.

A conjecture

The following conjecture is a particular case of a more general conjecture concerning blocks with abelian defect groups (see [Br1]).

Conjecture 2.4. Let G be a finite group whose Sylow p-subgroups are abelian. Let H be the normalizer of a Sylow p-subgroup. Then there is a normalized isotypy between $B_p(G)$ and $B_p(H)$.

The preceding conjecture is true if G is p-solvable, by 1.4 and 1.5, (2), above.

It has also been checked in the following cases.

- G is a symmetric group (Rouquier, [Rou1]), or an alternating group (Fong),
- G is a sporadic non abelian simple group (Rouquier, [Rou1]),

- G is a "finite reductive group" in non-describing characteristic (Broué-Malle-Michel, [BMM] and [BrMi]),
- $p = 2$ – and G is any finite group with abelian Sylow 2-subgroups (Fong-Harris, [FoHa]).

3. Rickard equivalences

We shall explain now why the existence of an isotypy between the principal blocks of two finite groups must be the "shadow" of a much deeper connection between the two blocks, which we call a "Rickard equivalence".

p-permutation modules and Rickard complexes

p-permutation modules and the Brauer functor

Let R denote either $\mathcal{O}$ or k. We call p-permutation RG-modules the summands of the permutation G-modules over R. The following characterization of p-permutation RG-modules is well known (see for example [Br3]).

Proposition 3.1. *The p-permutation RG-modules are the modules which, once restricted to a Sylow p-subgroup of G, are permutation modules.*

Let us denote by ${}_{RG}\mathbf{perm}$ the category of all p-permutation RG-modules. For P a p-subgroup of G, we set $\overline{N}_G(P) := N_G(P)/P$.

For Ω a finite G-set, we denote by Ω^P the set of fixed points of Ω under P, viewed as a $\overline{N}_G(P)$-set.

Proposition 3.2. *There is a functor*

$$\mathrm{Br}_P : {}_{\mathcal{O}G}\mathbf{perm} \longrightarrow {}_{k\bar{N}_G(P)}\mathbf{perm}$$

which "induces" the "fixed points" functor, i.e., *which is such that the diagram of natural transformations*

$$\begin{array}{ccc} {}_G\mathbf{sets} & \xrightarrow{.^P} & {}_{\bar{N}_G(P)}\mathbf{sets} \\ \downarrow & & \downarrow \\ {}_{\mathcal{O}G}\mathbf{perm} & \xrightarrow{\mathrm{Br}_P} & {}_{k\bar{N}_G(P)}\mathbf{perm} \end{array}$$

is commutative.

SKETCH OF PROOF. For X a p-permutation $\mathcal{O}G$-module and Q a p-subgroup of G, we define

$$\mathrm{Tr}_Q^P\, X^Q \to X^P \quad \text{by}$$
$$\mathrm{Tr}_Q^P(x) := \textstyle\sum_{g \in P/Q} g(x)\,.$$

We set

$$\mathrm{Br}_P(X) := (X/\mathfrak{p}X)^P / \sum_{Q \subsetneq P} \mathrm{Tr}_Q^P((X/\mathfrak{p}X)^Q)\,.$$

□

For V any kG-module, and g a p'-element of G, we denote by $\mathrm{Brtr}(g\,;\,V)$ the value at g of the Brauer character of V. Recall that we view the Brauer character $\mathrm{Brtr}(\cdot\,;\,V)$ as a class function on G vanishing outside the set $G_{p'}$ of p-regular elements of G. The following proposition generalizes to p-permutation modules a result which is well known for actual permutation modules.

Proposition 3.3. *Let X be a p-permutation $\mathcal{O}G$-module. If $g = g_p g_{p'}$ where g_p is a p-element, $g_{p'}$ is p-regular and $g_p g_{p'} = g_{p'} g_p$, then*

$$\mathrm{tr}(g\,;\,X) = \mathrm{Brtr}(g_{p'}\,;\,\mathrm{Br}_{\langle g_p \rangle}(X))\,.$$

Rickard complexes

Let us start with some notation related to complexes of modules.

Let

$$\Gamma := \left(\cdots \to 0 \to \Gamma^m \xrightarrow{d^m} \Gamma^{m+1} \xrightarrow{d^{m+1}} \cdots \xrightarrow{d^{m+a-1}} \Gamma^{m+a} \to 0 \to \cdots\right)$$

be a complex of modules on some $\mathcal{O}$-algebra. The $\mathcal{O}$-dual of Γ is by definition the complex

$$\Gamma^* := \left(\cdots \to 0 \to \mathrm{Hom}_{\mathcal{O}}(\Gamma^{m+a}, \mathcal{O}) \xrightarrow{{}^t d^{m+a-1}} \cdots \xrightarrow{{}^t d^m} \mathrm{Hom}_{\mathcal{O}}(\Gamma^m, \mathcal{O}) \to 0 \to \cdots\right).$$

From now on, the following hypothesis and notation will be in force:

> *We denote by G a finite group whose Sylow p-subgroups are* abelian.
> *We denote by S_p one of the Sylow p-subgroups, and we set $H := N_G(S_p)$.*

Definition 3.4. A Rickard complex for the principal blocks $B_p(G)$ and $B_p(H)$ is a bounded complex of $(B_p(G), B_p(H))$-bimodules

$$\Gamma := \left(\cdots \to 0 \to \Gamma^m \xrightarrow{d^m} \Gamma^{m+1} \xrightarrow{d^{m+1}} \cdots \xrightarrow{d^{m+a-1}} \Gamma^{m+a} \to 0 \to \cdots\right)$$

with the following properties:

(1) Each constituent Γ^n of Γ, viewed as an $\mathcal{O}[G \times H]$-module, is a p-permutation module with vertex contained in $\Delta_{G\times H^\circ}(S_p)$ (where $\Delta_{G\times H^\circ} : S_p \to G \times H$ is defined by $\Delta_{G\times H^\circ}(x) := (x, x^{-1})$).

(2) We have homotopy equivalences:

$$\Gamma \bigotimes_{\mathcal{O}H} \Gamma^* \simeq B_p(G) \quad \text{as complexes of } (B_p(G), B_p(G))\text{-bimodules,}$$
$$\Gamma^* \bigotimes_{\mathcal{O}G} \Gamma \simeq B_p(H) \quad \text{as complexes of } (B_p(H), B_p(H))\text{-bimodules.}$$

One of the main properties of Rickard complexes is that they automatically define Rickard complexes at the "local level" as well, as shown by the following result.

Theorem 3.5. (J. Rickard) *Let Γ be a Rickard complex for $B_p(G)$ and $B_p(H)$. Then, for every subgroup P of S_p, there is a finite complex Γ_P of $(B_p(C_G(P)), B_p(C_H(P)))$-bimodules, unique up to isomorphism, such that*

(1) Γ_P is a Rickard complex for $B_p(C_G(P))$ and $B_p(C_H(P))$,

(2) we have

$$\mathrm{Br}_{\Delta_{G\times H^\circ}(P)}(\Gamma) = k \otimes_{\mathcal{O}} \Gamma_P\,.$$

Rickard complexes and derived equivalences

For A an $\mathcal{O}$-algebra (finitely generated as an $\mathcal{O}$-module), we denote by $\mathcal{D}^b(A)$ the *derived bounded category of the module category* ${}_A\mathbf{mod}$, *i.e.* the triangulated category whose

- objects are the complexes

$$X := \left(\cdots \to X^n \xrightarrow{d^n} X^{n+1} \xrightarrow{d^{n+1}} \cdots \xrightarrow{d^{a-1}} X^a \to 0 \to \cdots\right)$$

  of finitely generated projective A-modules, bounded on the right, and exact almost everywhere,
- morphisms are chain maps modulo homotopy.

If Γ is a Rickard complex for $B_p(G)$ and $B_p(H)$, it is easy to see that the functor $Y \mapsto \Gamma \bigotimes_{B_p(H)} Y$ defines an equivalence of triangulated categories from $\mathcal{D}^b(B_p(H))$ to $\mathcal{D}^b(B_p(G))$ (see for example [Br5] for more details).

Thus the datum of a Rickard complex for $B_p(G)$ and $B_p(H)$ induces a consistent family of derived equivalences between $B_p(C_G(P))$ and $B_p(C_H(P))$, where P runs over the set of subgroups of S_p.

Rickard complexes and isotypies

• Let

$$\Gamma := \left(\cdots \to 0 \to \Gamma^m \xrightarrow{d^m} \Gamma^{m+1} \xrightarrow{d^{m+1}} \cdots \xrightarrow{d^{m+a-1}} \Gamma^{m+a} \to 0 \to \cdots \right)$$

be a Rickard complex for $B_p(G)$ and $B_p(H)$. We denote by μ_Γ the character of Γ as a complex of $(\mathcal{O}G, \mathcal{O}H)$-bimodules, *i.e.* $\mu_\Gamma := \sum_n (-1)^n \mathrm{tr}(\cdot\,; \Gamma^n)$.

• For every subgroup P of S_p, let Γ_P be the complex of $(B_p(C_G(P)), B_p(C_H(P)))$-bimodules defined as in 3.5 above. We denote by μ_{Γ_P} the character of Γ_P as a complex of $(\mathcal{O}C_G(P), C_H(P))$-bimodules.

The following result is a consequence of the definition of a Rickard complex, of 3.3 and of 3.5. It shows that a Rickard complex for $B_p(G)$ and $B_p(H)$ provides a natural isotypy between $B_p(G)$ and $B_p(H)$.

Theorem 3.6. *There is an isotypy $I = (I(P))_{P \subseteq S_p}$ such that, for each subgroup P of S_p, we have*

$$\mu_{\Gamma_P} = \sum_{\zeta \in \mathrm{Irr}(B_p(C_H(P)))} \zeta \otimes I(P)(\zeta)\,.$$

A conjecture

The following conjecture (one of J. Rickard and the author's dreams) makes more precise earlier conjectures about derived equivalences between blocks (see [Br1] and [Br5]).

Conjecture 3.7. Let G be a finite group with abelian Sylow p-subgroups, and let H be the normalizer of one of the Sylow p-subgroups. Then there exists a Rickard complex for $B_p(G)$ and $B_p(H)$.

It follows from 3.6 that the preceding conjecture implies the Conjecture 2.4.

Notice that Conjecture 3.7 holds if G is p-solvable by 1.4. Indeed, in this case the $(B_p(G), B_p(H))$-bimodule $\mathcal{O}(G/O_{p'}(G))$ is a Rickard complex for $B_p(G)$ and $B_p(H)$ (in this case, the algebras $B_p(G)$ and $B_p(H)$ are actually isomorphic).

Conjecture 3.7 is known to hold only in a few cases if G is not p-solvable.

- Some examples are provided below for the case $G = \mathfrak{A}_5$, and others are provided in §4 for the case where G is a finite reductive group.
- It follows from recent work of Rouquier ([Rou2]) that Conjecture 3.7 holds when S_p is cyclic.

The case of $G = \mathfrak{A}_5$

The case $p = 2$

The following explanation of the isotypy I_2 described in §1 is due to Rickard (*c.f.* [Ri4]).

We set $H := N_G(S_2)$, where S_2 denotes a Sylow 2-subgroup of G.

We view $B_2(G)$ as a $(B_2(G), B_2(H))$-bimodule, where $B_2(G)$ acts by left multiplication, while $B_2(H)$ acts by right multiplication.

Let us denote by $IB_2(G)$ the kernel of the augmentation map $B_2(G) \to \mathcal{O}$. Thus $IB_2(G)$ is a $(B_2(G), B_2(H))$-sub-bimodule of $B_2(G)$. Let C denote a projective cover of the bimodule $IB_2(G)$.

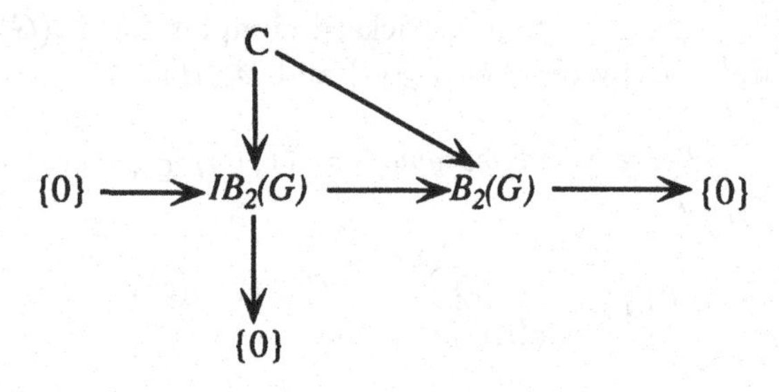

We denote by

$$\Gamma_2 := (\{0\} \to C \to B_2(G) \to \{0\})$$

the complex of $(B_2(G), B_2(H))$-bimodules defined by the preceding diagonal arrow, where $B_2(G)$ is in degree 0 (and C in degree -1).

We denote by $K\Gamma_2$ the complex of $(KB_2(G), KB_2(H))$-bimodules deduced by extension of scalars up to K. Let $H^0(K\Gamma_2)$ and $H^{-1}(K\Gamma_2)$ be the corresponding homology groups, viewed as (KG, KH)-bimodules.

It is clear that $H^0(K\Gamma_2)$ is the trivial (KG, KH)-bimodule, hence its character is $1 \otimes 1$.

Theorem 3.8. (J. Rickard)

(1) The character of $H^{-1}(K\Gamma_2)$ (with suitables choices of $\sqrt{5}$ and $\sqrt{-3}$ in the field K) is

$$(\chi_5 \otimes \alpha_3) + (\chi_3 \otimes \alpha_1) + (\chi_3' \otimes \alpha_1') .$$

In particular the character of $\sum_n (-1)^n H^n(K\Gamma_2)$ is

$$(1 \otimes 1) - ((\chi_5 \otimes \alpha_3) + (\chi_3 \otimes \alpha_1) + (\chi_3' \otimes \alpha_1')) .$$

(2) *We have homotopy equivalences:*

$$\Gamma_2 \bigotimes_{\mathcal{O}H} \Gamma_2^* \simeq B_2(G) \quad \textit{as complexes of } (B_2(G), B_2(G))\textit{-bimodules,}$$
$$\Gamma_2^* \bigotimes_{\mathcal{O}G} \Gamma_2 \simeq B_2(H) \quad \textit{as complexes of } (B_2(H), B_2(H))\textit{-bimodules.}$$

Remark. The above Theorem 3.8 may be viewed now as a particular consequence of Rouquier's recent theorem (*c.f.* [Rou2]) which provides Rickard complexes from certain stable equivalences.

The case $p = 3$

Now we view G as $\mathrm{SL}_2(4)$. We then denote by T the group of diagonal matrices in G (the split torus, which is also a Sylow 3-subgroup of G), and by U the Sylow 2-subgroup of G, consisting of unipotent uppertriangular matrices.

We set $H := N_G(T)$, and $\sigma := \begin{pmatrix} 0 & -1 \\ 1 & 0 \end{pmatrix}$. We have $H = T \rtimes \langle\sigma\rangle$.

Let Γ_3 denote the free $\mathbb{Z}_3$-module with basis G/U. Then Γ_3 is a $(\mathbb{Z}_3G, \mathbb{Z}_3T)$-bimodule, where G acts by left multiplication while T acts by right mutiplication.

Theorem 3.9.

(1) *The $(\mathbb{Z}_3G, \mathbb{Z}_3T)$-bimodule Γ_3 extends to a $(\mathbb{Z}_3G, \mathbb{Z}_3H)$-bimodule, whose character as a (KG, KH)-bimodule is*

$$(1 \otimes 1) + (\chi_4 \otimes \beta_1) + (\chi_5 \otimes \beta_2).$$

(2) *We have isomorphisms*

$$\Gamma_3 \bigotimes_{\mathcal{O}H} \Gamma_3^* \simeq B_3(G) \quad \textit{as } (B_3(G), B_3(G))\textit{-bimodules,}$$
$$\Gamma_3^* \bigotimes_{\mathcal{O}G} \Gamma_3 \simeq B_3(H) \quad \textit{as } (B_3(H), B_3(H))\textit{-bimodules.}$$

Remark. The previous example is a particular case of a much more general situation, as will be shown in §4.

The case $p = 5$

Now we set $H := N_G(S_5)$, where S_5 is a Sylow 5-subgroup of G. Since S_5 is cyclic, we can apply Rouquier's constructions as in [Rou2].

The Brauer trees of the blocks $B_5(G)$ and $B_5(H)$ are

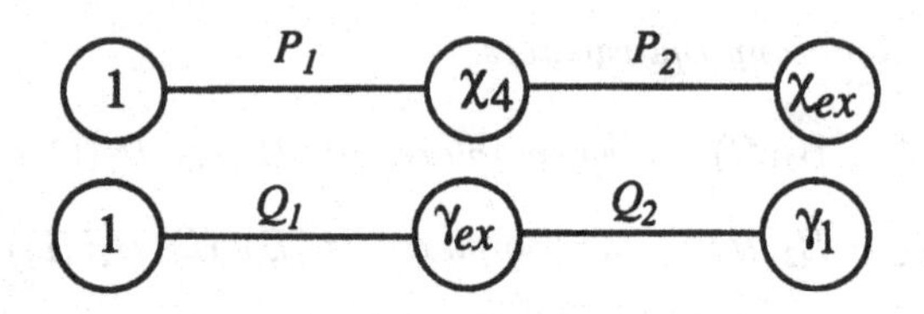

for G and H respectively, where the exceptional vertices correspond to characters

$$\chi_{ex} := \chi_3 + \chi_3', \ \gamma_{ex} := \gamma_2 + \gamma_2' .$$

In other words, we have minimal projective resolutions of the trivial representations, periodic of period 4, of the following shapes

$$\begin{array}{c} \ldots \to P_1 \to P_2 \to P_2 \to P_1 \to \mathcal{O} \to \{0\} \\ \ldots \to Q_1 \to Q_2 \to Q_2 \to Q_1 \to \mathcal{O} \to \{0\} . \end{array}$$

The $(B_5(G), B_5(H))$-bimodule $B_5(G)$ is indecomposable and its projective cover has the shape (*c.f.* [Rou2])

$$(P_1 \otimes \mathrm{Hom}_{\mathcal{O}}(Q_1, \mathcal{O})) \oplus (P_2 \otimes \mathrm{Hom}_{\mathcal{O}}(Q_2, \mathcal{O})) \xrightarrow{\pi} B_5(G) .$$

Following [Rou2], we then define

$$\Gamma_5 := \Big(\{0\} \to P_2 \otimes \mathrm{Hom}_{\mathcal{O}}(Q_2, \mathcal{O}) \xrightarrow{\pi} B_5(G) \to \{0\} \Big) .$$

Theorem 3.10. (R. Rouquier)

(1) The character of $H^{-1}(K\Gamma_5)$ is

$$(\chi_4 \otimes \gamma_1) + (\chi_3 \otimes \gamma_2) + (\chi_3' \otimes \gamma_2') .$$

In particular the character of $\sum_n (-1)^n H^n(K\Gamma_5)$ is

$$(1 \otimes 1) - ((\chi_4 \otimes \gamma_1) + (\chi_3 \otimes \gamma_2) + (\chi_3' \otimes \gamma_2')) .$$

(2) We have homotopy equivalences:

$$\begin{array}{ll} \Gamma_5 \otimes_{\mathcal{O}H} \Gamma_5^* \simeq B_5(G) & \textit{as complexes of } (B_5(G), B_5(G))\textit{-bimodules,} \\ \Gamma_5^* \otimes_{\mathcal{O}G} \Gamma_5 \simeq B_5(H) & \textit{as complexes of } (B_5(H), B_5(H))\textit{-bimodules.} \end{array}$$

4. The case of the finite reductive groups

In the case where G is a "finite reductive group", the Conjecture 3.7 can be made more precise and closely linked with the underlying algebraic geometry (for more details, see [BrMa]).

In this paragraph, we change our notation to fit with the usual notation of finite reductive groups: our prime p (the characteristic of our field $k := \mathcal{O}/\mathfrak{p}$) is now denoted by ℓ, and q denotes a power of another prime $p \neq \ell$.

From now on, we denote by $\mathbf{G}$ a connected reductive algebraic group over $\overline{\mathbb{F}}_q$, endowed with a Frobenius endomorphism F which defines a rational structure on $\mathbb{F}_q$. The finite group $\mathbf{G}^F$ of fixed points of $\mathbf{G}$ under F is called a finite reductive group.

The Deligne–Lusztig variety and its ℓ–adic cohomology

The Deligne–Lusztig variety

Let $\mathbf{P}$ be a parabolic subgroup of $\mathbf{G}$, with unipotent radical $\mathbf{U}$, and with F–stable Levi subgroup $\mathbf{L}$.

We denote by $\mathrm{Y}(\mathbf{U})$ the associated Deligne–Lusztig variety defined (*c.f.* [DeLu] and [Lu]) by

$$\mathrm{Y}(\mathbf{U}) := \{g(\mathbf{U} \cap F(\mathbf{U})) \in \mathbf{G}/\mathbf{U} \cap F(\mathbf{U})\,;\, g^{-1}F(g) \in F(\mathbf{U})\}\,.$$

It is clear that $\mathbf{G}^F$ acts on $\mathrm{Y}(\mathbf{U})$ by left multiplication while $\mathbf{L}^F$ acts on $\mathrm{Y}(\mathbf{U})$ by right multiplication.

It is known (*c.f.* [Lu]) that $\mathrm{Y}(\mathbf{U})$ is an $\mathbf{L}^F$–torsor on a variety $\mathrm{X}(\mathbf{U})$, which is smooth of pure dimension equal to $\dim(\mathbf{U}/\mathbf{U} \cap F(\mathbf{U}))$, and which is affine (at least if q is large enough). In particular $\mathrm{X}(\mathbf{U})$ is endowed with a left action of $\mathbf{G}^F$. If R is a commutative ring, the image of the constant sheaf R on $\mathrm{Y}(\mathbf{U})$ through the finite morphism $\pi : \mathrm{Y}(\mathbf{U}) \to \mathrm{X}(\mathbf{U})$ is a locally constant sheaf $\pi_*(R)$ on $\mathrm{X}(\mathbf{U})$. We denote this sheaf by $\mathcal{F}_{R\mathbf{L}^F}$.

A consequence of yet another theorem of J. Rickard

From now on, we denote by ℓ a prime number which does not divide q, and which is *good for* $\mathbf{G}$.

The following theorem is a consequence of the main result of [Ri5] (for a "character theoretic approach" of this result, see [Br1], §2.A).

Theorem 4.1. (J. Rickard) *There exists a bounded complex*

$$\Lambda_c(\mathrm{X}(\mathbf{U}), \mathcal{F}_{\mathbb{Z}_\ell\mathbf{L}^F}) = \left(\cdots \to 0 \to \Lambda^m \xrightarrow{d^m} \Lambda^{m+1} \xrightarrow{d^{m+1}} \cdots \xrightarrow{d^{m+a-1}} \Gamma^{m+a} \to 0 \to \cdots\right)$$

of $(\mathbb{Z}_\ell\mathbf{G}^F, \mathbb{Z}_\ell\mathbf{L}^F)$*–bimodules, with the following properties:*

(1) *For each positive integer* n, $(\mathbb{Z}_\ell/\ell^n\mathbb{Z}_\ell) \otimes \Lambda_c(\mathrm{X}(\mathbf{U}), \mathcal{F}_{\mathbb{Z}_\ell\mathbf{L}^F})$ *is a representative, in the derived bounded category of* $((\mathbb{Z}_\ell/\ell^n\mathbb{Z}_\ell)\mathbf{G}^F, (\mathbb{Z}_\ell/\ell^n\mathbb{Z}_\ell)\mathbf{L}^F)$*–bimodules, of the "$\ell$–adic cohomology complex"* $\mathrm{R}\Gamma_c(\mathrm{X}(\mathbf{U}), \mathcal{F}_{\mathbb{Z}_\ell/\ell^n\mathbb{Z}_\ell\mathbf{L}^F})$.

(2) For each integer n, the $\mathbb{Z}_\ell[\mathbf{G}^F \times \mathbf{L}^F]$–module Λ^n is an ℓ–permutation module, such that each of its indecomposable constituents has a vertex contained in

$$\Delta_{\mathbf{G}^F\times(\mathbf{L}^F)^\circ}(\mathbf{L}^F)$$

If $\mathcal{O}$ is the ring of integers of a finite extension K of $\mathbb{Q}_\ell$, we set

$$\Lambda_c(\mathrm{X}(\mathbf{U}), \mathcal{F}_{\mathcal{O}\mathbf{L}^F}) := \mathcal{O} \otimes_{\mathbb{Z}_\ell} \Lambda_c(\mathrm{X}(\mathbf{U}), \mathcal{F}_{\mathbb{Z}_\ell\mathbf{L}^F}).$$

A conjecture

Notation

From now on, we assume that ℓ is a prime number, $\ell \neq p$, which is good for $\mathbf{G}$, and such that the Sylow ℓ–subgroups of $\mathbf{G}^F$ are *abelian.*

• Let $\mathcal{O}$ be the ring of integers of a finite unramified extension k of the field of ℓ–adic numbers $\mathbb{Q}_\ell$, with residue field k, such that the finite group algebra $k\mathbf{G}^F$ is split.

• Let e be the principal block idempotent of $\mathcal{O}\mathbf{G}^F$, so that $\mathcal{O}\mathbf{G}^F e$ is the principal block $B_\ell(\mathbf{G}^F)$ of $\mathcal{O}\mathbf{G}^F$.

• Let S be a Sylow ℓ–subgroup of $\mathbf{G}^F$ and let $\mathbf{L} := C_{\mathbf{G}}(S)$. The group $\mathbf{L}$ is a rational Levi subgroup of $\mathbf{G}$.

We have $N_{\mathbf{G}^F}(S) = N_{\mathbf{G}^F}(\mathbf{L})$. The group S is a Sylow ℓ–subgroup of $Z(\mathbf{L})^F$, and ℓ does not divide $|N_{\mathbf{G}^F}(\mathbf{L})/\mathbf{L}^F|$.

Let f be the principal block idempotent of $\mathcal{O}\mathbf{L}^F$, so that $\mathcal{O}\mathbf{L}^F f$ is the principal block $B_\ell(\mathbf{L}^F)$ of $\mathcal{O}\mathbf{L}^F$.

Conjecture 4.2. There exists a parabolic subgroup of $\mathbf{G}$ with unipotent radical $\mathbf{U}$ and Levi complement $\mathbf{L}$, such that

(C1) the idempotent e acts as the identity on the complex $\Lambda_c(\mathrm{X}(\mathbf{U}), \mathcal{F}_{\mathcal{O}\mathbf{L}^F}).f$,

(C2) the structure of the complex of $(\mathcal{O}\mathbf{G}^F e, \mathcal{O}\mathbf{L}^F f)$–bimodules of $\Lambda_c(\mathrm{X}(\mathbf{U}), \mathcal{F}_{\mathcal{O}\mathbf{L}^F}).f$ extends to a structure of the complex of $(\mathcal{O}\mathbf{G}^F e, \mathcal{O}N_{\mathbf{G}^F}(\mathbf{L})f)$–bimodules,

(C3) we have homotopy equivalences:

$$\Lambda_c(\mathrm{X}(\mathbf{U}), \mathcal{F}_{\mathcal{O}\mathbf{L}^F}).f \bigotimes_{\mathcal{O}N_{\mathbf{G}^F}(\mathbf{L})f} f.\Lambda_c(\mathrm{X}(\mathbf{U}), \mathcal{F}_{\mathcal{O}\mathbf{L}^F})^* \simeq \mathcal{O}\mathbf{G}^F e$$

as complexes of $(\mathcal{O}\mathbf{G}^F e, \mathcal{O}\mathbf{G}^F e)$–bimodules,

$$f.\Lambda_c(\mathrm{X}(\mathbf{U}), \mathcal{F}_{\mathcal{O}\mathbf{L}^F})^* \bigotimes_{\mathcal{O}\mathbf{G}^F e} \Lambda_c(\mathrm{X}(\mathbf{U}), \mathcal{F}_{\mathcal{O}\mathbf{L}^F}).f \simeq \mathcal{O}N_{\mathbf{G}^F}(\mathbf{L})f$$

as complexes of $(\mathcal{O}N_{\mathbf{G}^F}(\mathbf{L})f, \mathcal{O}N_{\mathbf{G}^F}(\mathbf{L})f)$–bimodules.

By 4.1, one sees that if the above conjecture is true, the complex $\Lambda_c(\mathrm{X}(\mathbf{U}), \mathcal{F}_{\mathcal{O}\mathbf{L}^F})$ is indeed a Rickard complex for $\mathcal{O}\mathbf{G}^F e$ and $\mathcal{O}N_{\mathbf{G}^F}(\mathbf{L})f$.

Although some evidence in favour of Conjecture 4.2 is indeed available (see [BrMa]), it is actually known to be true in very few cases.

The particular case where ℓ divides $(q-1)$ (which had been conjectured and almost proved by Hiß) may be deduced from some results of Puig ([Pu]).

Theorem 4.3. *Assume that ℓ divides $q-1$ and does not divide the order of the Weyl group of $\mathbf{G}$.*

(1) We have $\mathbf{L} = \mathbf{T}$, a quasi-split maximal torus of $\mathbf{G}$. For $\mathbf{P} = \mathbf{B}$, a rational Borel subgroup of $\mathbf{G}$ containing $\mathbf{T}$ with unipotent radical $\mathbf{U}$, we have

$$\Lambda_c(\mathrm{X}(\mathbf{U}), \mathcal{F}_{\mathcal{O}\mathbf{L}^F}) \simeq \mathcal{O}[\mathbf{G}^F/\mathbf{U}^F].$$

(2) (L. Puig) We have isomorphims:

$$\mathcal{O}[\mathbf{G}^F/\mathbf{U}^F].f \bigotimes_{\mathcal{O}N_{\mathbf{G}^F}(\mathbf{T})f} f.\mathcal{O}[\mathbf{U}^F\backslash\mathbf{G}^F] \simeq \mathcal{O}\mathbf{G}^F e$$

as $(\mathcal{O}\mathbf{G}^F e, \mathcal{O}\mathbf{G}^F e)$–bimodules,

$$f.\mathcal{O}[\mathbf{U}^F\backslash\mathbf{G}^F] \bigotimes_{\mathcal{O}\mathbf{G}^F e} \mathcal{O}[\mathbf{G}^F/\mathbf{U}^F].f \simeq \mathcal{O}N_{\mathbf{G}^F}(\mathbf{T})f$$

as $(\mathcal{O}N_{\mathbf{G}^F}(\mathbf{T})f, \mathcal{O}N_{\mathbf{G}^F}(\mathbf{T})f)$–bimodules,

providing a Morita equivalence between $\mathcal{O}\mathbf{G}^F e$ and $\mathcal{O}N_{\mathbf{G}^F}(\mathbf{T})f$.

Remark. The case where $\mathbf{G}^F = \mathrm{SL}_2(4)$ and $\ell = 3$ (see Theorem 3.9 above) is a particular case of the preceding theorem.

References

[Al1] J.L. Alperin, Weights for finite groups, in *The Arcata Conference on Representations of Finite Groups* (Proc. Symp. Pure Math. **47**, Amer. Math. Soc., Providence, 1987), 369–379.

[Al2] J.L. Alperin *Local representation theory* (Cambridge studies in advanced mathematics **11**, Cambridge University Press, Cambridge, 1986).

[AlBr] J.L. Alperin and M. Broué, Local Methods in Block Theory, *Ann. Math.* **110**(1979), 143–157.

[Br1] M. Broué, Isométries parfaites, types de blocs, catégories dérivées, *Astérisque* **181–182**(1990), 61–92.

[Br2] M. Broué, Isométries de caractères et équivalences de Morita ou dérivées, *Publ. Math. I.H.E.S.* **71**(1990), 45–63.

[Br3] M. Broué, On Scott modules and p–permutation modules, *Proc. Amer. Math. Soc.* **93**(1985), 401–408.

[Br4] M. Broué, Théorie locale des blocs, in *Proceedings of the International Congress of Mathematicians, Berkeley, 1986* (I.C.M.), 360–368.

[Br5] M. Broué, Equivalences of blocks of group algebras (in *Proceedings of the International Conference on Representations of Algebras, Ottawa, 1992*, 1993).

[BrMa] M. Broué and G. Malle, Zyklotomische Heckealgebren, *Astérisque* **212** (1993), 119–190.

[BMM] M. Broué, G. Malle and J. Michel, Generic blocks of finite reductive groups, *Astérisque* **212**(1993), 7–92.

[BrMi] M. Broué and J. Michel, Blocs à groupes de défaut abéliens des groupes réductifs finis, *Astérisque* **212**(1993), 93–118.

[DeLu] P. Deligne and G. Lusztig Representations of reductive groups over finite fields, *Ann. Math.* **103**(1976), 103–161.

[Fe] W. Feit, *The representation theory of finite groups* (North–Holland, Amsterdam, 1982).

[FoHa] P. Fong and M. Harris, On perfect isometries and isotypies in finite groups, *Invent. Math.* (1993).

[Gl] G. Glauberman, Central elements in core–free groups *J. Algebra* **4**(1966), 403–420.

[Gro] A. Grothendieck, Groupes des classes des catégories abéliennes et triangulées, complexes parfaits, in *Cohomologie ℓ-adique et fonctions L (SGA 5)* (Springer–Verlag L.N. **589**, 1977), 351–371.

[Lu] G. Lusztig, Green functions and character sheaves, *Ann. Math.* **131**(1990) 355–408.

[Pu] L. Puig Algèbres de source de certains blocs des groupes de Chevalley, *Astérisque* **181–182**(1990), 221–236.

[Ri1] J. Rickard, Morita Theory for Derived Categories, *J. London Math. Soc.* **39**(1989), 436–456.

[Ri2] J. Rickard, Derived categories and stable equivalences, *J. Pure Appl. Alg.* **61**(1989), 307–317.

[Ri3] J. Rickard, Derived equivalences as derived functors, *J. London Math. Soc.* **43**(1991), 37–48.

[Ri4] J. Rickard, Derived equivalences for the principal blocks of $\mathfrak{A}_4$ and $\mathfrak{A}_5$, preprint.

[Ri5] J. Rickard, Finite group actions and étale cohomology, preprint.

[Ro] G.R. Robinson, The Z_p^*–theorem and units in blocks, *J. Algebra* **134**(1990) 353–355.

[Rou1] R. Rouquier, Sur les blocs à groupe de défaut abélien dans les groupes symétriques et sporadiques, *J. Algebra*, to appear.

[Rou2] R. Rouquier, From stable equivalences to Rickard equivalences for blocks with cyclic defect, these Proceedings.

[Se] J.-P. Serre, *Représentations linéaires des groupes finis* 3ème édition (Hermann, Paris, 1978).

COMPUTING THE CONJUGACY CLASSES OF ELEMENTS OF A FINITE GROUP[1]

GREG BUTLER

Centre Interuniversitaire en Calcul Mathématique Algébrique, Department of Computer Science, Concordia University, Montreal, Quebec, H3G 1M8 Canada

Abstract

There are several problems associated with the computation of the conjugacy classes of elements of a finite group, for example, the computation of centralizers and the determination of conjugacy of two elements. The known algorithms for solving these associated problems and the main problem of determining the conjugacy classes are presented. The effectiveness of the algorithms is illustrated by examples. Some applications are briefly discussed.

1. Introduction

Let G be a finite group. The *conjugate*, x^g, of an element $x \in G$ by an element $g \in G$ is the element $g^{-1}xg$. Two elements x and y are *conjugate* in G if there exists $g \in G$ such that $x^g = y$. The relation of being conjugate in G is an equivalence relation on the set of elements of the group G. The equivalence classes are called *conjugacy classes* of elements. For each element $g \in G$, the map $\Phi_g : G \to G$ defined by $x \mapsto x^g$ is an automorphism of the group G. The maps are called *inner* automorphisms and they form a normal subgroup of the automorphism group, $Aut(G)$, of G. Therefore, the elements in a conjugacy class are in the same orbit of $Aut(G)$ in its action on G, and have similar properties. Hence, a set of class representatives contains one of each distinct "type" of element in the group.

The conjugacy classes of elements are an important piece of information about the structure of a group. The information is a prerequisite for many algorithms such as those for computing the lattice of normal subgroups, the automorphism group, and the character table. Characters and representations play an important role in quantum physics, coding theory, the classification of finite simple groups, and image processing.

There are several distinct problems related to the computation of conjugacy classes:

1. **Class elements**: Given an element $x \in G$, determine the elements in the class, $K_G(x) = \{ x^g \mid g \in G \}$, of x in G.

[1]This work was supported in part by the Australian Research Council, and FCAR of Quebec.

2. **Fusion of powers of an element**: Given an element $x \in G$, determine the *rational class*, $Kr_G(x) = \bigcup_{<x^m>=<x>} K_G(x^m)$, as a disjoint union of classes $K_G(x^m)$ of powers of x.
3. **Centralizer problem**: Given an element $x \in G$, determine the centralizer $C_G(x) = \{ g \in G \mid x^g = x \}$ of x in G.
4. **Conjugacy problem**: Given two elements $x, y \in G$, determine whether they are conjugate in G; and if they are conjugate then find an element $g \in G$ such that $x^g = y$.
5. **Classes**: Determine a set of representatives for the classes (or rational classes) of elements of G.
6. **Class map problem**: Given representatives of the classes of G and an element $x \in G$, determine the class in which x lies.

We will concentrate on algorithms for determining a set of representatives of the (rational) classes of elements of a finite group G. These algorithms will be presented in historical order:

- Classes as orbits under the action of conjugation for groups of moderate order;
- First element of each class of a permutation group from the first tuple of each orbit of k-tuples of points (Sims,1971; Butler, 1979);
- Random method for permutation and matrix groups (Butler, 1979);
- Orbit-stabiliser algorithm for p-groups (Felsch and Neubüser, 1979);
- Orbit-stabiliser algorithm for soluble groups (Laue et al, 1984); and
- Inductive approach for permutation groups using homomorphisms (Butler, 1990b).

The subproblems of determining centralizers and conjugacy of pairs of elements are discussed in the background section, while the class map problem and the problem of representing all the elements of a class are discussed together with relevant applications.

The effectiveness of the algorithms is illustrated by examples computed using Cayley, v3.7, of Cannon(1984) running on a Decstation 5000.

2. Computational background

This section discusses the computational description of groups, especially permutation groups, p-groups and soluble groups. As well as the fundamental tasks of representing elements, subgroups, quotients, and homomorphisms, this section also discusses some algorithms for computing centralizers and determining conjugacy. The reader is referred to Hall(1959), Huppert(1967) and Wielandt(1964) for elementary definitions and results from group theory.

2.1. Permutation groups

For computational purposes, a permutation group G acting on the points Ω is described by a *base and strong generating set.* A base for G is a sequence of points $B = [\beta_1, \beta_2, ..., \beta_k]$ in Ω such that the only element in G fixing every point β_i is the identity element. A strong generating set relative to B is a set T of elements of G which contains a set of generators for each stabiliser $G^{(i)} = G_{\beta_1,\beta_2,...,\beta_{i-1}}$, $1 \leq i < k$. A Schreier system of a transversal $U^{(i)}$ of $G^{(i+1)}$ in $G^{(i)}$ is stored in a *Schreier vector* $v^{(i)}$. For a point $\gamma \neq \beta_i$ in the orbit $\beta_i^{G^{(i)}}$, there is a unique element $u^{(i)}(\gamma)$ in the transversal $U^{(i)}$ which maps β_i to γ. The entry $v^{(i)}[\gamma]$ is the generator $s \in T$ which is the last factor of the word for $u^{(i)}(\gamma)$ in the Schreier system. Leon(1980) describes algorithms for constructing these data structures and for their use in answering basic questions such as:

PERM1: *Given a permutation group by a set of generating permutations, compute a base, a set of strong generators, and a Schreier vector for each level in the stabiliser chain.*

PERM2: *Given a base and strong generating set for a group G perform the following tasks:*

(a) Given a permutation π, test whether $\pi \in G$.

(b) Given a sequence of points $[\gamma_1, \gamma_2, \ldots \gamma_l]$, with $1 \leq l \leq k$, determine whether there exists an element $g \in G$ mapping β_i to γ_i for all i, $1 \leq i \leq l$.

If an element $g \in G$ exists in **PERM2**(b), the sequence of points is called a *partial base image* and the coset $G^{(l+1)}g$ is the set of all elements of the group which map the initial segment of the base to the partial base image.

Let us begin by describing some of the more expensive computations which might arise as subproblems in determining the conjugacy classes.

PERM3: *Given a base and strong generating set for a group G, and an element $z \in G$, compute a base and strong generating set for $C_G(z)$.*

PERM4: *Given a base and strong generating set for a group G, and two elements $z_1, z_2 \in G$, determine whether z_1 is conjugate to z_2 in G, and if so, determine an element $g \in G$ such that $z_1^g = z_2$.*

The backtrack search algorithms enumerate the search tree of partial base images, and prune the tree using known subgroups and properties. For example, an element g in $C_G(z)$ must permute the cycles of z, and $\langle z \rangle$ is known to be a subgroup of $C_G(z)$. Conjugacy testing could use the double cosets of $C_G(z_1)$ and $C_G(z_2)$ in G to prune the search tree, but more simply could choose the image γ_1 of β_1 to be first point in its $C_G(z_2)$-orbit. It is beneficial to arrange that the base points be chosen from long cycles (of z or z_1) first.

Both of the computations **PERM3** and **PERM4** are generally quite efficient. However, the algorithms of Butler(1982) are backtrack searches and are subject to combinatorial explosion of the size of the search tree. So in bad cases, the cost could be two or three orders of magnitude worse than the average cost. Furthermore, if the determination of the conjugacy classes requires thousands of conjugacy tests, then the cost accumulates, and the chance of a bad case increases. Hence, a major aim of any approach to determining the conjugacy classes should be to minimize the number of conjugacy tests in the permutation group G.

There are other backtrack algorithms which compute normalisers and set stabiliser.

Recent work by Leon(1991) has produced significantly more efficient backtrack algorithms. These improvements are not reflected in the running times given here.

PERM5: *Given a base and strong generating set for a group G, and a prime p dividing the order of G, compute a base and strong generating set for a Sylow p-subgroup of G.*

The algorithm of Butler and Cannon(1991) for **PERM5** requires a small number of centralizer computations, and its total cost is essentially the cost of these computations.

The following computations can be done very efficiently.

PERM6: *Given a base and strong generating set for a group G acting on Ω, set up any of the following homomorphisms: the action of G on an invariant subset Δ of Ω; the action of G on an invariant partition Υ of Ω; for a p-group G, the isomorphism between G and the group defined by a power-commutator (pc) presentation of G; for a soluble group G, the isomorphism between G and the group defined by a conditioned pc presentation of G. Allow the computation of the image, kernel, image of an element, preimage of an element, image of a subgroup, and preimage of a subgroup.* (Butler, 1985, 1988, 1990a; Butler and Cannon, 1993)

2.2. Matrix groups

For finite matrix groups, such as groups of matrices over finite fields, there are analogues of most of the techniques for permutation groups which are founded on base and strong generating set. The group is considered to act on the vectors and the one-dimensional subspaces of the vector space. A base is chosen from amongst both vectors and one-dimensional subspaces. There is no normalizer algorithm, nor are there algorithms for group homomorphisms (other than obtaining the permutation action on the vector space). See Butler and Cannon(1982, 1989); Butler(1982).

MAT1: *Given a group by a set of generating matrices, over a finite field, compute a base and a set of strong generators.*

MAT2: *Given a base and strong generating set for a group G perform the following tasks:*

(a) Given a matrix π, test whether $\pi \in G$.

(b) Given a sequence of vectors/one-dimensional subspaces $[\gamma_1, \gamma_2, \ldots \gamma_l]$, with $1 \leq l \leq k$, determine whether there exists an element $g \in G$ mapping β_i to γ_i for all i, $1 \leq i \leq l$.

MAT3: *Given a base and strong generating set for a group G, and an element $z \in G$, compute a base and strong generating set for $C_G(z)$.*

MAT4: *Given a base and strong generating set for a group G, and two elements $z_1, z_2 \in G$, determine whether z_1 is conjugate to z_2 in G, and if so, determine an element $g \in G$ such that $z_1^g = z_2$.*

MAT5: *Given a base and strong generating set for a group G, and a prime p dividing the order of G, compute a base and strong generating set for a Sylow p-subgroup of G.*

2.3. p-groups

For computational purposes, a p-group is described by a *power-commutator presentation*, or pc presentation. This is a presentation of the form

$$G = \langle\, a_1, a_2, ..., a_n \mid a_i{}^p = u_i, \text{ for } 1 \leq i \leq n,\ [a_j, a_i] = v_{ij}, \text{ for } 1 \leq i < j \leq n \,\rangle$$

where the u_i are words in $\{a_{i+1}, a_{i+2}, ..., a_n\}$, and the v_{ij} are words in $\{a_{j+1}, a_{j+2}, ..., a_n\}$. This defines a descending central series

$$G = G(1) \rhd G(2) \rhd \ldots \rhd G(r) \rhd G(n+1) =< identity >$$

where $G(i) = \langle a_i, a_{i+1}, ..., a_n \rangle$. Every descending central series is a *subnormal* series; that is $G(c+1)$ is normal in $G(c)$. Every descending central series is a *normal* series; that is $G(c+1)$ is normal in G. Every descending central series is a *central* series; that is $G(c)/G(c+1)$ is contained in the centre of $G/G(c+1)$. Each of the quotients $G(c)/G(c+1)$ of a descending central series is abelian.

The relations of a pc presentation allow any word in the generators to be reduced to a *normal word*, $a_1^{k_1} a_2^{k_2} \ldots a_n^{k_n}$, where the generators come in order and the exponents lie in the restricted range $0 \leq k_i < p$. The process of reducing a word to normal form is called *collection*.

The *lower exponent-p-central* chain for a prime p is

$$G = P_0(G) \rhd P_1(G) \rhd \ldots \rhd P_c(G) \rhd P_{c+1}(G) \rhd \ldots$$

where $P_{c+1}(G) = [P_c(G), G](P_c(G))^p$, for each c. Each of the quotients $P_c(G)/P_{c+1}(G)$ of the lower exponent-p-central series is an elementary abelian p-group.

A pc presentation of a finite p-group G is *conditioned* if it is prime-step and defines a refinement of the lower exponent-p-central series. That is, it has the form

$$\langle\, a_1, a_2, ..., a_n \mid a_i{}^p = u_i, \text{ for } 1 \leq i \leq n, \ [a_j, a_i] = v_{ij}^*, \text{ for } 1 \leq i < j \leq n \,\rangle$$

where the u_i are words in $\{a_{i+1}, a_{i+2}, ..., a_n\}$, and the v_{ij}^* are normal words in $\{a_{j+1}, a_{j+2}, ..., a_n\}$. Furthermore, there are integers $w(c)$ such that $P_c(G) = \langle a_{w(c)}, a_{w(c)+1}, \ldots, a_n \rangle$, for each c. The *weight* of a generator a_i (denoted $wt(a_i)$) is the value c such that $w(c) \leq i < w(c+1)$. The generators of weight zero are the *defining generators*. There are d ($= w(1) - 1$) of them. The defining generators are important because any algorithm which needs to compute an action by G only needs to use the action of the defining generators and not the action of all n generators.

Relative to a chosen generating sequence $\{a_1, a_2, ..., a_n\}$ for a group G, each subgroup H has a canonical generating sequence. The non-commutative Gaussian algorithm (Laue et al, 1984, pp.109–110) will find the canonical generating sequence (gs) given a set of generators for H. Each element g may be represented by a vector $exp(g) = [k_1 k_2 \ldots k_n]$ of the exponents of the normal word $a_1^{k_1} a_2^{k_2} \ldots a_n^{k_n}$ for g. Define the *leading coefficient* of an element g to be the first non-zero entry of its corresponding vector, and define the *leading index* to be the index of the leading coefficient in the vector. Denote these by $lc(g)$ and $li(g)$ respectively. A subgroup H of G has an induced series defined by $H(c) = H \cap G(c) = H \cap \langle a_c, a_{c+1}, \ldots, a_n \rangle$ where there may be duplicate terms in the series. We can eliminate these terms and determine a canonical set of generators for a subgroup H as follows. A sequence $[h_1, h_2, ..., h_r]$ of elements of H is called a *canonical generating sequence* (cgs) relative to a fixed pc presentation of G if and only if

1. the sequence $[h_1, h_2, ..., h_r]$ of elements defines a prime-step subnormal series of H;
2. $li(h_i) > li(h_j)$ for $i > j$;
3. $lc(h_i) = 1$ for $i = 1, 2, ..., r$;
4. $\exp(h_j)[li(h_i)] = 0$ for $i \neq j$.

(If only conditions (i) and (ii) hold, then the sequence is called a *generating sequence* (gs).)

P1: *For a p-group G defined by a pc presentation, compute any of the following: conjugacy classes of elements; centralizer of an element; determine if two elements are conjugate, and if so, determine a conjugating element; normalizers, centre, upper central series, chief series.* (Felsch and Neubüser, 1979; Laue et al 1984)

P2: *Given a presentation of a p-group G, find a pc presentation of G.* (Havas and Newman 1980)

P3: *Given a permutation group G which is a p-group, find a pc presentation of G.* (Holt, 1984; Butler 1988; Butler and Cannon 1993)

2.4. Soluble groups

A *soluble* group G is a group whose composition series has only cyclic factors. Hence, subgroups and quotients of soluble groups are soluble. In particular, a soluble group G has a *prime-step subnormal* series

$$G = G(1) \triangleright G(2) \triangleright \ldots \triangleright G(r) \triangleright G(r+1) = < identity > .$$

That is, for each value of i, $1 \leq i \leq r$, the group $G(i+1)$ is normal in $G(i)$, and the index of $G(i+1)$ in $G(i)$ is a prime $p(i)$. Let $g(i) \in G(i) \setminus G(i+1)$, for each i. Then every element of G can be uniquely expressed as a *normal word* $g(1)^{\varepsilon(1)} g(2)^{\varepsilon(2)} \ldots g(r)^{\varepsilon(r)}$, where $0 \leq \varepsilon(i) < p(i)$, for each i. Note that $G(i) = < g(i), g(i+1), \ldots, g(r) >$. The sequence $[g(1), g(2), \ldots, g(r)]$ is called a *series-generating sequence* (or simply a generating sequence (gs)).

A *power-commutation presentation* (pc presentation) of a finite soluble group G has the form

$$\langle\, a_1, a_2, ..., a_n \mid a_i{}^{\rho_i} = u_i, \text{ for } 1 \leq i \leq n,\ a_j a_i = a_i a_j v_{ij}, \text{ for } 1 \leq i < j \leq n\, \rangle$$

where each ρ_i is a positive integer, and where the u_i and v_{ij} are words in $\{a_{i+1}, a_{i+2}, ..., a_n\}$. (These are also called *AG systems* or *power-conjugate presentations.*) A pc presentation defines a subnormal series of subgroups where $G(i) = \langle a_i, a_{i+1}, \ldots, a_n \rangle$. The quotient group $G(i)/G(i+1)$ is generated by $a_i G(i+1)$ and the order of the quotient divides ρ_i. When the order is precisely ρ_i, we say the pc presentations is *confluent* or *consistent.* If each ρ_i is prime, then we say the series is *prime-step.* We can always refine a pc presentation to a prime-step series by introducing new generators which correspond to powers of the generators a_i when ρ_i is composite.

The *derived subgroup* (or commutator subgroup) of a group G is the subgroup generated by all commutators $[a,b]$ $(= a^{-1}b^{-1}ab)$ where $a, b \in G$. It is written $[G,G]$ or $D(G)$. It is the smallest normal subgroup of G such that the quotient of G by the normal subgroup is abelian. The *derived series* of G is defined by

1. $D^0(G) = G$, and
2. $D^{c+1}(G) = [D^c(G), D^c(G)]$, for each $c \geq 0$.

A group G is defined to be *soluble* if and only if some term of the derived series of G is the identity subgroup. The derived series is a subnormal series since $D(G)$ is normal in G. It is also a *normal series* since each term $D^c(G)$ is normal in G. Furthermore, each quotient $D^c(G)/D^{c+1}(G)$ is abelian. Each abelian group H can be written as a direct product of cyclic groups of prime power order. Hence, H has a normal series

$$H = H(1) \triangleright H(2) \triangleright \ldots \triangleright H(r) \triangleright H(m+1) = < identity >,$$

where the quotients $H(i)/H(i+1)$ are *elementary abelian*. Therefore, the derived series of G can be refined to a normal series with elementary abelian quotients, and this *normal* series can be further refined to a prime-step *subnormal* series. This leads to the following definition:

A *conditioned* pc presentation of a finite soluble group G is a confluent pc presentation of the form

$$\langle\, a_1, a_2, \ldots, a_n \mid a_i^{p_i} = u_i, \text{ for } 1 \leq i \leq n,\ a_j a_i = a_i a_j v_{ij}^*, \text{ for } 1 \leq i < j \leq n \,\rangle$$

where

1. p_i is a prime,
2. u_i, v_{ij}^* are normal words in $\{a_{i+1}, a_{i+2}, \ldots, a_n\}$, and
3. there exists a *normal* series

   $$G = N(1) \rhd N(2) \rhd \ldots \rhd N(r) \rhd N(r+1) = < identity >$$

   of G where each quotient $N(i)/N(i+1)$ is elementary abelian, and each $N(i)$ has the form $\langle a_{n(i)}, a_{n(i)+1}, \ldots, a_n \rangle$ for some integer $n(i)$.

Every finite soluble group has a conditioned pc presentation, and the extra properties are very useful when computing in a finite soluble group.

SOL1: *For a soluble group G defined by a pc presentation, compute any of the following: conjugacy classes of elements; centralizer of an element; determine if two elements are conjugate, and if so, determine a conjugating element; Sylow subgroups, Hall subgroups, normalizers, centre.* (Laue et al 1984; Glasby 1988; Mecky and Neubüser 1989; Glasby and Slattery 1990)

Relative to a chosen generating sequence for a group G, each subgroup H has a canonical generating sequence. The non-commutative Gaussian algorithm will find the canonical gs given a set of generators for H.

For a soluble permutation group G, given a base B for G, we can choose a series-generating sequence $[g(1), g(2), \ldots, g(r)]$ such that each generating set $\{g(i), g(i+1), \ldots, g(r)\}$ is also a strong generating set of $G(i)$ relative to B. Such a sequence of elements is called a *B-strong series-generating sequence* (or B-ssgs). As an example, the symmetric group S_4 of degree 4 is soluble. It has a base B=[1,2,3], a B-ssgs [$g(1)$=(3,4), $g(2)$=(2,3,4), $g(3)$=(1,2)(3,4), $g(4)$=(1,3)(2,4)].

A common task is to construct a cyclic extension $K = \langle H, z \rangle$ of a group H by a normalizing permutation z. The algorithm *normalizing_generator* of Sims(1990) constructs a base and strong generating set of K given them for H.

SOL2: *Given a soluble permutation group G, find a base B and a B-strong series generating sequence, and a corresponding conditioned pc presentation of G.* (Butler 1990a, 1991)

3. Theory of conjugacy classes

Lemma 3.1. *The length of the class $K_G(z)$ is equal to the index $|G : C_G(z)|$.*

The next few results concern the lifting of the conjugacy classes of a quotient group G/N to obtain information about the classes of G.

Proposition 3.1. *Suppose N is a normal subgroup of G and let $\overline{G} = G/N$. Let $g \in G$, let $\overline{C}$ be the centralizer of gN in $\overline{G}$, and let C be the complete preimage of $\overline{C}$. Then*

(a) *the preimage L of the class $K_{\overline{G}}(gN)$, that is, the union of the cosets zN which are conjugate in $\overline{G}$ to gN, is a union of G-conjugacy classes $L_1, L_2, \ldots, L_r$;*

(b) *C acts of the coset gN by conjugation;*

(c) *C has r orbits in its action on gN, and the orbit representatives $g_1, g_2, \ldots, g_r$ are representatives of the G-classes $L_1, L_2, \ldots, L_r$;*

(d) *the centralizer $C_G(g_i)$ is the stabiliser of g_i in C in its action on gN.*

The algorithms make use of the fact that N is central, is elementary abelian, or has prime order when calculating orbits and stabilisers in the action of C on the coset gN.

While the centralizer of an element g determines the length of the class, the normalizer determines the classes within the rational class of g. Define the *Galois group*, $Gal_G(g)$, of $g \in G$ to be the subgroup of $Aut(\langle g \rangle)$ isomorphic to $N_G(\langle g \rangle)/C_G(g)$. The classes within the rational class $Kr_G(g)$ are in 1-1 correspondence with the cosets of the Galois group, $Gal_G(g)$, in the automorphism group, $Aut(\langle g \rangle)$.

Proposition 3.2. ([Huppert, 1967, Satz 13.19, page 84])

(1) *$Aut(Z_n)$ is isomorphic to the multiplicative group of $\mathbb{Z}/n\mathbb{Z}$.*

(2) *If $n = n_1 n_2 \ldots n_l$ with the n_i pairwise coprime then*

$$Aut(Z_n) \simeq Aut(Z_{n_1}) \times Aut(Z_{n_2}) \times \ldots \times Aut(Z_{n_l}).$$

(3) *For $p > 2$, $Aut(Z_{p^r})$ is a cyclic group of order $p^{r-1}(p-1)$.*

(4) *For $r \geq 3$, $Aut(Z_{2^r}) \simeq Z_{2^{r-2}} \times Z_2$ and we can take 5 mod 2^r and -1 mod 2^r as generators of the respective direct factors.*

(5) *$Aut(Z_4) \simeq Z_2$, and $Aut(Z_2)$ is the trivial group.*

Let ε_g denote the obvious isomorphism between $Aut(\langle g \rangle)$ and the multiplicative group of $\mathbb{Z}/n\mathbb{Z}$, where $n = |g|$. Hence, $\varepsilon_g(Gal_G(g))$ is a subgroup of the multiplicative group of $\mathbb{Z}/n\mathbb{Z}$. If $x \in N_G(\langle g \rangle)$ or x is a coset in $N_G(\langle g \rangle)/C_G(g)$ then $\varepsilon_g(x)$ will denote the corresponding element of $\varepsilon_g(Gal_G(g))$.

If $n = qt$, where q and t are coprime, then $Aut(Z_n) \simeq Aut(Z_q) \times Aut(Z_t)$. Hence, there is a projection $\downarrow_t^n$ from the multiplicative group of $\mathbb{Z}/n\mathbb{Z}$ to the multiplicative group of $\mathbb{Z}/t\mathbb{Z}$. We denote the inverse embedding by $\uparrow_t^n$.

Corollary 3.1.

Let S be a Sylow p-subgroup of G and let $g \in G$ have order p^r.

(1) *For $p > 2$ and $r > 1$, regard $Aut(Z_{p^r})$ as $Z_{p^{r-1}} \times Z_{(p-1)}$. There exists $g_1 \in S \cap K_G(g)$ such that $Gal_S(g_1)$ is the $Z_{p^{r-1}}$-component of $Gal_G(g_1)$.*

(2) *For $p > 2$ and $r > 1$, the $Z_{(p-1)}$-component of $Gal_G(g)$ projects faithfully to the $Z_{(p-1)}$-component of $Gal_G(g^p)$.*

(3) *For $p = 2$, there exists $g_1 \in S \cap K_G(g)$ such that $Gal_S(g_1) = Gal_G(g_1)$.*

For cases (1) and (3) of the corollary, we may take $g_1 \in S \cap K_G(g)$ such that $C_S(g_1)$ is a Sylow p-subgroup of $C_G(g_1)$ and the normalizer $N_S(\langle g_1 \rangle)$ is as large as possible.

In case (2) of the corollary, if $C_G(g) = C_G(g^p)$, then take an element $x \in G$ which normalizes g^p and (modulo the centralizer) generates the $Z_{(p-1)}$-component of $Gal_G(g^p)$. The generator of the $Z_{(p-1)}$-component of $Gal_G(g)$ is a power of x.

Proposition 3.3. *If $H \leq G$ and $h \in H$ then $Gal_H(h) \leq Gal_G(h)$.*

Corollary 3.2. *Let $H \leq G$, $h \in H$ and $g \in G$ such that $|h| = |g| = n$. If $\varepsilon_h(Gal_H(h)) \not\leq \varepsilon_g(Gal_G(g))$ then h is not conjugate in G to g.*

The following properties can be used to decide that two elements are not conjugate. Hence an explicit conjugacy test in a group is only performed when all the conditions (for which the information is readily available) have been checked.

1. If z_1 and z_2 have distinct invariants (including the order of the element) then they are not conjugate in G.
2. If, for some integer t, the powers z_1^t, z_2^t are not conjugate in G then z_1, z_2 are not conjugate in G.
3. If $z_1 \in H$ and $z_2 \in G$, where $H \leq G$, such that the order of the centralizer $C_H(z_1)$ does not divide the order of $C_G(z_2)$ then z_1, z_2 are not conjugate in G.
4. If $z_1 \in H$ and $z_2 \in G$, where $H \leq G$, such that $Gal_H(z_1)$ is not isomorphic to a subgroup of $Gal_G(z_2)$ then z_1, z_2 are not conjugate in G.

Invariants include the order of the element in general; cycle shapes for permutations; and for matrices A invariants include the characteristic polynomial of A and the rank of $A + I$.

4. Classes as orbits under conjugation

For a small group G, where one can store a list of the elements, the classes can be directly computed as orbits under the action of conjugation. This obvious approach to the problem has been in use since the 60's. A list of elements provides a bijection between the elements of G and the integers $1,2,\ldots,|G|$. A subset of G, such as a conjugacy class, can be stored as a bitstring representing the characteristic function of the set, or as a list of elements. If the list is hashed then both the bitstring and list data structures provide fast methods to test membership of an element in the set. The list has the advantage that it can also queue the elements of the set in the order in which they were added to the set.

An orbit of an element can be computed by closing the set of images under the action of a set of generators for the group. Algorithm 1 computes each class this way. While the class Γ_m is under construction, it is stored as a hashed list of elements, which also acts as a queue for the closure algorithm. The elements in the list could be represented by an integer in the range 1 to $|G|$. The set Γ is the union of the known classes and is stored as a bitstring. The algorithm performs $|S| \times (|G| - 1)$ conjugations in total.

Algorithm 1
Input : a finite group G with generating set S;
Output : the classes $\Gamma_0, \ldots, \Gamma_m$ of G;
begin
    $\Gamma_0 := \{$ identity $\}$; $m := 0$;
    $\Gamma := \Gamma_0$; "set of all elements in known classes"
    for $i := 2$ **to** $|G|$ **do**
        $x := G[i]$; "the i-th element of G"
        if $x \notin \Gamma$ **then** "form next class"
            $m := m + 1$; $\Gamma_m := [x]$;
            $j := 1$;
            while $j \leq length(\Gamma_m)$ **do**
                $y := \Gamma_m[j]$;
                for each s **in** $S \mid y^s \notin \Gamma_m$ **do**
                    append y^s to Γ_m;
                end for;
            end while;
            $\Gamma := \Gamma \cup \Gamma_m$;
        end if;
    end for;
end.

The algorithm can be modified to examine the powers of x, the new class representative. If x^i is a new class representative then $K_G(x^i) = \{y^i \mid y \in K_G(x)\}$, and if x and x^i have the same order then their classes have the same length. The cost of conjugation and the cost of powering an element are similar, so in practice the class of x^{-1} is formed by inverting the elements of $K_G(x)$ but the classes of other powers are formed as the orbit of x^i under conjugation.

Examples: (1) The group $D_4 wr S_4$ of degree 16 and order $98304 = 2^{15}3^1$ has 190 classes and rational classes. The algorithm takes 77 seconds.

(2) The Mathieu group M_{11} has order 7920, has 10 classes and 8 rational classes. As a permutation group of degree 11, the algorithm takes 2 seconds; while as a group of 5×5 matrices over GF(3) the algorithm takes 8 seconds.

(3) The symplectic group $Sp(4,4)$ of 4×4 matrices over GF(4) has order 979,200, has 27 classes and 18 rational classes. The algorithm takes 591 seconds.

(4) The special linear group $L(2,128)$ has order 2,097,024, has 129 classes and 6 rational classes. For the permutation representation of degree 129, the orbit algorithm takes 18,200 seconds.

5. Orbits of tuples for permutation groups

In 1970, C.C. Sims outlined a method for computing the conjugacy classes of a permutation group based on an ordering of the elements of the group and knowledge of the orbits of tuples of points. He used the method in a program to look for 2-elements when determining (by hand) the conjugacy classes of Suzuki's sporadic simple group and its automorphism group described by permutations of degree 1782. Butler(1979) implemented the approach and explored several variations. However, the approach only seems competitive for highly transitive groups like the Mathieu groups.

Let G be a group acting on the set of points Ω. Let g be an element of G. Define $cyc(g)$ to be the cycle decomposition of g where each cycle starts with the least possible integer, and the cycles are ordered in increasing order of their first point. Define $seq(g)$ to be the sequence of points obtained by removing the parentheses from $cyc(g)$. For example, consider the Mathieu group M_{11} generated by a=(1,10)(2,8)(3,11)(5,7) and b=(1,4,7,6)(2,11,10,9) of degree 11. Then the cycle decomposition $cyc(a)$ = (1,10)(2,8)(3,11)(4)(5,7)(6)(9), and hence $seq(a)$ = [1,10,2,8,3,11,4,5,7,6,9].

The elements of the group are ordered as follows: g is less than h if either

1. for some i and for all j, $1 \le j < i$, the j-th cycle of $cyc(g)$ has the same length as the j-th cycle of $cyc(h)$, but the i-th cycle of $cyc(g)$ is longer than the i-th cycle of $cyc(h)$.

2. for all i, the i-th cycle of $cyc(g)$ has the same length as the i-th cycle of $cyc(g)$ and $seq(g)$ is lexicographically less than $seq(h)$.

The representative of a conjugacy class is the first element in the class under this ordering of elements. The ordering is designed so that long cycles should be at the start of $cyc(g)$ if g is low in the order. This is sensible in light of the choice of an appropriate base for the computation of the centralizer of g.

Let $\lambda = (\alpha_1, \alpha_2, \ldots, \alpha_k)$ be a cycle of the element g. Then $\lambda^h = (\alpha_1^h, \alpha_2^h, \ldots, \alpha_k^h)$ is a cycle of g^h. Suppose τ is the first tuple in the orbit λ^G. Then there exists a conjugate g' of g such that τ is a cycle of g'. Define T_k to be the set of the first tuple of each G-orbit of k-tuples with distinct entries. Define

$$G[k] = \{\, g \in G \mid \text{for some } \tau \in T_k,\ \tau \text{ is a cycle of } g\}.$$

Then $G[k]$ contains the first element of every class of elements which have a cycle of length k.

Suppose $\tau = (\beta_1, \beta_2, \ldots, \beta_k) \in T_k$. Further suppose that $[\beta_1, \beta_2, \ldots, \beta_k, \ldots]$ is the base for G and that τ is a cycle of the element $g \in G$. Then the coset $G^{(k+1)}g$ is the set of elements of G which have τ as a cycle. Note that the cycle τ defines the initial segment $[\beta_2, \beta_3, \ldots, \beta_k, \beta_1]$ of the base image of g. Hence, g can be easily constructed if there exists an element of G with τ as a cycle.

The set of prefixes of length $k-1$ of the tuples in T_k contains the set T_{k-1}. We regard $\tau \in T_k$ as the initial segment of $seq(g)$ for candidate class representatives g. The initial part of $cyc(g)$ is obtained by introducing parentheses consistent with the ordering of elements that has long cycles first in $cyc(g)$. We call each such initial part a *bracketing* of τ. For cycles of length greater than k, or incomplete cycles at the end of the initial part of $cyc(g)$ in a bracketing, the cycle will be open rather than closed. The elements associated with a particular tuple and bracketing form a coset of $G^{(k+1)}$, or a coset of $G^{(k)}$ if the last cycle is open, though not every bracketing necessarily gives rise to elements of G. The union of all the cosets associated with the tuples and their bracketings contains the first element of each class.

For example, consider the Mathieu group M_{11}. The group is sharply 4-transitive, so $T_4 = \{(1,2,3,4)\}$. The bracketings are

$$(1)(2)(3)(4)\ \ (1,2)(3)(4)\ \ (1,2)(3,4)\ \ (1,2,3)(4)$$
$$(1,2,3)(4,\ \ (1,2,3,4)\ \ (1,2,3,4,$$

and the union of the cosets is

$g_0 =$ identity

$g_1 = (1,2)(5,8)(7,10)(9,11)$

$g_2 = (1,2)(3,4)(5,10)(7,8)$

$g_3 = (1,2,3)(5,9,7)(8,10,11)$

$g_4 = (1,2,3)(4,5,10)(6,11,7)$
$g_5 = (1,2,3)(4,6)(5,8,7,11,9,10)$
$g_6 = (1,2,3)(4,7,8)(6,10,9)$
$g_7 = (1,2,3)(4,8,11,6,9,5)(7,10)$
$g_8 = (1,2,3)(4,9,11)(5,6,8)$
$g_9 = (1,2,3)(4,10,8,6,7,9)(5,11)$
$g_{10} = (1,2,3)(4,11,10,6,5,7)(8,9)$

$g_{11} = (1,2,3,4)(7,9,8,11)$

$g_{12} = (1,2,3,4,11,9)(5,10,7)(6,8)$
$g_{13} = (1,2,3,4,8)(5,9,6,11,10)$
$g_{14} = (1,2,3,4,5,6,7,8,9,10,11)$
$g_{15} = (1,2,3,4,10,6,9,7)(5,8)$
$g_{16} = (1,2,3,4,6)(5,7,10,9,11)$
$g_{17} = (1,2,3,4,7,11,6,5)(8,10)$
$g_{18} = (1,2,3,4,9,5,11,8,7,6,10)$

Since long cycles should come first, the elements g_1, g_3, g_5, g_7, g_9, and g_{10} are not first in their class. Moreover, conjugacy tests will confirm that g_4 and g_6 are conjugate to g_8, and that g_{13} is conjugate to g_{16}. Hence, the first elements in the classes of M_{11} are

$$\{g_0, g_2, g_4, g_{11}, g_{12}, g_{14}, g_{15}, g_{16}, g_{17}, g_{18}\}.$$

The first tuples in the orbits of k-tuples are computed by extending each tuple $\tau = [\gamma_1, \gamma_2, \ldots, \gamma_{k-1}]$, which is first in its orbit, by the first points in the orbits of the stabiliser $G^{(k)} = G_{\gamma_1,\gamma_2,\ldots,\gamma_{k-1}}$ on $\Omega \setminus \{\gamma_1, \gamma_2, \ldots, \gamma_{k-1}\}$. As the stabiliser becomes smaller, the number of orbits becomes larger, especially if the degree is large. However, a small stabiliser reduces the size of the cosets $G^{(k)}g$ and $G^{(k-1)}g$ and their contribution to the size of $G[k]$. This trade-off as the length of tuples changes is further complicated by the combinatorial explosion of the number of bracketings as the tuple length k increases.

Experiments with variations that controlled tuple length, and examined the powers of newly found class representatives were performed on a CDC Cyber 72. When powers are examined we do not require the powers to be first in their class to be used as a class representative. All variations had problems with moderately sized groups that were not highly transitive, such as $L_5(2)$ of degree 31 and order $2^{10}3^2 5\ 7\ 31$; and G isomorphic to $2^6 : 3.S_6$ of degree 24 and order $2^{10}3^3 5$. The experiments allowed k to vary between 3 and 6 with a length less than 6 used if either $|G^{(k)}| \leq 50$ or the number of orbits of $G^{(k)}$ greater than 10. When no powers were examined, the classes of $L_5(2)$ took 1500 seconds, and the classes of G took 1000 seconds. When powers were examined, the two groups took 500 and 1000 seconds respectively. Hence, this approach cannot be recommended, especially when compared to the random method and the inductive schema.

6. Random method

After discovering the limitations of Sims' method based on orbits of tuples, Butler and Cannon developed a probabilistic approach for permutation and matrix groups — essentially an implementation of ad hoc hand methods used to find class representatives of simple groups. This was part of my thesis work (Butler,1979). Conservatism, and the speed of computers at the time, led us to regard 100-300 random elements as a large number to examine. However, Schneider(1990) had no hesitation in examining several thousand random elements when determining the classes in preparation for the calculation of character tables.

The random algorithm for permutation groups and groups of matrices defined over a finite field is based on two premises:

- The probability that a random element z of the group lies in a given class, or that a power of z lies in the class, is reasonably large, say greater than 0.1%.
- There are easily computed invariants of the elements in a class that distinguish most conjugacy classes of elements.

The second is not essential, but it provides a cheap means of determining that two elements definitely belong to different classes, and is useful in reducing the number of conjugacy tests via backtrack searches. When conjugacy is decided purely on the basis of invariants, then we call it *weak conjugacy testing*, whereas when a backtrack search is used to definitely determine whether two elements with the same invariants are conjugate we call it *strong conjugacy testing*.

For simple or near-simple groups the probabilities seem reasonably large from our experience. The third sporadic simple group $Co3$ of Conway of order 495,766,656,000 has precisely one class of elements of order 30. The centralizer is Z_{30} and the probability of finding an element in the class at random is 0.33%. This is the smallest probability for a class in $Co3$. On the other hand, the probability that a random element or a power of a random element lies in a class of involutions is 58% and 16% respectively for the two classes of involutions.

As an example of a bad case for the random method consider the Frobenius group G of degree p and order $p(p-1)$. The $p-1$ elements of order p in G form a single conjugacy class K, and there are no elements outside this class which have a power in K. Hence, the probability that an element or a power of an element lies in K is $1/p$, which can be made arbitrarily small.

The algorithm consists of three stages, and is parameterized by two integers, n_1 and n_2, which determine the number of random elements considered. For a large near-simple group, I would typically use $n_1 = 5000$ and $n_2 = 3000$.

Stage 1: Generate n_1 random elements of G and save one element with each distinct set of invariants encountered (that is, use weak conjugacy testing). Compute the centralizer of each element saved in order to determine the length of its class.

Stage 2: Examine the powers of the elements saved in Stage 1 for new class representatives. (In particular, determine the rational class of each element saved in Stage 1.) Use strong conjugacy testing to decide conjugacy. Compute the centralizer and determine the class length for each new class representative.

Stage 3: Generate n_2 further random elements and use strong conjugacy testing to determine conjugacy of the random element with the known class representatives having the same invariants. For each new class representative z, compute its centralizer, determine the length of its class, and examine its powers for further class representatives.

The algorithm maintains a running count of the number of elements in the known conjugacy classes, and will terminate once the count equals the order of G. The algorithm is unsuccessful if it completes the three stages without accounting for each element of the group in its list of known classes.

Uniformly distributed random elements may be formed by choosing a random coset representative $u^{(i)} \in U^{(i)}$, for each i, $1 \leq i \leq k$. The choice of each $u^{(i)}$ follows a uniform distribution, and is independent of the choices of the other $u^{(j)}$, $j \neq i$. The product $u^{(k)}u^{(k-1)} \ldots u^{(1)}$ is a random element distributed uniformly.

The strong conjugacy tests between a random element z and a known class representative x are speeded up by use of the orbits of the centralizer of x.

One can reduce the number of conjugacy tests when determining the fusion within a rational class and when examining the powers of an element z by noting that

$$z \sim_G z^i \;\Rightarrow\; z^m \sim_G z^{im}$$

for all integers m. This is a simple use of positive conjugacy tests. In Butler(1979), the algorithm maintains a subset Γ_j of the integers i less than $|z|$ with $\gcd(i,|z|) = j$, for each j dividing $|z|$. The subset stores the candidates for the class representatives in the rational class of the power of z. The algorithm runs through Γ_1 selecting the next integer i, tests conjugacy between z and z^i, and if they are conjugate it discards entries from each Γ_j. The entries discarded are the cycles $l \times i^k \ mod \ |z|$, for each l in Γ_j. However, l itself, the first point in the cycle is not discarded. After Γ_1 has been processed, its entries are the class representatives of the rational class. Each Γ_j is treated similarly, but with two differences: first we check whether z^j is a new class representative - if it is not then we set each Γ_{jk} to the empty set; secondly, a positive conjugacy test between z^j and z^{ij} only affects the sets Γ_{jk}. As an example, consider the group M_{11}. Suppose Stage 1 finds elements with each distinct cycle shape. Let z be an element of order 8. Then z and z^5 are the class representatives in the rational class of z. Stage 2 would initialise the sets $\Gamma_1 = \{1,3,5,7\}$, $\Gamma_2 = \{2,6\}$, $\Gamma_4 = \{4\}$. The test $z \sim z^3$ is positive, so we would discard (non-first integers in) the cycles (1,3) and (5,7) from Γ_1, (2,6) from Γ_2, and (4) from Γ_4 to leave $\Gamma_1 = \{1,5\}$, $\Gamma_2 = \{2\}$, $\Gamma_4 = \{4\}$. The test $z \sim z^5$ is negative. When processing Γ_2, we find that z^2 is not a new class representative so both Γ_2 and Γ_4 are set to the empty set.

Examples: (1) The group $Co3$ has a permutation representation of degree 276. It has 38 rational classes and 42 classes. Only two rational classes have the same invariants - cycle shape $8^{30}4^7 2^3 1^2$ - and the probability of finding an element (or an element with a power) in the class is 4.7% and 3.1% respectively. The time taken by the random method is 280 seconds.

(2) The second sporadic simple group $Co2$ of Conway has order $2^{18}3^6 5^3 7\,11\,23$ and degree 2300. It has 56 rational classes and 60 classes. All rational classes have distinct cycle shape. The random method took 3800 seconds.

(3) The wreath product G of the dihedral group D_4 with the symmetric group S_4 has degree 16 and order 98 304 = $2^{15}3$. G has 190 rational classes and 190 classes. There are 51 distinct cycle shapes amongst its elements. There is a class of involutions which have no roots and the class has length 24. Hence the probability of finding an element in this class is less than 0.03%. Furthermore, there are at least 12 classes where the probability that a random element (or a power of it) lies in the class is less than 0.1%. The random method with $n_1 = 10000$ and $n_2 = 5000$ takes 120 seconds but fails to find a representative of each class. When the parameters are both increased by a factor of 10 the random method takes 593 seconds and is still not successful.

(4) For M_{11} as a group of 5×5 matrices over GF(3), the random method takes 2 seconds.

(5) For the symplectic group $Sp(4,4)$ as a group of 4×4 matrices over GF(4), the random method takes 21 seconds.

(6) The symplectic group $Sp(4,8)$ has order $2^{12}3^4 5\ 7^2 13$, 83 classes and 28 rational classes. As a group of 4×4 matrices over GF(8), the random method takes 7980 seconds. As a permutation group of degree 585, the random method takes 1830 seconds.

(7) The special linear group $L_5(2)$ has order 9,999,360 = $2^{10}3^2 5\ 7\ 31$, 27 classes and 18 rational classes. As a group of 5×5 matrices over GF(2), the random method takes 55 seconds. As a permutation group of degree 31, the random method takes 10 seconds.

7. Orbit-stabiliser algorithm for p-groups

Felsch and Neubüser(1979) devised a top-down approach for computing the conjugacy classes of elements of a p-group given by a pc presentation. The classes of $G/G(i)$ are extended to classes of $G/G(i+1)$ using the fact that $G(i)/G(i+1)$ is central of order p. Leedham-Green and Cannon improved the approach by extending by a central elementary abelian section at each step.

These are very powerful algorithms limited more by the number of conjugacy classes of the group than by the order of the group.

At each step of the algorithm the following information about the classes of $G/G(i)$ is stored:

- each class representative $gG(i)$, and
- the centralizer $C_{G/G(i)}(\, gG(i)\,)$.

A centralizer is represented by a gs, which is not necessarily canonical.

The algorithm begins by noting that a group of order p^2 is abelian. Hence, for $G/G(3)$, each element is a class representative with centralizer $G/G(3)$ and gs $[a_1G(3), a_2G(3)]$. The algorithm then proceeds top-down extending the classes of $G/G(i)$ to classes of $G/G(i+1)$ using the fact that $G(i)/G(i+1) = \langle a_iG(i+1)\rangle$ is central of order p. The extension is based on the results in Proposition 3.1 and the following theorem.

Theorem 7.1. ([Felsch and Neubüser, 1979, p. 455])
Let $N = \langle z\rangle$ be central of order p in G. Let C/N be centralizer of gN in G/N be given by a generating sequence $[c_1N, c_2N, \ldots, c_mN]$. Then there are two cases to consider:

Case I: $[g, C] = \langle id\rangle$ (i.e. p orbits of length 1)

Then each element of gN is a class representative with centralizer C, which has gs $[c_1, c_2, \ldots, c_m, z]$.

Case II: $[g, C] \neq \langle id\rangle$ (i.e. 1 orbit of length p)

Then g is the class representative with centralizer C^*, given by a gs $[c_1^*, c_2^*, \ldots, c_m^*]$, where the elements c_i^* are defined as follows: Let $j \in \{1, 2, \ldots, m\}$ be maximal with $[g, c_j] \neq id$. For each i, $1 \leq i < j$, there is an integer $k(i)$ with $0 \leq k(i) < p$ such that $[g, c_i] = [g, c_j]^{k(i)}$. Let

$$c_i^* = \begin{cases} c_i c_j^{-k(i)} & \text{for } 1 \leq i < j, \\ c_{i+1} & \text{for } j \leq i < m, \text{ and} \\ z & \text{for } i = m. \end{cases}$$

As an example of the algorithm, consider the group Q of order 2^6 defined by the pc presentation

$$Q = \langle a_1, a_2, a_3, a_4, a_5, a_6 \mid a_1^2 = 1, a_2^2 = a_4, a_3^2 = a_6, a_4^2 = 1, a_5^2 = a_6, a_6^2 = 1,$$

$$(a_2, a_1) = a_4, (a_3, a_2) = a_5, (a_4, a_3) = a_6, (a_5, a_1) = a_6, (a_5, a_3) = a_6,$$

$$\text{other commutators are trivial} \rangle$$

isomorphic to a Sylow 2-subgroup of the non-split extension $2^3L(3,2)$. The classes of each quotient $Q/Q(i)$, i = 3,4,5,6, and Q itself are displayed in Figure 1 — a gs for the centralizer in Q of the class representative is also given. The group $Q/Q(4)$ is elementary abelian, so only Case I of Theorem 7.1 is encountered when extending $Q/Q(3)$ to $Q/Q(4)$. However, mainly Case II is encountered in the other steps. As an example of Case II, consider $g = a_1a_2Q(4)$ when determining the classes of $Q/Q(5)$. The centralizer $C/Q(4)$ has gs $[a_1Q(4), a_2Q(4), a_3Q(4)]$, so j=2 and $[g, c_2] = a_4Q(5)$. For i = 1, $k(i)$ = 1 since $[g, c_1] = a_4Q(5)$, so $c_1^* = a_1a_2^{-1}Q(5) = a_1a_2a_4Q(5)$. Therefore a gs of the centralizer C^* is $[a_1a_2a_4Q(5), a_3Q(5), a_4Q(5)]$. The classes of Q take less than 0.05 seconds to compute on a Decstation 5000.

$Q/Q(3)$	$Q/Q(4)$	$Q/Q(5)$	$Q/Q(6)$	Q	*Centralizer*
id	*id*	*id*	*id*	*id*	$[a_1, a_2, a_3, a_4, a_5, a_6]$
				a_6	$[a_1, a_2, a_3, a_4, a_5, a_6]$
			$a_5Q(6)$	a_5	$[a_1a_3a_6, a_2, a_4, a_5, a_6]$
		$a_4Q(5)$	$a_4Q(6)$	a_4	$[a_1, a_2, a_4, a_5, a_6]$
			$a_4a_5Q(6)$	a_4a_5	$[a_1, a_2, a_4, a_5, a_6]$
	$a_3Q(4)$	$a_3Q(5)$	$a_3Q(6)$	a_3	$[a_1, a_3, a_4a_5a_6, a_6]$
		$a_3a_4Q(5)$	$a_3a_4Q(6)$	a_3a_4	$[a_1, a_3a_5a_6, a_4a_5a_6, a_6]$
$a_1Q(3)$	$a_1Q(4)$	$a_1Q(5)$	$a_1Q(6)$	a_1	$[a_1, a_3, a_4, a_6]$
			$a_1a_5Q(6)$	a_1a_5	$[a_1a_5a_6, a_3a_5a_6, a_4, a_6]$
	$a_1a_3Q(4)$	$a_1a_3Q(5)$	$a_1a_3Q(6)$	a_1a_3	$[a_1, a_3, a_5, a_6]$
			$a_1a_3a_5Q(6)$	$a_1a_3a_5$	$[a_1, a_3, a_5, a_6]$
$a_2Q(3)$	$a_2Q(4)$	$a_2Q(5)$	$a_2Q(6)$	a_2	$[a_2a_3, a_4a_5a_6, a_6]$
				a_2a_6	$[a_2, a_4, a_5, a_6]$
	$a_2a_3Q(4)$	$a_2a_3Q(5)$	$a_2a_3Q(6)$	a_2a_3	$[a_2a_3, a_4a_5a_6, a_6]$
$a_1a_2Q(3)$	$a_1a_2Q(4)$	$a_1a_2Q(5)$	$a_1a_2Q(6)$	a_1a_2	$[a_1a_2a_4, a_4, a_6]$
	$a_1a_2a_3Q(4)$	$a_1a_2a_3Q(5)$	$a_1a_2a_3Q(6)$	$a_1a_2a_3$	$[a_1a_2a_3, a_5, a_6]$

Figure 1. Class representatives of example Q

The nilpotent quotient algorithm (NQA) is a common way in which pc presentations of p-groups are created. The NQA defines a lower exponent-p-central series, which has central elementary abelian sections. Cannon and Leedham-Green improved the top-down approach of Felsch and Neubüser to compute the conjugacy classes of elements by extending by a central elementary abelian group N, rather than only by a central group of order p. The foundations for the improved approach are given in the following theorem.

Theorem 7.2. ([Leedham-Green, see (Mecky and Neubüser, 1989, p284)]) *Let N be a central, elementary abelian subgroup of G. Let $[n_1, n_2, \ldots, n_d]$ be a gs of N. Let the centralizer C/N of gN in G/N be given by gs $[c_1N, c_2N, \ldots, c_mN]$. Then there are two cases to consider:*

Case I: $[g, C] = \langle id \rangle$

Then each element of gN is a class representative with centralizer C, which has a generating sequence $[c_1, c_2, \ldots, c_m, n_1, n_2, \ldots, n_d]$.

Case II: $[g, C] \neq \langle id \rangle$

Let K be a complement of $[g,C]$ in N. Then each element of gK is a class representative. Their centralizer is given by a gs $[c_1^*, c_2^*, \ldots, c_{m-r}^*, n_1, n_2, \ldots, n_d]$ defined as follows: Let r be the rank of $[g,C]$. Define a basis $B = \{b_1, b_2, \ldots, b_r\}$ of $[g, C]$ and the elements c_j^* by

> *$d := 0$;*
> **for** *$i := m$* **downto** *1* **do**
>   *$b := [g,c_i]$;*
>   **if** *b is independent of $\{b_1, b_2, \ldots, b_d\}$* **then**
>   *$d := d + 1$; $b_d := b$; $c_{i_d} := c_i$;*
>   **else**
>   *let $b = b_1^{e_1} b_2^{e_2} \ldots b_d^{e_d}$;*
>   *$c_{i-r+d}^* := c_i \left(c_{i_1}^{e_1} c_{i_2}^{e_2} \ldots c_{i_d}^{e_d}\right)^{-1}$;*
>   **end if**;
> **end for**;

We repeat the computation of the classes of our example Q, but this time we go down by elementary abelian sections. The top section is $Q/Q(4)$ which is elementary abelian of order 8. The group $Q(4)/Q(6)$ is elementary abelian of order 4 and central in $Q/Q(6)$. It is the section used for the next step of the algorithm. Figure 2, like Figure 1 lists the class representatives of $Q/Q(4)$, $Q/Q(6)$, and Q, but Figure 2 also lists the bases B of $[g, C]$ for each class representative of $Q/Q(4)$, from which we can easily deduce the classes of $Q/Q(6)$. The final step, extending the classes of $Q/Q(6)$ to the classes of Q is exactly the same as the algorithm of Felsch and Neubüser.

Let us now consider the performance of the improved algorithm on some larger examples. In their paper, Felsch and Neubüser(1979) define a space group S and two quotients of order 2^{34} and 2^{42} respectively of S. The first

$Q/Q(4)$	$[g,C]$	$Q/Q(6)$	Q	*Centralizer*
id	*id*	*id*	*id*	$[a_1, a_2, a_3, a_4, a_5, a_6]$
			a_6	$[a_1, a_2, a_3, a_4, a_5, a_6]$
		$a_4Q(6)$	a_4	$[a_1, a_2, a_4, a_5, a_6]$
		$a_5Q(6)$	a_5	$[a_1a_3a_6, a_2, a_4, a_5, a_6]$
		$a_4a_5Q(6)$	a_4a_5	$[a_1, a_2, a_4, a_5, a_6]$
$a_1Q(4)$	$[a_4Q(6)]$	$a_1Q(6)$	a_1	$[a_1, a_3, a_4, a_6]$
		$a_1a_5Q(6)$	a_1a_5	$[a_1a_5a_6, a_3a_5a_6, a_4, a_6]$
$a_2Q(4)$	$[a_4Q(6), a_5Q(6)]$	$a_2Q(6)$	a_2	$[a_2a_3, a_4a_5a_6, a_6]$
			a_2a_6	$[a_2, a_4, a_5, a_6]$
$a_3Q(4)$	$[a_5Q(6)]$	$a_3Q(6)$	a_3	$[a_1, a_3, a_4a_5a_6, a_6]$
		$a_3a_4Q(6)$	a_3a_4	$[a_1, a_3a_5a_6, a_4a_5a_6, a_6]$
$a_1a_2Q(4)$	$[a_4Q(6), a_5Q(6)]$	$a_1a_2Q(6)$	a_1a_2	$[a_1a_2a_4, a_4, a_6]$
$a_1a_3Q(4)$	$[a_4a_5Q(6)]$	$a_1a_3Q(6)$	a_1a_3	$[a_1, a_3, a_5, a_6]$
		$a_1a_3a_5Q(6)$	$a_1a_3a_5$	$[a_1, a_3, a_5, a_6]$
$a_2a_3Q(4)$	$[a_4Q(6), a_5Q(6)]$	$a_2a_3Q(6)$	a_2a_3	$[a_2a_3, a_4a_5a_6, a_6]$
$a_1a_2a_3Q(4)$	$[a_4Q(6), a_5Q(6)]$	$a_1a_2a_3Q(6)$	$a_1a_2a_3$	$[a_1a_2a_3, a_5, a_6]$

Figure 2. Improved algorithm applied to example Q

quotient has 17896 classes, which require 1400 seconds to compute. The second quotient has so many classes that it exhausts 32 Mbytes of workspace after 20000 seconds. A Sylow 3-subgroup of the projective special linear group $L(5,9)$ of dimension 5 over $GF(9)$ has order 3^{20}. It has 32401 classes, which requires 580 seconds to compute when the Sylow subgroup is described by a pc presentation.

The above algorithms can be adapted to solve related problems such as compute only those classes larger than a certain size; determine a gs of the centralizer of a given element; given an element g, determine a representative of the class containing g which is canonical relative to the given pc presentation of the group (and determine an element which conjugates g to the canonical class representative); and determine whether two elements of the group are conjugate.

8. Orbit-stabiliser algorithm for soluble groups

Laue, Neubüser, and Schoenwaelder(1984) recognised that the orbit-stabiliser approach for p-groups would work for soluble groups given by a pc presentation. The approach was implemented by Mecky in 1986 as part of the SOGOS system using Fortran, and independently by Slattery and Cannon in 1987 as part of Cayley using C. The use of affine transformations by Mecky and Neubüser(1989) improved the performance of the algorithm considerably.

The soluble group G is given by a conditioned pc presentation, which defines a subnormal chain $G(i)$, $i = 1, 2, \ldots, n+1$ which refines a normal chain $N(i)$, $i = 1, 2, \ldots, r+1$. The top-down approach applies Proposition 3.1, a general result about lifting classes of G/N to G, to the case where the classes of $G/N(i)$ are lifted to classes of $G/N(i+1)$. Here N is $N(i)/N(i+1)$. At each step of the algorithm the information stored about the classes of $G/N(i)$ is (1) each class representative $gN(i)$, and (2) the centralizer $C_{G/N(i)}(\, gN(i)\,)$. A centralizer is represented by a gs, which is not necessarily canonical. For each class of $G/N(i)$, the algorithm determines the orbits of the preimage C of its centralizer on the coset $gN(i)$ to find the orbit representatives and their stabilisers (which define the class representatives and their centralizer respectively). The algorithm begins with the classes of the trivial group $G/N(1)$.

The orbits are calculated by working up the subnormal series given by the gs of the preimage C, since the orbits of a normal subgroup are blocks of the group. The following algorithm (Laue et al, 1984, 3.1) forms an orbit and the stabiliser simultaneously.

Algorithm 2
Input : a finite soluble group G
 with prime-step generating sequence $[g_1, g_2, \ldots, g_n]$
 and which acts on a set Ω;
 $\omega \in \Omega$;
Output : the orbit $\Delta = \omega^G$;
 a prime-step generating sequence S of the stabiliser of ω in G;
begin
 $\Delta := \{\ \omega\ \}$; $S :=$ empty sequence;
 for $i := n$ **downto** 1 **do**
  if $\omega^{g_i} \notin \Delta$ **then**
   set Δ to be the disjoint union of $\Delta^{g_i^j}$, for $j = 0, 1, 2, \ldots, p_i - 1$;
  else
   find $x \in G(i+1)$ such that $\omega^{g_i} = \omega^x$;
   append $g_i x^{-1}$ to S;
  end if;
 end for;
end.

As an example of the algorithm, consider the group Q of order $2^6 3$ defined by the pc presentation

$$Q = \langle a_1, a_2, a_3, a_4, a_5, a_6, a_7 \mid a_1^2 = a_5, a_2^3 = 1, a_3^2 = a_7, a_4^2 = a_7, a_5^2 = 1,$$
$$a_6^2 = 1, a_7^2 = 1, (a_2, a_1) = a_2 a_5 a_6, (a_3, a_1) = a_3 a_4 a_7, (a_4, a_1) = a_3 a_4 a_7,$$

$$(a_6, a_1) = a_5, (a_3, a_2) = a_4, (a_4, a_2) = a_3a_4a_7, (a_5, a_2) = a_5a_6,$$
$$(a_6, a_2) = a_5, (a_4, a_3) = a_7, (a_5, a_3) = a_7, (a_6, a_3) = a_7, (a_5, a_4) = a_7,$$
$$\text{other commutators are trivial}\rangle$$

isomorphic to a subgroup 2^3S_4 of the non-split extension $2^3L(3,2)$. The classes of each quotient $Q/Q(i)$, i = 2,3,5, and Q itself are displayed in Figure 3 — a gs for the centralizer in Q of the class representative is also given.

$Q/Q(2)$	$Q/Q(3)$	$Q/Q(5)$	Q	*Centralizer*
id	id	id	id	$[a_1, a_2, a_3, a_4, a_5, a_6, a_7]$
			a_6	$[a_1a_2^2, a_4, a_5, a_6, a_7]$
			a_7	$[a_1, a_2, a_3, a_4, a_5, a_6, a_7]$
		$a_4Q(5)$	a_4	$[a_1a_2^2, a_3a_5, a_4, a_6, a_7]$
			a_4a_5	$[a_3, a_4a_5, a_6, a_7]$
			a_4a_6	$[a_1a_2^2, a_3, a_4, a_6, a_7]$
	$a_2Q(3)$	$a_2Q(5)$	a_2	$[a_2, a_7]$
			a_2a_7	$[a_2, a_7]$
$a_1Q(2)$	$a_1Q(3)$	$a_1Q(5)$	a_1	$[a_1, a_3a_4, a_5, a_7]$
			a_1a_6	$[a_1a_6, a_5, a_7]$
			a_1a_7	$[a_1, a_3a_4, a_5, a_7]$
		$a_1a_4Q(5)$	a_1a_4	$[a_1a_4, a_3a_4a_5, a_7]$
			$a_1a_4a_6$	$[a_1a_4a_6, a_3a_4, a_7]$

Figure 3. Class representatives of example Q

The normal series is $Q(i)$, for $i = 1, 2, 3, 5, 8$. The quotient $Q/Q(5)$ is isomorphic to S_4, whose classes are well-known, so we will look at extending the classes of $Q/Q(5)$ to those of $Q/Q(8)$, which is Q. Consider the class of $a_1Q(5)$ with centralizer generated by $[a_1Q(5), a_3a_4Q(5)]$. The complete preimage C is generated by $[a_1, a_3a_4, a_5, a_6, a_7]$. The action of C on the coset $a_1Q(5)$ of size 8 is required. The application of Algorithm 2 for $\omega = a_1$ proceeds by appending a_7 to S since $a_1^{a_7} = a_1$; setting $\Delta = \{a_1, a_1a_5\}$ since $a_1^{a_6} = a_1a_5$; appending a_5 to S since $a_1^{a_5} = a_1$; appending a_3a_4 to S since $a_1^{a_3a_4} = a_1$; appending a_1 to S since $a_1^{a_1} = a_1$. The application of Algorithm 2 to $\omega = a_1a_7$ proceeds similarly, giving the orbit $\{a_1a_7, a_1a_5a_7\}$ and stabiliser generated by $[a_1, a_3a_4, a_5, a_7]$. The application of Algorithm 2 for $\omega = a_1a_6$ proceeds by appending a_7 to S since $\omega^{a_7} = \omega$; setting $\Delta = \{a_1a_6, a_1a_5a_6\}$ since $\omega^{a_6} = a_1a_5a_6$; appending a_5 to S since $\omega^{a_5} = \omega$; setting $\Delta = \{a_1a_6, a_1a_5a_6, a_1a_6a_7, a_1a_5a_6a_7\}$ since $\omega^{a_3a_4} = a_1a_6a_7$; appending a_1 to S since $\omega^{a_1} = \omega$.

In a common case where the normal subgroup N is elementary abelian group of order p^t, the use of affine transformations is very beneficial. Consider lifting the class gN of G/N by determining the orbits of the preimage C of its centralizer on the coset gN. Let $c \in C$ and $gn \in gN$. Since N is elementary abelian and $[g, c] \in N$,

$$(gn)^c = gn^c[g, c].$$

Hence, the action of C on gN is equivalent to the action of the group $A = \{\alpha_{c,g} \mid c \in C\}$ on N, where

$$\alpha_{c,g} \; : \; n \; \mapsto n^c[g, c].$$

The equivalence is given by the bijection $n \mapsto gn$ between N and gN. The map $cnj_{c,g} \; : \; n \; \mapsto n^c$ is a linear transformation of the vector space defined by N, while the map $mlt_{c,g} \; : \; n \; \mapsto n[g, c]$ is a translation. So their composition $\alpha_{c,g}$ is called an *affine transformation.* Rather than use collection, which can be expensive, to determine the action of C on gN, the action of the affine group A on N is used.

Consider our previous example group Q and the task of lifting the class $a_1Q(5)$. So $N = Q(5)$ and $g = a_1$. Figure 4 presents the affine transformation associated with each generator in the gs of C relative to the "basis" $[a_5, a_6, a_7]$ of the vector space N.

c	$[g, c]$	$cnj_{c,g}$	orbits of $\alpha_{c,g}$ on N
a_1	id	$\begin{pmatrix} 1 & 0 & 0 \\ 1 & 1 & 0 \\ 0 & 0 & 1 \end{pmatrix}$	$\{id\}, \{a_5\}, \{a_6, a_5a_6\}, \{a_7\}, \{a_5a_7\}, \{a_6a_7, a_5a_6a_7\}$
a_3a_4	id	$\begin{pmatrix} 1 & 0 & 0 \\ 0 & 1 & 1 \\ 0 & 0 & 1 \end{pmatrix}$	$\{id\}, \{a_5\}, \{a_6, a_6a_7\}, \{a_7\}, \{a_5a_7\}, \{a_5a_6, a_5a_6a_7\}$
a_5	id	I	discrete orbits
a_6	a_5	I	$\{id, a_5\}, \{a_6, a_5a_6\}, \{a_7, a_5a_7\}, \{a_6a_7, a_5a_6a_7\}$
a_7	id	I	discrete orbits

Figure 4. Affine transformations on $Q(5)$.

Hence, the A-orbits on N are

$$\{id, a_5\}, \;\; \{a_6, a_5a_6, a_5a_6a_7, a_6a_7\}, \;\; \{a_7, a_5a_7\}.$$

Hence, the C-orbits on gN are as in the above example.

Since $N \triangleleft C$ the N orbits are blocks of the action of C on the coset gN. Furthermore, for $m \in N$,

$$\alpha_{m,g} \; : \; n \; \mapsto n[g, m]$$

which is a translation. The translating elements form a subgroup (or subspace) $[g, N]$ of N and the orbits of A on N can be deduced from the orbits of $\overline{A}$ on $N/[g, N]$. Here one defines

$$\overline{\alpha_{c,g}} \; : \; n[g, N] \; \mapsto n^c[g, c][g, N].$$

To return to our example of lifting $a_1Q(5)$ we note that $[g, N] = \langle a_5 \rangle$ and $N/[g, N]$ has "basis" $\overline{a_6}, \overline{a_7}$. The $\overline{A}$-orbits on $N/[g, N]$ are

$$\{\overline{id}\}, \quad \{\overline{a_6}, \overline{a_6a_7}\}, \quad \{\overline{a_7}\}.$$

These, when expanded by the subgroup $[g, N] = \{id, a_5\}$, give the A-orbits on N and the corresponding C-orbits on gN.

Certain special cases are useful to recognize as an aid to improved efficiency. When N is central then one can use the result of Leedham-Green (Theorem 7.2). When N is cyclic and non-central of order p then calculations with the affine transformations reduce to calculations in $\mathbb{Z}/p\mathbb{Z}$.

Examples: (1) The wreath product $D_4 wr S_4$ order 98 304 $= 2^{15}3$. This group has 190 rational classes and 190 classes. The orbit-stabiliser algorithm computes the classes in 18 seconds.

(2) The Cayley library `imp768` defines a soluble permutation group of degree 768 and order 1,088,391,168 $= 2^{11}3^{12}$. One can obtain a conditioned pc presentation for the group using the *pcrepresentation* function of Cayley. Then the orbit-stabiliser algorithm computes the 117 classes in 186300 seconds. In this group collection is often very time-consuming. Almost all the time of the orbit-stabiliser algorithm is spent forming the orbits of the last section $N(9)/N(10)$ of size $3^8 = 6561$.

The effect of affine transformations must be estimated, since the algorithm implemented in Cayley v3.7 does not use them. Mecky and Neubüser(1989) report that for a group G_1, where `imp768` is a quotient G_1/Z_3 (and on a computer other than the Decstation 5000), the use of affine transformations reduced the total time from 2496 seconds to 101 seconds. Hence, we could estimate that on a Decstation 5000 the total time of 186300 seconds would be reduced to 7600 seconds by the use of affine transformations.

9. Inductive schema for permutation groups

The ideas behind the inductive schema are inherent in theoretical work on the classification of simple groups. Sims in private discussions with Butler and Cannon in 1976 suggested treating elements of prime-power order and composite order separately, and finding the elements of composite order inside centralizers of elements of prime order. In 1984, Peter Neumann pointed out that the kernel of the action of the centralizer of an element of prime order on the cycles of the element is elementary abelian — giving an effective algorithm for Sylow subgroups, and a route to finding the elements of composite order. The computation of a Sylow subgroup and the algorithm of Felsch and Neubüser(1979) would lead to the classes of elements of prime power order if one could compute a pc presentation of the Sylow subgroup.

In 1985, Butler and Cannon completed the task, started by Holt(1984), of computing a pc presentation and setting up the isomorphism between the permutation representation of the Sylow subgroup and the p-group (see Butler and Cannon(1993)). In 1987, Butler implemented a first prototype of the inductive schema in Cayley and introduced several important ideas such as the order in which to process the primes dividing $|G|$, and discarding roots of a class representative of a Sylow subgroup if it fuses in G to an earlier representative. With Cheryl Praeger and John Cannon, he developed and implemented a method to lift classes of elements in the quotient of the centralizer to the classes of elements of composite order. Butler observed that each class of the quotient lifted to a single class of elements of composite order, and in 1990 he proved this fact and re-implemented the prototype, computing rational classes and fully utilising the theory of fusion of classes within a rational class.

The main ideas behind the approach are

- For a given prime p, a representative of each class of p-elements can be found in a fixed Sylow p-subgroup S of G.
- Representatives of the classes of elements of order $p^r t$, where t is coprime to p, can be found in the centralizers of the class representatives of elements of order p^r.
- The class representatives of elements of order $p^r t$ can be found by taking preimages of the class representatives of elements of order t in the quotient of the centralizer given by its action on the cycles of the element of order p^r.

Let $z \in G$ have order p^r and let $C = C_G(z)$. Then C permutes the cycles of z and the kernel of the action is an abelian p-group whose elements all have order dividing p^r.

The approach can treat the primes p dividing the order of G in terms of "increasing difficulty", so that for the last prime it is not necessary to analyse the classes of the centralizers or their quotients. It is outlined in Algorithm 3. Rational classes should be determined rather than classes, as the latter entails duplication of effort in computing centralizers and analysing the classes of their quotients.

Consider the Mathieu group M_{11} of degree 11 and order $7920 = 2^4 3^2 5\ 11$. For $p = 11$, the Sylow subgroup is cyclic of order p and therefore has one rational class of elements of order 11 and it consists of 10 classes. A representative is $d = (1,2,3,4,5,6,7,8,9,10,11)$. $Aut(Z_{11}) \simeq Z_{10}$ is generated by $2\ mod\ 11$ and its subgroups are $Z_{10} = \langle 2\ mod\ 11 \rangle$, $Z_5 = \langle 4\ mod\ 11 \rangle$, $Z_2 = \langle 10\ mod\ 11 \rangle$, and the identity. To determine the fusion within the G-rational class of d, we test whether d is conjugate to d^2 — it is not — and then whether d is conjugate to d^4 — it is. Hence, we know that $Gal_{M_{11}}(z) = Z_5$

Algorithm 3
Input : a finite permutation group G of order $p_1^{n_1} p_2^{n_2} \dots p_s^{n_s}$;
Output : the classes of G;
begin
for each prime p_i dividing the order of G **do**
$S :=$ Sylow p_i-subgroup of G;
let $f_1 : S \to \bar{S}$, where $\bar{S}$ is defined by a pc presentation of S;
compute the classes of $\bar{S}$ (and hence, of S);
determine their fusion in G; "the classes of p_i-elements of G"
for each class $K_G(z)$ of p_i-elements **do** "(some) classes of roots of z"
$C := C_G(z)$;
let $f_2 : C \to \bar{C}$, the action of C on the cycles of z;
determine the classes of $\bar{C}$ of elements whose order only involves the primes $p_{i+1}, p_{i+2}, \dots, p_s$;
lift the classes of $\bar{C}$ to roots of z and determine their fusion in G;
end for;
end for;
end.

and that the rational class consists of two classes with representatives z and z^2. The Sylow 5-subgroup is treated similarly, requiring a single conjugacy test. However, the Sylow 3-subgroup S is elementary abelian of order 3^2 with 4 rational classes of elements of order 3 each consisting of 2 classes. In M_{11}, the rational classes are all conjugate, and there is only one class per rational class. Four conjugacy tests are needed. Let $z = (1,6,2)(3,7,11)(4,5,9)$ be the representative. The centralizer $C = C_G(z)$ has order $2\,3^2$ and is generated by z, x=(1,4,3)(2,9,11)(5,7,6), and y=(1,2,6)(3,9,7,4,11,5)(8,10). The only possible composite order is 6. The class of elements of order 6 is found by forming a Sylow 2-subgroup $\langle \bar{y} \rangle$ of the quotient $\bar{C}$ of C acting on the cycles of z. Its unique rational class of elements of order 2 lifts to a single rational class of elements of order 6 which must consist of a single class. A Sylow 2-subgroup of M_{11} has 6 rational classes and 7 classes. The representatives have order 1, 2, 2, 4, 4, and 8 and centralizers (in the Sylow subgroup) of order 2^4, 2^4, 2^2, 2^3, 2^2, and 2^3. The last rational class consists of two classes. Elements of the same order have identical cycle shape. Choosing the first class of each cycle shape and computing the centralizer in M_{11} of the representative, and the fact that the fusion in M_{11} within a rational class is determined by the fusion in the Sylow subgroup, shows that we have accounted for all the elements of the group. No conjugacy tests in M_{11} are required to determine the classes of 2-elements. Hence our list of rational classes is complete. A total of 7 conjugacy tests is needed.

There are several major subproblems to be solved. Some have technical solutions, but others require a strategy to be developed.

Problem 1. Fusion within a rational class of p-elements: Determine the classes within each rational class $Kr_G(z)$, where z has prime power order.

The solution uses the theory of Galois groups to minimise the number of conjugacy tests performed. We have information about $F = Gal_S(z)$, where S is the Sylow subgroup, and we search the lattice of subgroups between F and a conservative approximation to $Aut(\langle z \rangle) \cap G$. The examination of each subgroup requires one conjugacy test. We stop once a conjugacy test returns a positive result.

Careful ordering of the classes of S implies, by Corollary 3.1, that we know the p-part of the Galois group. Moreover, for $p = 2$, this is the whole Galois group.

Problem 2. Classes of elements of composite order: Given an element z of prime power order p^r, determine the classes of elements y where $y^t \in Kr_G(z)$, and t is coprime to p.

Problem 3. Fusion within a rational class of elements of composite order: Determine the classes within each rational class $Kr_G(y)$, where y has composite order.

For elements y of composite order, we assume we know the fusion of the elements z of prime power order, and use the following results to determine conjugacy and fusion in G from information about the quotient $\overline{C}$ of the centralizer $C = C_G(z)$ in its action on the cycles of z. Suppose that

$$y \in G \text{ has order } p^r t, \ t \text{ is coprime to } p, \text{ and } y^t = z. \qquad (\ddagger)$$

Let Υ be the partition of Ω determined by the cycles of z. Let $f : C \rightarrow C\downarrow_\Upsilon$, $N = ker(f)$, and $\bar{C} = C/N$. Then $\bar{y} = f(y) \in \bar{C}$ has order t.

Theorem 9.1. ([Butler,1990]) *Suppose y_1, y_2 satisfy (‡). Then $y_1 \sim_G y_2$ if and only if $\bar{y_1} \sim_{\bar{C}} \bar{y_2}$.*

Theorem 9.2. ([Butler,1990]) *Suppose y_1, y_2 satisfy (‡). Let $g \in N_G(z)$ be an element conjugating z to z^m, $m \neq 1$. Then $y_1 \sim_G {y_2}^m$ if and only if $\overline{{y_1}^g} \sim_{\bar{C}} \bar{y_2}^m$.*

Problem 4. Classes of p-elements: Determine the classes in G of elements of prime power order.

This is really the problem of determining the fusion of the classes of a Sylow subgroup S in G. It is the bottleneck for the inductive schema, as a Sylow subgroup may have many thousand classes. The strategy used is to analyse and sort the rational classes of S, and then to examine their fusion in G according to the sorted order. The analysis categorizes the classes of the Sylow subgroup by the order of the elements, and their cycle structure. Within each category, the analysis sorts the classes in decreasing order of their centralizer order (so that the roots of a discarded class can be safely discarded), and within a centralizer order it sorts the classes in decreasing order of their normalizer order (so that the p-part of the Galois group in G is known). The first class with each distinct cycle structure clearly represents a new G-class and they are listed first. The remaining classes are then listed. The classes of elements of order p are listed using the sorted order. After a class of elements of order p — with representative z— and before the next class of elements of order p, we list the roots of z in a recursive fashion. (If z is discarded, then so will its roots be discarded, but if z represents a new class of G then there will also be some new classes amongst its roots, so they should be considered at an early stage.)

Problem 5. Ordering the primes: Determine the order in which the primes p_i dividing $|G|$ should be processed.

One strategy is to process the primes in increasing order of their exponent, and within the primes of equal exponent to process them in decreasing order. The rationale is that the p-elements, for the smaller primes p such as 2 and 3, have larger, more complex centralizers and so the recursive treatment of the quotients of their centralizers should be avoided, or attention restricted to a small set of element orders t. The primes with high exponent generally have corresponding Sylow subgroups with a large number of classes, and so the fusion of p-elements in G may be difficult to determine. The prime 2 is always last, in order to avoid p-elements with non-cyclic Galois groups.

Examples: (1) The simple group $Co3$ of Conway has order $2^{10}3^{7}5^{3}7\ 11\ 23$ and degree 276. The time taken by the inductive schema is 223 seconds.

(2) The second sporadic simple group $Co2$ of Conway has order $2^{18}3^{6}5^{3}7\,11\,23$ and degree 2300. The inductive schema took 4781 seconds.

(3) The wreath product $D_4 wr S_4$ has degree 16 and order 98 304 $= 2^{15}3$. The inductive schema took 409 seconds and found all rational classes.

(4) The Cayley library `imp768` defines a soluble group of degree 768 of order 1,088,391,168 $= 2^{11}3^{12}$. The inductive approach fails to complete because the fusion on the classes of a Sylow 3-subgroup in G takes too long. At the time the job was terminated there had been at least five conjugacy tests that each take over 45000 seconds.

(5) The symplectic group $Sp(4,8)$ has order $2^{12}3^{4}5\ 7^{2}13$, degree 585, 83 classes and 28 rational classes. The inductive schema takes 978 seconds.

(6) The special linear group $L_5(2)$ has order 9,999,360 = $2^{10}3^{2}5\ 7\ 31$, degree 31, 27 classes and 18 rational classes. The inductive schema takes 38 seconds.

10. Applications

This section discusses the use of classes in the construction of the normal subgroup lattice, the character table, and the automorphism group of a group G.

A normal subgroup is a union of rational classes, so it may be represented as a characteristic function (bitstring) of the set of all rational classes of elements. This gives a very space-efficient representation of a normal subgroup.

One approach to constructing the character tables (Schneider, 1990) uses the class structure constants $\#(K_1, K_2, K_3)$ which counts the number of products k_1k_2 that lie in the class K_3, where the elements k_1 and k_2 run over the classes K_1 and K_2 respectively. The classes may be large, so the number of products may be very large. For each product, we need to quickly determine which class it lies in. Note that an element may occur several times as a product. A *class map* is a map from G to the natural numbers, where the image of an element g gives the index of the class containing g. For small groups, one can form a table indexed by $1 \ldots |G|$ listing the corresponding class. For larger groups this is not possible, so one attempts to identify the class based on invariants of the element, and invariants of powers of the element. All the classes are known (and one can determine the power maps as well) so it is easy to decide if this information uniquely determines the class. If it does not uniquely determine the class then one must resort to conjugacy testing.

The automorphism group of G is represented as a permutation group acting faithfully on a union of classes of elements of G. This union is called the *support*, and one attempts to choose a support as small as possible. An automorphism is found by considering the possible images of the generators of G. Once one is found the automorphism must be expanded from its action on the generators to its action on the whole support. Robertz(1976) reported that most of the time in constructing $Aut(G)$ is spent determining the full permutation action on the support. In his implementation, each element in the support was given as a word in the generators, and its image under the automorphism was determined by evaluating the word using the images of the generators. Thus each image required several group operations to evaluate the word. Using a Schreier system for each class in the support, it is possible to use fewer group operations to determine these images. Essentially one conjugation per element in the support is required. Each class is an orbit of the group acting by conjugation. The Schreier system stores a spanning tree of this orbit. The Schreier system stores a word in the generators for the class representative, and for every other element x in the class it stores a (previous) element $parent[x]$ of the class and a generator $gen[x]$ such that x is obtained from its parent through conjugation by the generator. The image of the class representative is found by evaluating the word using the images of the generators, and the image of x is found from the image of $parent[x]$ by conjugating by the image of the generator $gen[x]$.

11. Conclusion

There have been several algorithms developed for determining the conjugacy classes of elements of a finite group. They are surveyed here. In general they heavily utilise the properties of the special family of groups and the specific computer representations that apply to the family of groups. There are still problems to be solved however.

(1) For permutation and matrix groups we need better centralizer and conjugacy algorithms. The recent work of Leon(1991) will greatly help alleviate this problem, but until we have some experience with the effect of his implementations on the computation of conjugacy classes we will not know if the problem has been solved completely.

(2) For p-groups and soluble groups the problem is the sheer number of classes. One needs to rephrase the question and determine only those classes necessary to answer the particular theoretical or computational question at hand, or one must be satisfied with only counting the number of classes.

(3) For soluble groups, despite the improvements through the use of affine transformations, there is still a problem if the size of the section $N(i)/N(i+1)$ is large. This may require one to represent very large orbits.

(4) The inductive schema for permutation groups may flounder not only because of the cost of centralizers and conjugacy tests but also when the fusion in G of S-classes for a Sylow subgroup S of G is attempted in the case where S has very many classes. More group theory may be useful here.

The progress in computing the composition factors of a permutation group suggests that one might be able to exploit them, in a manner similar to exploiting the chief series of a soluble group, using Proposition 3.1 to lift classes.

References

Butler, G. (1979), *Computational Approaches to Certain Problems in the Theory of Finite Groups*, Ph.D. Thesis, University of Sydney.

Butler, G. (1982), Computing in permutation and matrix groups II : backtrack algorithm, *Math. Comp.* **39**, 671–680.

Butler, G. (1983), Computing normalizers in permutation groups, *J. Algorithms* **4**, 163–175.

Butler, G. (1985), Effective computation with group homomorphisms, *J. Symb. Comp.* **1**, 143–157.

Butler, G. (1988), A proof of Holt's algorithm, *J. Symb. Comp.* **5**, 275–283.

Butler, G. (1990a), Computing a conditioned pc presentation of a soluble permutation group, TR 392, Basser Department of Computer Science, University of Sydney.

Butler, G. (1990b), An inductive schema for computing conjugacy classes in permutation groups, TR 394, Basser Department of Computer Science, University of Sydney (to appear in *Math. Comp.*).

Butler, G. (1991a), Implementing some algorithms of Kantor, (in *AAECC-9*, H.F. Mattson, T. Mora, T.R.N. Rao (eds.), Springer LNCS **539**), 82–93.

Butler, G. (1991b), *Fundamental Algorithms for Permutation Groups* (Lecture Notes in Computer Science **559**, Springer-Verlag, Heidelberg).

Butler, G., and Cannon, J.J. (1982), Computing in permutation and matrix groups I : normal closure, commutator subgroup, series, *Math. Comp.* **39**, 663–670.

Butler, G., and Cannon, J.J. (1989), Computing in permutation and matrix groups III : Sylow subgroups, *J. Symb. Comp.* **8**, 241–252.

Butler, G., and Cannon, J.J. (1991), Computing Sylow subgroups of permutation groups using homomorphic images of centralizers, *J. Symb. Comp.* **12**, 443–457.

Butler, G., and Cannon, J.J. (1993), On Holt's algorithm, *J. Symb. Comp.* **15**, 229–233.

Cannon, J.J. (1984), An introduction to the group theory language, Cayley, in *Computational Group Theory* (M.D. Atkinson (ed.), Academic Press, London), 145–183.

Felsch, V., and Neubüser, J. (1979), An algorithm for the computation of conjugacy classes and centralizers in p-groups, in *EUROSAM '79* (E.W. Ng (ed.), Lecture Notes in Computer Science **72**, Springer-Verlag, Berlin), 452–465.

Glasby, S.P. (1988), Constructing normalisers in finite soluble groups, *J. Symb. Comp.* **5**, 285–294.

Glasby, S.P. and Slattery, M.C. (1990), Computing intersections and normalizers in soluble groups, *J. Symb. Comp.* **9**, 637–651.

Hall, Jr, M. (1959) *The Theory of Groups* (Macmillan, New York).

Hoffman, C.M. (1982), *Group Theoretic Algorithms and Graph Isomorphism* (Lecture Notes in Computer Science **136**, Springer-Verlag, Berlin).

Holt, D.F. (1984), The calculation of the Schur multiplier of a permutation group, in *Computational Group Theory* (M.D. Atkinson (ed.), Academic Press, London), 307–319.

Holt, D.F. (1991), Computing normalizers in permutation groups, *J. Symb. Comp.* **12**, 599–516.

Huppert, B. (1967), *Endliche Gruppen I* (Springer-Verlag, Berlin).

Laue, R., Neubüser, J. and Schoenwaelder, U. (1984), Algorithms for finite soluble groups and the SOGOS system, in *Computational Group Theory* (M.D. Atkinson (ed.), Academic Press, New York), 105–135.

Leon, J.S. (1980), On an algorithm for finding a base and strong generating set for a group given by generating permutations, *Math. Comp.* **35**, 941–974.

Leon, J.S. (1991), Permutation group algorithms based on partitions, I : theory and algorithms, *J. Symb. Comp.* **12**, 533–583.

Mecky, M. and Neubüser, J. (1989), Some remarks on the computation of conjugacy classes of soluble groups, *Bull. Austral. Math. Soc.* **40**, 281–292.

Robertz, H. (1976), *Eine Methode zur Berechnung der Automorphismengruppe einer endliche Gruppe*, Diplomarbeit, R.W.T.H. Aachen.

Schneider, G.J.A. (1990), Dixon's character table algorithm revisited, *J. Symb. Comp.* **9**, 601–606.

Sims, C.C. (1971), Determining the conjugacy classes of a permutation group, in *Computers in Algebra and Number Theory* (G. Birkhoff and M. Hall, Jr (eds.), SIAM-AMS Proc. 4, Amer. Math. Soc., Providence, R.I.), 191–195.

Sims, C.C. (1990), Computing the order of a solvable permutation group, *J. Symb. Comp.* **9**, 699–705.

Wielandt, H. (1964), *Finite Permutation Groups* (Academic Press, New York).

QUOTIENT CATEGORIES OF MODULES OVER GROUP ALGEBRAS

JON F. CARLSON[1]

Department of Mathematics, University of Georgia, Athens, Georgia 30602, U.S.A.

1. Introduction

The following paper is a report on recent progress in the study of quotient categories of modules over modular group algebras. The investigation began with joint work by the author, Peter Donovan and Wayne Wheeler [CDW]. Continuations have involved collaborations with Wheeler [CW] and Geoff Robinson [CR]. Extensions of the theory to more general groups and coefficient rings are undoubtedly possible, but the investigation to this point has been restricted to the case in which G is a finite group and k is an algebraically closed field of characteristic $p > 0$. This case will be assumed throughout this paper. Also we assume that all kG-modules are finitely generated.

The quotient category construction is essentially a localization process, in the sense that the groups of morphisms between objects in the quotients are localized versions of the ext groups in the ordinary cohomology. In this way the quotient construction is a means by which we can highlight certain properties of the module theory and of the cohomology. In addition, the investigation requires the adoption of an entirely new point of view. In the past few months the different viewpoint has led to new results in the study of modules with vanishing cohomology. Also many questions have been raised concerning the module theory and the role of the homological techniques. In reality, the study of the quotient categories is still in its beginning stages and many more applications should be expected.

What follows is a bare sketch of the results in the area. Included is some speculation on the direction of future research. Very few proofs are given and even the details of the quotient category construction are not made precise. The interested reader is referred to [CDW] or its references for authoritative details. Most of the background material is already written into the texts by Benson [B] and Evens [E]. The paper of Happel [H] provides a readable exposition on triangulated categories and quotient categories.

2. Basic definitions

We begin by considering the stable category of kG-modulo projectives. Call this Stmod-kG. The objects in the stable category are kG-modules. The

[1]Partially supported by a grant from NSF.

group of morphisms from an object M to an object N is the set

$$\underline{\mathrm{Hom}}_{kG}(M,N) = \mathrm{Hom}_{kG}(M,N)/\mathrm{PHom}_{kG}(M,N)$$

where PHom denotes the set of all homomorphisms which factor through projective modules. It's important to note here that kG is a self injective ring, so that projective modules are injective and vice versa.

The crucial property of the stable category is that it is triangulated, it has triangles, and, in fact, every morphism $\alpha : M \to N$ fits into a unique triangle:

$$L \longrightarrow M \xrightarrow{\alpha} N \longrightarrow L^{[1]}$$

where $L \to L^{[1]}$ is a translation functor (see [H]). In the case of kG-modules the translation functor is $\Omega^{-1}(\ )$, the inverse loop space operator. So if L is a kG-module and if Q is an injective hull of L, then $\Omega^{-1}(L)$ is the cokernel of the injection $L \to Q$. Now, to fit α into a triangle, we take the direct sum of M and a projective module P which is large enough so that there is a surjection $\alpha' : M \oplus P \to N$ coinciding with α on M. Thus we have an exact sequence

$$0 \longrightarrow L \longrightarrow M \xrightarrow{\alpha'} N \longrightarrow 0$$

where the kernel of α', L, is the third object in the triangle containing α. Suppose on the other hand that we take the direct sum of N with a projective module Q which is large enough so that there is an injection $\alpha'' : M \to N \oplus Q$ whose composition with the projection onto N is α. The reader may check that the cokernel of such an α'' must be isomorphic to $\Omega^{-1}(L)$, at least up to projective direct summand. Hence we also have an exact sequence

$$0 \longrightarrow M \xrightarrow{\alpha''} N \oplus Q \longrightarrow \Omega^{-1}(L) \longrightarrow 0.$$

Now in the stable category, P and Q are equivalent to zero and also α' and α'' are equivalent to α. Thus the triangle is well defined in the stable category. In order to prove that Stmod–kG is triangulated it would be necessary to show that several other axioms are satisfied.

To describe the quotients we need to introduce several notations (see [B] or [E]). The cohomology ring $H^*(G,k) \cong \mathrm{Ext}^*_{kG}(k,k)$ is a finitely generated graded-commutative k-algebra. Its maximal ideal spectrum $V_G(k)$ is a finite dimensional homogeneous affine variety. If M and N kG-modules then $\mathrm{Ext}^*_{kG}(M,N)$ is a finitely generated module over $H^*(G,k)$. Let $J(M)$ be the annihilator in $H^*(G,k)$ of $\mathrm{Ext}^*_{kG}(M,M)$ and let $V_G(M) \subseteq V_G(k)$ be the subvariety consisting of all maximal ideals which contain $J(M)$. Finally, the complexity of M is the dimension of the variety $V_G(M)$. It can also be described as the polynomial rate of growth of the cohomology ring $\mathrm{Ext}^*_{kG}(M,M)$ or of the dimensions of the terms of a minimal projective resolution of M. A fundamental theorem of Quillen (see [B], (5.6.1)) tells us that the complexity of the trivial module k is equal to the p-rank, r, of G. Here r is the largest integer such that G has an elementary abelian p-subgroup, $E \cong (\mathbb{Z}/p)^r$.

Now suppose that c is a positive integer and $\mathcal{M}_c$ is the full subcategory of Stmod–kG whose objects are all kG-modules of complexity at most c. Thus by the theorem mentioned above $\mathcal{M}_r = \text{Stmod}-kG$. It can be shown that if M and N are in $\mathcal{M}_c$, then any triangle with M and N as two of the terms has the third term also in $\mathcal{M}_c$. Hence $\mathcal{M}_c$ is a triangulated subcategory of $\mathcal{M}_r$ or of any $\mathcal{M}_b$ with $b > c$.

Definition. Suppose that d is a positive integer and that M and N are kG-modules. A morphism $\alpha \in \underline{\text{Hom}}_{kG}(M, N)$ is said to be a mod-d-complexity isomorphism if the third object L in the triangle

$$L \longrightarrow M \xrightarrow{\alpha} N \longrightarrow \Omega^{-1}(L)$$

is in $\mathcal{M}_d$.

With this definition we are prepared to define the quotient category $\mathcal{M}_c/\mathcal{M}_d$, assuming $d < c$ (see [H]). The objects in the quotient are the same as those in $\mathcal{M}_c$. If M and N are in $\mathcal{M}_c$, then a morphism from M to N is an equivalence class of triples (s, U, f), where $U \in \mathcal{M}_c$, $s \in \underline{\text{Hom}}_{kG}(U, M)$ is a mod-d-complexity isomorphism and $f \in \underline{\text{Hom}}_{kG}(U, N)$. If $V \in \mathcal{M}_c$ and $t : V \to U$ is a mod-d-complexity isomorphism then the triple (st, V, ft) is equivalent to (s, U, f). The equivalence relation defining the morphism is the minimal equivalence relation on such triples that includes equivalences of the type given above.

Now given a morphism from M to N represented by (s, U, f), we may think of the morphism as having the form $fs^{-1} : M \to N$. That is to say, the quotient category construction inverts the mod-d-complexity isomorphisms. The equivalence says that $(ft)(st)^{-1} = fs^{-1}$, which is exactly what should be expected. We should also mention that the subcategory $\mathcal{M}_d$ is a thick subcategory and hence the quotient is also triangulated.

3. The difference one case

To this point most of the work on complexity quotient categories has focused on the case of the quotients $\mathcal{Q}_c = \mathcal{M}_c/\mathcal{M}_{c-1}$. In any additive category it's a natural question as to what might be the groups of morphisms between objects. In the case of $\mathcal{Q}_c$, the answer is particularly nice. In $\mathcal{M}_c$ we shall say that a morphism is a complexity isomorphism if it is a mod-$(c-1)$-complexity isomorphism. For notation, we define $\Omega^n(M)$, for M a kG-module, to be the kernel of ∂_{n-1} in a minimal projective resolution

$$\cdots \xrightarrow{\partial_2} P_1 \xrightarrow{\partial_1} P_0 \longrightarrow M \longrightarrow 0$$

of M, $n > 0$.

The main principal in the characterization of the Hom's in $\mathcal{Q}_c$ is the following. Suppose that U and M are in $\mathcal{M}_c$ and $s : U \to M$ is a complexity isomorphism. Then there is a positive integer n and a homomorphism $t : \Omega^n(M) \to U$ such that t is also a complexity isomorphism. Hence $\zeta = st \in \underline{\mathrm{Hom}}_{kG}(\Omega^n(M), M)$ is invertible in $\mathcal{Q}_c$. Moreover we know that

$$\underline{\mathrm{Hom}}_{kG}(\Omega^n(M), M) \cong \mathrm{Ext}^n_{kG}(M, M)$$

and ζ is a non-zero cohomology class. The trick of the proof of the principal is to choose ζ in such a way that it annihilates the cohomology of the third object in the triangle containing s, U and M. Specifically the theory comes out as follows.

Theorem 3.1. ([CDW]) *Suppose that M and N are in $\mathcal{M}_c$. Then*

$$Hom_{\mathcal{Q}_c}(M, N) = [Ext^*_{kG}(M, N) \cdot S^{-1}]_0$$

where $S \subseteq Ext^*_{kG}(M, M)$ is the multiplicative set

$$S = \{\zeta \cup Id_M \mid \zeta \in H^*(G, k), \dim(V_G(\zeta) \cap V_G(M)) < c\}.$$

Here the symbol $[\ ]_0$ indicates the zero grading. That is both $\mathrm{Ext}^*_{kG}(M, N)$ and S are graded and the grading on the product is given by $\deg(fs^{-1}) = \deg(f) - \deg(s)$.

An immediate consequence of the theorem is that all objects in $\mathcal{Q}_c$ are periodic. That is, for $M \in \mathcal{Q}_c$, there exists some $\zeta \in H^*(G, k)$ such that the morphism $\zeta \cup \mathrm{Id}_M \in \underline{\mathrm{Hom}}_{kG}(\Omega^n(M), M)$ is a complexity isomorphism. The period of M (least such positive n) will be determined by the degrees of the generators of $H^*(G, k)$ whose varieties intersect transversely with $V_G(M)$.

An item of particular interest is the structure of the endomorphism rings in $\mathcal{Q}_c$ of modules. In the ordinary module category, a kG-module is indecomposable if and only if its endomorphism ring is local. In the quotient categories, $\mathcal{Q}_c$, every object is a finite direct sum of indecomposable objects [CDW]. But endomorphism ring of even the indecomposable objects need not be local. The problem, simply stated, is that the quotient categories have no Krull-Schmidt theorem. The structure theorem for the Hom's indicates that the endomorphism ring of a non-zero object must be, at least, a sum of rings, one for each component of dimension c of the variety of the module. This has several consequences which we illustrate with a couple of examples.

First consider the endomorphism ring of the trivial kG-module k in $\mathcal{M}_r/\mathcal{M}_{r-1} = \mathcal{Q}_r$. The theorem of Quillen, quoted earlier, is actually much more specific about the variety $V_G(k)$. It says that $V_G(k) = \cup V_E$, where the union is over a set of representatives of the conjugacy classes of maximal elementary abelian p-subgroups E of G. The V_E are the irreducible components of $V_G(k)$ and for E a maximal elementary abelian p-subgroup of G, $V_E = \text{Image}(\text{res}^*_{G,E})$ where $\text{res}^*_{G,E}$ is the map induced on varieties by the restriction map $H^*(G,k) \to H^*(E,k)$. Moreover the dimension of V_E is the same as the dimension of $V_E(k) = k^t$, which is the p-rank, t, of E. So in $H^*(G,k) = \text{Ext}^*_{kG}(k,k)$ we can choose cohomology elements ζ_E, all of the same degree, whose varieties correspond to the components, and whose sum, ζ, is a complexity isomorphism. Then the elements $\zeta_E\zeta^{-1} \in \text{Hom}_{\mathcal{Q}_r}(k,k)$ are orthogonal idempotents. Of course, in the present context it is only necessary to consider those E of maximal rank r. So let $\mathcal{A}$ be a set of representatives of the conjugacy classes of elementary abelian p-subgroups of rank r. The result [CDW] is that

$$\text{Hom}_{\mathcal{Q}_r}(k,k) = \coprod_{E \in \mathcal{A}} R_E$$

where R_E is a local ring. The quotient $R_E/\text{Rad } R_E$ is a transcendental extension (not pure in general) of degree $r-1$ over k.

A similar thing can be done for a general module. The endomorphism ring will decompose according to the components of the variety. It seems likely that the transcendence degree for a module in $\mathcal{M}_c$ should by $c-1$. But this is not yet proved. What can be seen by other means is that for any module M in $\mathcal{M}_c$,

$$M \oplus \Omega(M) \cong U_1 \oplus \cdots \oplus U_s$$

(complexity isomorphism) where the modules $U_1, \ldots, U_s$ correspond to the components of the variety of M of dimension c.

The trivial module is actually indecomposable in $\mathcal{Q}_r$. However a suitable multiple of the trivial module has a decomposition along the lines predicted by the decomposition of its endomorphism ring. The construction of the triangle giving the decomposition is accomplished by means which are similar to the known techniques for creating transfers in cohomology. If E is an elementary abelian P-subgroup of G, let $D = D_G(E)$ be the set of all elements in the normalizer $N_G(E)$ whose conjugation action on $E \cong \mathbb{F}_p^r$ is by a scalar multiplication. That is, if $x \in D$ then there an integer a such for any $y \in E$ we have $xyx^{-1} = y^a$.

Theorem 3.2 ([C]) *Let m be the least common multiple of the indices $|N_G(E) : D_G(E)|$ for $E \in \mathcal{A}$. Let $m_E = m/|N_G(E) : D_G(E)|$. Then in $\mathcal{Q}_r$*

$$k^m \cong \bigoplus_{E \in \mathcal{A}} \left(k^{\uparrow G}_{D_G(E)}\right)^{m_E}.$$

The theorem indicates that modulo a certain complexity factor ($\mathcal{M}_{r-1}$), the module theory for kG is controlled at the level of the "diagonalizers", D, of the elementary abelian p-subgroups of maximal rank by induction (Frobenius reciprocity). Of course, if either $p = 2$ or G is a p-group then $D \cong C_G(E)$.

The idea has found an application in the study of modules with vanishing cohomology. It can be shown that if $p > 2$ and if M is a module in the principal kG-block with $H^*(G, M) = 0$ then the complexity of M must be strictly less than r [CR]. Something like this had been suspected at the time that the work on [BCR] was done, but the viewpoint of the quotient categories seems essential for the proof. The main point in the proof is that if $p > 2$, then the centralizers of the maximal elementary abelian p-subgroups must be p-nilpotent. Hence the cohomology or complexity of a module in the principal block can be analyzed as if the group were a p-group. The actual quotient category construction is not really necessary but the triangle (exact sequence) giving the decomposition of the trivial module seems to be essential.

Numerous variations can be made on the construction. One such variation is to fix a component of the variety $V_E \subseteq V_G(k)$. Let $\mathcal{M}(E)$ be the full subcategory of all modules M with $V_G(M) \cap V_E < V_E$. Let $\mathcal{Q}(G, E) =$ Stmod–$kG/\mathcal{M}(E)$. In $\mathcal{Q}(G, E)$ the trivial module has a local endomorphism ring. The theorem that can be proved in this setting is that if $N = N_G(E)$ contains a Sylow p-subgroup of G then $\mathcal{Q}(G, E)$ and $\mathcal{Q}(N, E)$ are equivalent as triangulated categories [CW]. There should be other such equivalences in other circumstances.

4. Higher differences and open problems

Consider now the quotients $\mathcal{M}_c/\mathcal{M}_d$ for fixed d. Each is a full subcategory of $\mathcal{M}_r/\mathcal{M}_d$. In the difference one case, $\mathcal{M}_r/\mathcal{M}_{r-1}$, the endomorphism ring of any module is finitely generated as a module over the endomorphism ring of the trivial module [CW]. It seems likely that the same would hold in the higher difference cases also. Thus initial efforts might best be devoted to characterizing the endomorphism ring of the trivial module.

In the difference one case, the endomorphism ring of k was obtained by showing that it was only necessary to invert cohomology elements ζ such that $\dim V_G(\zeta) < r$. This certainly will not suffice if $r - d \geq 2$. The problem is that $\dim V_G(\zeta) \geq r - 1$ for all $\zeta \in H^*(G, k)$. Hence if $d < r - 1$, then no homomorphism $\zeta' : \Omega^n(k) \to k$ is ever a mod-d-complexity isomorphism. Consequently it seems appropriate to find a universal set of modules to substitute for the $\Omega^n(k)$'s in the theory. Such a collection exists and can be described as follows. For convenience of notation we shall consider only the case that $d = r - 2$.

Suppose that $\alpha : U \to k$ is a mod-$(r-2)$-complexity isomorphism so that the kernel, W, of α has $\dim V_G(W) \leq r - 2$. It is possible to choose elements $\zeta_1, \zeta_2 \in H^*(G, k)$ which are homogeneous and have the properties that $\dim(V_G(\zeta_1) \cap V_G(\zeta_2)) = r - 2$ and ζ_1, ζ_2 both annihilate the cohomology of W, $\text{Ext}^*_{kG}(W,W)$. Now ζ_1 and ζ_2 are represented by homomorphisms:

$$\hat{\zeta}_1 : \Omega^{n_1}(k) \to k, \quad \hat{\zeta}_2 : \Omega^{n_2}(k) \to k,$$

where $n_i = \deg(\zeta_i)$. Taking tensor products over k we get a complex:

$$\Omega^{n_1}(k) \otimes \Omega^{n_2}(k) \xrightarrow{\gamma} \Omega^{n_1} \oplus \Omega^{n_2}(k) \longrightarrow k.$$

Let $L(\zeta_1, \zeta_2)$ be the third object in the triangle containing γ. We want this object in the position of the cokernel of γ, i.e. of γ made injective by the addition of a projective module to the middle term of the complex. Then it can be shown that there is a complexity isomorphism $\varphi : L(\zeta_1, \zeta_2) \to U$. Thus it would seem that we must invert pairs of elements of $H^*(G,k)$, or more generally we seem to be localizing at all subvarieties of dimension $r-2$. However we do not understand the structure of the localized ring in even the simplest cases.

In general, the Hom's and Ext's in the quotient categories give new cohomology theories of a sort. It should be possible to define complexities and varieties of modules in the quotients. A first step in this direction should be to show that the Hom's and Ext's are finitely generated over the endomorphism ring of the trivial module and the cohomology ring. In another direction it might be possible to approximate Morita equivalences of blocks of kG-modules by looking at successive quotients of the module category.

References

[B] D. J. Benson, *Representations and Cohomology II: Cohomology of Groups and Modules* (Cambridge University Press, Cambridge, 1991).

[BCR] D. J. Benson, J. F. Carlson and G. R. Robinson, On the vanishing of group cohomology, *J. Algebra* **131**(1990), 40–73.

[C] J. F. Carlson, Decomposition of the trivial module in the complexity quotient category, to appear.

[CDW] J. F. Carlson, P. W. Donovan and W. W. Wheeler, Complexity and quotient categories for group algebras, *J. Pure Appl. Algebra*, to appear.

[CW] J. F. Carlson and W. W. Wheeler, Varieties and localization of module categories, to appear.

[CR] J. F. Carlson and G. R. Robinson, Varieties and modules with vanishing cohomology, *Math. Proc. Cambridge Philos. Soc.*, to appear.

[E] L. Evens, *The Cohomology of Groups* (Oxford University Press, New York, 1991).

[H] D. Happel, *Triangulated Categories in the Representation Theory of Finite-Dimensional Algebras* (Cambridge University Press, Cambridge, 1988).

WEAK CHAIN CONDITIONS FOR NON-ALMOST NORMAL SUBGROUPS

GIOVANNI CUTOLO* and LEONID A. KURDACHENKO†

*Università di Napoli 'Federico II', Dipartimento di Matematica e Applicazioni, Via Cintia, Monte S. Angelo, I-80126 Napoli, Italy
†Ukraine, Dnepropetrovsk, Gagarin-Prospekt 72, University- Dept. of Algebra

Abstract

We describe locally (soluble-by-finite) groups in which the set of all subgroups with infinitely many conjugates satisfies some weak chain condition.

AMS subject classification: 20E15, 20F16, 20F24.

1. Introduction

A subgroup H of a group G is said to be almost normal in G if it has finitely many conjugates in G or, equivalently, if its normalizer $N_G(H)$ has finite index in G. A celebrated theorem by B.H. Neumann [8] states that every subgroup of a group G is almost normal if and only if the factor $G/Z(G)$ is finite.

It is natural to ask what information on the structure of the group G can be obtained if the condition of being almost normal is imposed only on a large set of subgroups of G. For instance, I.I. Eremin [4] proved that every subgroup of G is almost normal (and so $G/Z(G)$ is finite), provided every abelian subgroup is. Problems of this type have been considered in various papers (see [5] and references quoted therein).

A way of ensuring the existence of many almost normal subgroups is to impose some chain condition on the set of the subgroups which are not almost normal. The second author and V.V. Pylaev [7] studied groups satisfying the minimal condition on non-almost normal subgroups. Here we consider groups in which non-almost normal subgroups satisfy some weaker chain condition.

Following Zaĭtsev [12], we say that the group G satisfies the *Min-Max-*∞ *condition* if there exists no family $(H_n)_{n\in\mathbb{Z}}$ of subgroups of G such that H_n is a subgroup of infinite index of H_{n+1} for all integers n. This condition generalizes both the weak minimal (Min-∞) and the weak maximal condition (Max-∞), which are defined by excluding the existence of an infinite descending (resp. ascending) chain of subgroups of G with the property that $|H : K|$ is infinite for any pair of subgroups H, K belonging to the chain and such that $K < H$.

The weak minimal and the weak maximal conditions had been introduced by R. Baer [2] and D.I. Zaĭtsev [11]. They showed that for soluble groups both conditions are equivalent to the weak minimal and to the weak maximal condition on abelian subgroups and also to the property of being a minimax group. A general reference for the many investigations on groups satisfying weak chain conditions on relevant systems of subgroups is the survey paper [6].

Theorem 2 of [12] states that a locally (soluble-by-finite) group satisfies the Min-Max-∞ condition if and only if it is a (soluble-by-finite) minimax group; thus conditions Min-∞, Max-∞ and Min-Max-∞ are equivalent for locally (soluble-by-finite) groups.

We shall study groups satisfying the Min-Max-∞ condition on the set of non-almost normal subgroups, which we denote by Min-Max-∞-$\overline{an}$. In particular our results hold for groups satisfying the weak minimal condition (Min-∞-$\overline{an}$) or the weak maximal condition (Max-∞-$\overline{an}$) on non-almost normal subgroups. Indeed, it turns out that Min-Max-∞-$\overline{an}$, Min-∞-$\overline{an}$ and Max-∞-$\overline{an}$ are equivalent for locally (soluble-by-finite) groups.

The starting point (and the motivation) of this paper is the observation that the proof of B.H. Neumann's theorem can be adapted to show that an FC-group satisfying the Min-Max-∞ condition on non-almost normal subgroups is centre-by-finite (see Theorem 4 below). This also generalizes the easy fact that minimax FC-groups are centre-by-finite. On the other hand, it will be shown that there exist nilpotent groups which are neither minimax nor centre-by-finite, and satisfy Min-Max-∞-$\overline{an}$. It will be proved that locally (soluble-by-finite) groups satisfying Min-Max-∞-$\overline{an}$ are soluble-by-finite and a description of such groups will be given.

Notation used throughout the paper is standard. We refer to the books [9] and [10] for general terminology. Recall that an $\mathfrak{S}_1$-group is a group having a finite series whose factors are abelian, either torsion-free of finite rank or satisfying the minimal condition. An $\mathfrak{S}_1\mathfrak{F}$-group is a group having a subgroup of finite index which is an $\mathfrak{S}_1$-group.

2. Results and proofs

Our first lemma will be used mainly to restrict ranks of (abelian) sections of groups satisfying Min-Max-∞-$\overline{an}$.

Lemma 1. *Let the group G satisfy Min-Max-∞-$\overline{an}$. If G has a section H/K which is a direct product of an infinite family $(U_\lambda/K)_{\lambda\in\Lambda}$ of nontrivial groups, then:*

(i) For every subset Δ of Λ, $H_\Delta = \langle U_\lambda \mid \lambda \in \Delta\rangle$ is an almost normal subgroup of G. In particular, H and K are almost normal in G.

(ii) If $K = 1$ and x is an element of G which normalizes U_λ for all $\lambda \in \Lambda$, then $x \in F_1(G)$, the FC-centre of G.

PROOF. (i) Assume first that $\Lambda \setminus \Delta$ is infinite. Choose two infinite disjoint subsets Λ_1 and Λ_2 of $\Lambda \setminus \Delta$ and let $H_i = \langle U_\lambda \mid \lambda \in \Lambda_i \rangle$, for $i = 1$ and 2. It is easily seen, by using the condition Min-Max-∞-$\overline{an}$, that there exists a subgroup $X_i \leq H_i$ containing K such that $H_\Delta X_i$ is almost normal in G. Then $H_\Delta = H_\Delta X_1 \cap H_\Delta X_2$ is almost normal in G. If $\Lambda \setminus \Delta$ is finite, Δ is the union of two subsets with infinite complement in Λ and it follows from the previous case that H_Δ is still almost normal in G.

ii) There exists a finite subset Δ of Λ such that $\langle x \rangle \cap \mathrm{Dr}_{\lambda \in \Lambda \setminus \Delta} U_\lambda = 1$. Let Λ_1 and Λ_2 be two infinite disjoint subsets of $\Lambda \setminus \Delta$. By the Min-Max-∞-$\overline{an}$ condition, for $i = 1$ and 2, a subgroup X_i of $\mathrm{Dr}_{\lambda \in \Lambda_i} U_\lambda$ can be found such that $\langle x \rangle X_i$ is almost normal in G. Then $\langle x \rangle = \langle x \rangle X_1 \cap \langle x \rangle X_2$ is almost normal, thus $x \in F_1(G)$. □

The next aim is to prove that FC-groups satisfying Min-Max-∞-$\overline{an}$ are centre-by-finite. This will be done by adapting the proof of the B.H. Neumann-Eremin theorem, as reported in [10], Chapter 7. We begin with a lemma that is useful to rule out the case of groups with finite commutator subgroup.

Lemma 2. *Let the FC-group G satisfy Min-Max-∞-$\overline{an}$and let $H \leq G$. Then H/H_G is a minimax group.*

PROOF. It is easily seen that U/U_G is finite for every almost normal subgroup U of G (see [10], Lemma 7.13). It follows that if U/H_G is an infinite subgroup of H/H_G then U is not almost normal in G. In particular the group $\bar{H} = H/H_G$ satisfies the Min-Max-∞ condition on infinite subgroups, that is the Min-Max-∞ condition. It is clear from Zaĭtsev's theorem ([12]) quoted in §1, that an FC-group satisfying the Min-Max-∞ condition is a (soluble-by-finite) minimax group (for instance, because its commutator subgroup is locally finite). Thus $\bar{H}$ is a minimax group, as we wanted to show. □

Lemma 3. *Let the group G satisfy Min-Max-∞-$\overline{an}$. If G' is finite then $G/Z(G)$ is finite.*

PROOF. Let A be a maximal abelian subgroup of G and let $B = A \cap G'$. For every $b \in B \setminus 1$, let M_b be a subgroup of A, maximal with respect to the condition $b \notin M_b$. Then A/M_b is monolithic, hence it is either a cyclic or a Prüfer p-group (p prime). Let $C = \bigcap_{b \in B \setminus 1} M_b$. Since B is finite, A/C satisfies the minimal condition. Moreover $C \cap G' = 1$, hence $C_G = C \cap Z(G)$. By Lemma 1, $C/C \cap Z(G)$ is minimax. Thus $A/Z(G)$ is minimax. Since $G/Z(G)$ is periodic and residually finite, $A/Z(G)$ is finite. Then $A = FZ(G)$, where F is a finitely generated subgroup. Thus $A = C_G(A) = C_G(F)$ has finite index in G. It follows that $G/Z(G)$ is finite. □

Theorem 4. *Let G be an FC-group. Then G satisfies Min-Max-∞-$\overline{an}$ if and only if $G/Z(G)$ is finite.*

PROOF. Assume by contradiction that G satisfies Min-Max-∞-$\overline{an}$ and that $G/Z(G)$ is infinite. If L is a torsion-free subgroup of $Z(G)$ such that $Z(G)/L$ is periodic, then $Z(G/L) = Z(G)/L$ has infinite index in G/L and G/L is periodic. This shows that G can assumed to be periodic. By Lemma 3, G' is infinite. Hence there exist two sequences $(a_n)_{n\in\mathbb{N}}$ and $(b_n)_{n\in\mathbb{N}}$ of elements of G such that:

$$c_n = [a_n, b_n] \neq 1,\ c_n \neq c_m \ \text{ and } \ [a_n, a_m] = [b_n, b_m] = [a_n, b_m] = 1$$

for all $n, m \in \mathbb{N}$ with nm (see [10], Theorem 7.6 (ii)). For every subset X of $\mathbb{N}$, let $A(X) = \langle a_n \mid n \in X\rangle$ and $B(X) = \langle b_n \mid n \in X\rangle$. If $X \subset Y \subseteq \mathbb{N}$, then $A(X) < A(Y)$. Indeed, if $n \in Y \setminus X$, then b_n centralizes $A(X)$ but not $A(Y)$. Analogously $B(X) < B(Y)$. Furthermore, if $Y \setminus X$ is infinite, then $|A(Y) : A(X)|$ and $|B(Y) : B(X)|$ are infinite, as there exist infinitely many subgroups between $A(X)$ and $A(Y)$ (resp. $B(X)$ and $B(Y)$). There exists an infinite subset N_1 of $\mathbb{N}$ such that $A = A(N_1)$ and $B = B(N_1)$ are both almost normal in G. Indeed, let $\{X_n \mid n \in \mathbb{Z}\}$ be a chain of subsets of $\mathbb{N}$ such that $X_n \subset X_{n+1}$ and $X_{n+1} \setminus X_n$ is infinite for all $n \in \mathbb{Z}$. Since G satisfies Min-Max-∞-$\overline{an}$ there exists an index n such that $A(Y)$ is almost normal in G for all Y such that $X_n \subset Y \subset X_{n+1}$. Again by condition Min-Max-∞-$\overline{an}$ there exists N_1 such that $X_n \subset N_1 \subset X_{n+1}$ and $B(N_1)$ is almost normal in G.

Let $C = \langle c_n \mid n \in N_1\rangle$. Then $C \leq A^G \cap B^G$. By Lemma 7.13 of [10] both $|A^G : A|$ and $|B^G : B|$ are finite, hence $|AC : A|$ and $|BC : B|$ are finite. Thus there exists an infinite subset N_2 of N_1 such that $Ac_i = Ac_j$ and $Bc_i = Bc_j$ for all i, $j \in N_2$. If $i \neq j$ are elements of N_2, then $c_i \in Ac_j \subseteq C_G(a_i)$. Hence $[c_i, A] = 1$ and by the same argument $[c_i, B] = 1$, for all $i \in N_2$. Let $G_1 = \langle a_n, b_n \mid n \in N_2\rangle$. Then $G_1' = \langle c_n \mid n \in N_2\rangle \leq Z(G_1)$ and G_1 is nilpotent of class 2. Let Λ be the set of all odd primes p such that the p-component of G_1 is not abelian. For any $p \in \Lambda$ let H_p be a non-normal p-subgroup of G_1. If Λ is infinite, then $\mathrm{Dr}_{p\in\Lambda} H_p$, is clearly not almost normal in G, contradicting Lemma 1 (i). Thus Λ is finite. This shows that there exists a prime p such that the p-component P of G_1 is not centre-by-finite. We claim that there exist two sequences $(a_n^*)_{n\in\mathbb{N}}$ and $(b_n^*)_{n\in\mathbb{N}}$ of elements of P such that for all n, $m \in \mathbb{N}$ with $n \neq m$ the following hold:

$$\begin{cases} [a_n^*, a_m^*] = [b_n^*, b_m^*] = [a_n^*, b_m^*] = 1; \\ c_n^* = [a_n^*, b_n^*] \text{ has order } p; \\ c_n^* \neq c_m^*. \end{cases}$$

Indeed, P' is infinite by Lemma 3 and hence, by Theorem 7.6 (ii) of [10], two sequences $(a_n^*)_{n\in\mathbb{N}}$ and $(b_n^*)_{n\in\mathbb{N}}$ of elements of P can be constructed, which satisfy the first and the third requirement above and such that $c_n^* = [a_n^*, b_n^*] \neq 1$ for all $n \in \mathbb{N}$. Let p^{k_n} be the order of c_n^*. Replacing each a_n^* with its $(p^{k_n} - 1)$-th power, we obtain the required sequences, as P has nilpotency class 2. It is clear that there exists an infinite subset X of $\mathbb{N}$ such that $C^* = \langle c_n^* \mid n \in \mathbb{N} \rangle = \mathrm{Dr}_{n \in X} \langle c_n^* \rangle$. Let $H = \langle a_n^*, b_n^* \mid n \in X \rangle$ and $K = \langle c_n^* (c_m^*)^{-1} \mid n, m \in X \rangle$. The commutator subgroup C^*/K of H/K has order p. If Z/K is the centre of H/K, then H/Z is finite by Lemma 3. Hence there exist $n, m \in X$ such that $n \neq m$ and $a_n^* Z = a_m^* Z$. But $c_n^* = [a_n^*, b_n^*] \notin K$ and $1 = [a_m^*, b_n^*] \in K$, which implies $a_n^* Z \neq a_m^* Z$. This contradiction proves the theorem. □

Theorem 4 and Lemma 1 can be used to restrict strongly the structure of abelian subgroups of a group G satisfying Min-Max-∞-$\overline{an}$ and such that $G/Z(G)$ is infinite. If G satisfies some solubility requirement, this information can be translated into global structure information about G.

Lemma 5. *Let the group G satisfy Min-Max-∞-$\overline{an}$. If G has a subgroup A which is a direct product of infinitely many nontrivial cyclic subgroups, then $G/Z(G)$ is finite.*

PROOF. By Theorem 4 it suffices to prove that G is an FC-group. Hence, as follows from Lemma 1 (*ii*), it will be enough to construct a subgroup H of G which is a direct product of infinitely many nontrivial normal subgroups of G.

By Lemma 1, $A \leq F = F_1(G)$. Since $C = Z(F)$ has finite index in F, it may be assumed without loss of generality $A \leq C$. Assume first that the torsion subgroup T of C does not satisfy the minimal condition. T is a direct product of its primary components, which are normal subgroups of G. Hence it can be assumed that T has finitely many nontrivial primary components. Thus there exists a prime p such that the p-component P of the socle of T is infinite. It is possible to construct a sequence of non-trivial finite G-invariant subgroups $(U_n)_{n \in \mathbb{N}}$ of P which generate their direct product. Indeed, let $n \in \mathbb{N}$ and suppose that $U_1, \ldots, U_{n-1}$ have been constructed. Then there exists V such that $P = U_1 \times \ldots \times U_{n-1} \times V$. By Lemma 1 (*i*), V is almost normal in G. Since P/V is finite, also $W = V_G$ has finite index in P. In particular $W \neq 1$. If a is a nontrivial element of W, then $U_n = \langle a \rangle^G$ is finite and $\langle U_1, \ldots, U_n \rangle = U_1 \times \ldots \times U_n$. Hence the subgroup $H = \mathrm{Dr}_{n \in \mathbb{N}} U_n$ can be constructed. This proves the proposition in this case.

Suppose now that T satisfies the minimal condition. In this case C has infinite torsion-free rank. Let U be a free abelian subgroup of C such that C/U is periodic. Then U is almost normal in G by Lemma 1 (*i*). In particular, if $B = U_G$, then C/B is periodic and $B \simeq U$. We can argue as above to construct a sequence of non-trivial finitely generated G-invariant subgroups $(U_n)_{n \in \mathbb{N}}$ of B which generate their direct product. Indeed, put $B = \mathrm{Dr}_{\lambda \in \Lambda} X_\lambda$, where each X_λ is infinite cyclic and assume that $U_1, \ldots, U_{n-1}$ have been constructed. Let $X = U_1 \times \ldots \times U_{n-1}$. There exists a finite subset Δ of Λ such that $X \cap V = 1$, where $V = \mathrm{Dr}_{\lambda \in \Lambda \setminus \Delta} X_\lambda$. Let $W = V_G$. Since V is almost normal in G and B/V is finitely generated, B/W is finitely generated, in particular $W \neq 1$. Let U_n be the normal closure in G of a nontrivial element of W. Then $\langle X, U_n \rangle = U_1 \times \ldots \times U_n$ and U_n is finitely generated, as W is contained in the FC-centre of G. The lemma is now completely proved. □

It is well-known that if G is a group such that $G/Z(G)$ is polycyclic-by-finite, then G' is also polycyclic-by-finite (see for instance [9], part I, p. 115). The converse does not hold in general, as the example of an infinite extra-special p-group shows. However, under some rank restrictions, the converse is also true.

Lemma 6. *Let G be a group. If G' is polycyclic-by-finite and $Z_2(G)/Z(G)$ has finite p-rank for every prime p, then $G/Z(G)$ is polycyclic-by-finite.*

PROOF. It is known that $G/Z_2(G)$ is polycyclic-by-finite if G' is (see [9], part I, p. 119). Thus we only have to prove that $Z_2/Z = Z_2(G)/Z(G)$ is finitely generated. Let H be a subgroup of Z_2 containing Z and such that Z_2/H has the maximum possible torsion-free rank subject to the condition of being finitely generated. For every $g \in G$, one has $H/C_H(g) \simeq [H,g] \leq G'$. The choice of H yields that $[H,g]$ is finite. There is a bound $n \in \mathbb{N}$ for the exponent of a finite subgroup of G'. Thus $[H^n, g] = [H,g]^n = 1$, for all $g \in G$, hence $H^n \leq Z$. It follows from the hypothesis that H/Z is finite. Hence Z_2/Z is finitely generated, as we wanted to prove. □

Proposition 7. *Let the group G satisfy Min-Max-∞-$\overline{an}$. Assume that G has an abelian subgroup which is not minimax. Then $G/Z(G)$ is polycyclic-by-finite.*

PROOF. It can be assumed that $G/Z(G)$ is infinite. Let A be an abelian subgroup of G which is not minimax. By Lemma 5, A has a finitely generated subgroup B such that A/B is periodic and does not satisfy the minimal condition. The socle of A/B is infinite, thus B is almost normal in G by Lemma 1 (*i*). Let $N = N_G(B)$. By Lemma 5, N/B is centre-by-finite. Since B^G is finitely generated and $|G : N|$ is finite, $B^G \cap N$ is finitely generated. Thus $B^G \cap N/B$ satisfies the maximal condition. It follows that B^G is polycyclic-by-finite. The periodic abelian subgroup AB^G/B^G of G/B^G does not satisfy the minimal condition, hence G/B^G is centre-by-finite. In particular $G'B^G/B^G$ is finite and G' is polycyclic-by-finite. Finally G is a soluble-by-finite group whose abelian subgroups have finite total rank, thus an $\mathfrak{S}_1\mathfrak{F}$-group (by a theorem of Čarin [3]). Therefore Lemma 6 can be applied to conclude that $G/Z(G)$ is polycyclic-by-finite. □

Proposition 8. *Let the group G satisfy Min-Max-∞-$\overline{an}$. If G has an ascending series with locally (soluble-by-finite) factors, then either G is an $\mathfrak{S}_1\mathfrak{F}$-group or $G/Z(G)$ is finite.*

PROOF. Assume that $G/Z(G)$ is infinite. Then (by Lemma 5) every every abelian subgroup of G has finite total rank. Thus every soluble subgroup of G is an $\mathfrak{S}_1$-group by the just quoted theorem of Čarin. Hence it is enough to show that G is soluble-by-finite. By Proposition 7, it can be assumed that every abelian subgroup of G is minimax. It follows from the Baer-Zaĭtsev theorem ([2,11], see §1) that every soluble subgroup of G is minimax.

Consider first the case that G is locally (soluble-by-finite). We claim that if G is not soluble-by-finite there exists an almost normal subgroup H of G such that neither H nor $N_G(H)/H$ is soluble-by-finite. Indeed, by the theorem of Zaĭtsev [12] quoted in §1, G' does not satisfy the Min-Max-∞ condition, hence there exists a family $(U_n)_{n\in\mathbb{Z}}$ of subgroups of G' such that U_n is a subgroup of infinite index of U_{n+1} for all $n \in \mathbb{Z}$. By condition Min-Max-∞-$\overline{an}$ there exists an integer n such that $H = U_n$ is almost normal in G. Clearly neither H nor G'/H_G is a minimax group. This implies that H is not soluble-by-finite. Let $N = N_G(H)$ and suppose by contradiction that N/H is soluble-by-finite. Then there exists a subgroup S of finite index in N (and hence in G) such that S/H is soluble, of derived length d, say. Let $K = S_G$. Then G/K is finite and the d-th term of the derived series of K is contained in H, hence in H_G. Thus $\bar{G} = G/H_G$ is soluble-by-finite. Since $\bar{G}'$ is not minimax, $\bar{G}/Z(\bar{G})$ is not polycyclic-by-finite. By Proposition 7 every abelian subgroup of $\bar{G}$ is minimax, hence $\bar{G}$ is minimax. This is a contradiction, as $\bar{G}'$ is not minimax. The claim is now proved.

Now let H_0 be an almost normal subgroup of G such that neither H_0 nor $N_G(H_0)/H_0$ is soluble-by-finite. Repeated use of the claim just proved shows that it is possible to construct an ascending series $H_0 \lhd H_1 \lhd \ldots \lhd H_n \lhd \ldots$ of almost normal subgroups of G and a descending chain $H_0 > H_{-1} > \ldots > H_{-n} > \ldots$ of subgroups such that H_{-n} is almost normal in H_{-n+1} for every $n \in \mathbb{N}$ and $F_n = N_{H_{n+1}}(H_n)/H_n$ is not soluble-by-finite for every $n \in \mathbb{Z}$. In particular F_n is not centre-by-finite. Hence F_n contains a subgroup K_n/H_n which is not almost normal. Then

$$\ldots < K_{-2n} < \ldots < K_{-2} < K_0 < K_2 < \ldots < K_{2n} < \ldots$$

is a chain of non-almost normal subgroups of G and the index $|K_{2n} : K_{2(n-1)}|$ is infinite for every $n \in \mathbb{Z}$. This is impossible, since G satisfies Min-Max-∞-$\overline{an}$. This contradiction proves that G is soluble-by-finite, provided it is locally (soluble-by-finite).

Consider now the general case. By the first part of the proof G has an ascending series whose factors are either abelian or finite. Since every abelian subgroup of G is minimax, G is a soluble-by-finite minimax group by a theorem of Amberg [1]. □

Now we have to describe the structure of $\mathfrak{S}_1\mathfrak{F}$-groups which are not centre-by-finite and satisfy Min-Max-∞-$\overline{an}$. Minimax groups obviously satisfy Min-Max-∞-$\overline{an}$, hence we only have to consider groups with these properties which are not minimax. Observe first that (nilpotent torsion-free) groups of this kind exist. Indeed, let C be the direct product of an infinite cyclic group $\langle x\rangle$ by a subgroup A of the additive group of rational numbers which is not minimax. Let $1 \neq a \in A$ and let G be the semidirect product of C by an infinite cyclic group $\langle y\rangle$, where the action of y on C is defined by $[A, y] = 1$ and $[x, y] = a$. Then $G' = \langle a\rangle \leq A = Z(G)$. Let H be a subgroup of G which is not minimax. Then $A \cap H \neq 1$ and so $B = \langle a\rangle \cap H \neq 1$. The group G/B has finite commutator subgroup and finite rank, hence it is centre-by-finite. Thus H is almost normal in G. Therefore every subgroup of G which is not minimax is almost normal. This yields that G satisfies Min-∞-$\overline{an}$ and hence Min-Max-∞-$\overline{an}$, but $G/Z(G)$ is infinite and G is not minimax.

The property that every subgroup which is not minimax is almost normal is not typical of this example, but holds in all soluble-by-finite group satisfying Min-Max-∞-$\overline{an}$.

Lemma 9. *Let the soluble-by-finite group G satisfy Min-Max-∞-$\overline{an}$ and let H be a subgroup of G which is not minimax. Then H is almost normal in G. Moreover, the following holds:*

(i) G/H_G is centre-by-finite and H/H_G is finite;

(ii) $|G' : G' \cap H|$ is finite.

PROOF. It can be assumed that $G/Z(G)$ is infinite. Since G is not minimax, then $G/Z(G)$ is polycyclic-by-finite, by Proposition 7. In particular $H' \leq G'$ is polycyclic-by-finite and H/H' is not minimax. Since G is an $\mathfrak{S}_1\mathfrak{F}$-group by Proposition 8, H has a periodic abelian quotient with infinitely many nontrivial primary components. It follows by Lemma 1 (i) that H is almost normal in G.

Since $G/Z(G)$ is polycyclic-by-finite and $H \cap Z(G) \leq H_G$, every subgroup of G containing H_G is not minimax and so is almost normal, by the first part of the proof. Therefore G/H_G is centre-by-finite. This implies that H/H_G and $G'H_G/H_G$ are finite. In particular $|G' : G' \cap H| \leq |G'/G' \cap H_G|$ is finite. □

Since $\mathfrak{S}_1\mathfrak{F}$-groups are locally minimax (see [9], part II, Theorem 10.38), Lemma 9 yields that the class of soluble-by-finite groups satisfying property Min-Max-∞-$\overline{an}$ contains the class of soluble-by-finite groups whose subgroups are either finitely generated or almost normal. The latter class has been studied in [5].

Let A be a torsion-free abelian group. We denote by $\bar{M}(A)$ the intersection of all pure subgroups of A which are not minimax. It is easily seen that $\bar{M}(A) = 1$ if A has infinite rank. Lemma 9 (*ii*) and the following lemma show that $\bar{M}(A)$ plays a rôle in the description of groups with Min-Max-∞-$\overline{an}$.

Lemma 10. *Let A be an abelian torsion-free group and let B be a finitely generated subgroup of A. Then $B \leq \bar{M}(A)$ if and only if $B/B \cap H$ is finite for all subgroups H of A which are not minimax.*

PROOF. If A has infinite rank $\bar{M}(A) = 1$ and it is clear that no subgroup of A has the property required for B in the second part of the statement. Thus it can be assumed that A has finite rank. Suppose $B \leq \bar{M}(A)$ and let H be a subgroup of A which is not minimax. Let H^* be the pure closure of H in A (i.e., H^*/H is the torsion subgroup of A/H). Then $B \leq \bar{M}(A) \leq H^*$ as H^* is pure in A and not minimax. Hence $B/B \cap H$ is isomorphic to a subgroup of the periodic group H^*/H and so is finite.

Conversely, assume that if $B/B \cap H$ is finite for every subgroup H of A which is not minimax and let C be a pure subgroup of A which is not minimax. Then $BC/C \simeq B/B \cap C$ is finite. Since A/C is torsion-free, this means that $B \leq C$. □

Lemma 11. *Let G be a soluble-by-finite group with finite abelian section rank. Let E be a finite normal subgroup of G. Then:*

(i) if H is a subgroup of G, then H is almost normal in G if and only if EH is almost normal in G;

(ii) G satisfies Max-∞-$\overline{an}$ if and only if G/E satisfies Max-∞-$\overline{an}$.

PROOF. (i) If H is almost normal, then clearly EH is almost normal. Conversely, assume that EH is almost normal in G. Let $n = |E|$. Then $(EH)^n \leq H \leq EH$. Since $EH/(EH)^n$ is finite and both EH and $(EH)^n$ are almost normal in G, it follows that H is almost normal in G.

(ii) follows easily from (*i*). □

Theorem 12. *Let the group G have an ascending series with locally (soluble-by-finite) factors. Then the following conditions are equivalent:*

(i) G satisfies Min-Max-∞-$\overline{an}$;

(ii) G satisfies Min-∞-$\overline{an}$;

(iii) G satisfies Max-∞-$\overline{an}$;

(iv) every subgroup of G which is not minimax is almost normal;

(v) either G is a minimax group or $G/Z(G)$ is finite or G has a finite normal subgroup E such that G/E is direct product of finitely many Prüfer groups and a nilpotent torsion-free group N of class 2 such that $N/Z(N)$ is finitely generated and $N' \leq \bar{M}(Z(N))$.

PROOF. Obviously both Min-∞-$\overline{an}$ and Max-∞-$\overline{an}$ imply Min-Max-∞-$\overline{an}$. By Lemma 9, (*i*) implies (*iv*) and (*iv*) clearly implies (*ii*). Hence it only remains to show that (*v*) holds if G satisfies Min-Max-∞-$\overline{an}$ and that if (*v*) holds then G satisfies Max-∞-$\overline{an}$.

Let G satisfy Min-Max-∞-$\overline{an}$ and assume that G is not minimax and $G/Z(G)$ is infinite. Then G is an $\mathfrak{S}_1\mathfrak{F}$-group and $G/Z(G)$ is polycyclic-by-finite by Proposition 7 and Proposition 8. In particular G' is polycyclic-by-finite and $Z(G)$ is not minimax, thus $G/Z_2(G)$ is finite by Lemma 9 and by B.H. Neumann's theorem ([8]). It follows that $\gamma_3(G)$ is finite (see for instance [9], part I, p. 113). Let R be the $\mathfrak{F}$-perfect radical of G. There exists $U \leq G$ such that $G = RU$ and $RG' \cap U = G'$. Since G is finite-by-nilpotent, the set of the periodic elements of U is a subgroup E. Since $R \cap U = R \cap G'$ is finite, E is finite. Let $N = U/E$. Then N is torsion-free and nilpotent of class 2 (clearly $\gamma_3(G) \leq E$ and if N were abelian, then G', and hence $G/Z(G)$, would be finite), G/E is direct product of N and RE/E, which is direct product of finitely many Prüfer groups. Clearly N is not minimax, thus $N/Z(N)$ is finitely generated by Proposition 7. Finally, it follows from Lemma 9 (*ii*) and Lemma 10 that N' is contained in $\bar{M}(Z(N))$. Thus (*v*) holds if G satisfies Min-Max-∞-$\overline{an}$.

Conversely, let (*v*) hold. In order to prove that G satisfies Max-∞-$\overline{an}$ it can be assumed that $E = 1$, as follows from Lemma 11 (*ii*). Since $N/Z(N)$ is finitely generated, $G' = N'$ is finitely generated. Let $(H_n)_{n\in\mathbb{N}}$ be an ascending sequence of subgroups of G such that $|H_{n+1} : H_n|$ is infinite for every index n. We have to prove that H_n is almost normal in G for some $n \in \mathbb{N}$. Let, for any $n \in \mathbb{N}$, $K_n = H_n \cap Z(N)$. Since $G/Z(N)$ is minimax, $|K_{n+1} : K_n|$ is infinite for all but finitely many $n \in \mathbb{N}$. Hence $K = \bigcup_{n\in\mathbb{N}} K_n$ is not minimax. Since $G' \leq \bar{M}(Z(N))$, Lemma 10 yields that $G'/K \cap G'$ is finite. Moreover there exists a positive integer n such that $H_n \cap G' = K_n \cap G' = K \cap G'$. Thus $\bar{G} = G/H_n \cap G'$ has finite commutator subgroup. Since $\bar{G}/Z(\bar{G})$ is finitely generated and $\bar{G}$ is nilpotent of class 2, $\bar{G}$ is centre-by-finite, hence H_n is almost normal in G. □

It is possible to obtain some further information on the structure of a soluble-by-finite group G satisfying Min-Max-∞-$\overline{an}$. Namely, if G is neither centre-by-finite nor minimax, then $Z(G)$ is an extension of a minimax group by a torsion-free abelian group of rank 1. In view of Theorem 12, this follows from the following proposition.

Proposition 13. *Let A be a torsion-free abelian group. If $\bar{M}(A) \neq 1$ then A has a pure minimax subgroup P such that A/P has rank* 1.

PROOF. Since $\bar{M}(A) \neq 1$, A has finite rank. Let X be a maximal $\mathbb{Z}$-independent subset of $\bar{M}(A)$. Then X is contained in a maximal $\mathbb{Z}$-independent subset Y of A. Put $B = \langle X \rangle$. Let $x \in X$ and $C = \langle Y \setminus \{x\} \rangle$. Then

$D = B \cap C = \langle X \setminus \{x\} \rangle$. Let P/C be the torsion subgroup of A/C. Then P is pure in A and A/P has rank 1. Since C/P is periodic $B \cap P = D$. Therefore P is minimax by Lemma 10. □

References

[1] B. Amberg, Fast-Polyminimaxgruppen, *Math. Ann* **175**(1968), 44–49.

[2] R. Baer, Polyminimaxgruppen, *Math. Ann.* **175**(1968), 1–43.

[3] V.S. Čarin, On soluble groups of type A_3, *Mat. Sb.* **54**(1961), 489–499.

[4] I.I. Eremin, Groups with finite classes of conjugate abelian subgroups, *Mat. Sb.* **47**(1959), 45–54.

[5] S. Franciosi, F. de Giovanni and L.A. Kurdachenko, On groups with many almost normal subgroups, *Ann. Mat. Pura Appl. (4)*, to appear.

[6] L.S. Kazarin and L.A. Kurdachenko, Finiteness conditions and factorizations in groups, *Uspekhi Mat. Nauk* **47**(3)(1992), 75–114; *Russian Math. Surveys* **47**(3)(1992), 81–126.

[7] L.A. Kurdachenko and V.V. Pylaev, Groups rich with almost-normal subgroups, *Ukrain. Mat. Ž.* **40**(1988), 326–330; *Ukrainian Math. J.* **40**(1988), 278–281.

[8] B.H. Neumann, Groups with finite classes of conjugate subgroups, *Math. Z.* **63**(1955), 76–96.

[9] D.J.S. Robinson, *Finiteness Conditions and Generalized Soluble Groups* (Springer, Berlin, 1972).

[10] M.J. Tomkinson, *FC-groups* (Pitman, London, 1984).

[11] D.I. Zaĭtsev, On groups satisfying a weak minimal condition, *Mat. Sb.* **78**(1969), 323–331; *Math. USSR Sb.* **7**(1969), 315–322.

[12] D.I. Zaĭtsev, Theory of minimax groups, *Ukrainian Mat. Ž.* **23**(1971), 652–660; *Ukrainian Math. J.* **23**(1971), 536–542.

COMPUTATION OF THE CHARACTER TABLE OF AFFINE GROUPS USING FISCHER MATRICES

M.R. DARAFSHEH* and A. IRANMANESH†

*Math. Dept., Tehran University, Tehran, Iran
†Math. Dept., Tarbiat Modarres University, Tehran, Iran & Center for Theo. Phys. & Mathematics, AEOI, Tehran, Iran

Abstract

Let $GL_n(q)$ be the general linear group and let $H_n = V_n(q).GL_n(q)$ denote the affine group of $V_n(q)$. In this paper we will compute the character tables of the groups H_2, H_3 and H_4 using Fischer's method. For this purpose, we use results from [1] and [2] to obtain Fischer matrices for each of the conjugacy classes of $GL_2(q), GL_3(q)$ and $GL_4(q)$. It turns out that the groups H_2, H_3 and H_4 have q^2+q-1, q^3+q^2-1 and $q^4+q^3+q^2-q-1$ irreducible characters respectively.

1990 *AMS Mathematics Subject Classification*: Primary 20C15.

1. Preliminary and notation

Let H be a group and $V \trianglelefteq H$. Then H acts on Irr(V) by conjugation. Let I_χ be the inertia group of χ in H. Since I_χ is the stabilizer of χ in H, it is clear that $V \trianglelefteq I_\chi$.

In [1], a method is presented for calculating the irreducible characters of the group extensions $H = V.G$ provided that every irreducible character of V extends to an irreducible character of its inertia group in H. Since the affine group has this property, we are able to compute the character table of this group by Fischer's method.

Let $\chi_1, \ldots, \chi_s$ be representatives of the orbits of $H = V.G$ acting on the irreducible characters of V and let $\mathcal{K}$ be a conjugacy class of G and $\mathcal{K}_1, \ldots, \mathcal{K}_t$ be all the conjugacy classes of H which map to $\mathcal{K}$. If we choose $\chi_1, \ldots, \chi_s$ such that each $I_{\chi_i}, 1 \leq i \leq s$,contains an element in $\mathcal{K}$, then the Fischer matrix of H at h is denoted by $F^{\mathcal{K}} = \begin{bmatrix} F^{\mathcal{K}}_{\chi_1} \\ \vdots \\ F^{\mathcal{K}}_{\chi_s} \end{bmatrix}$, where $F^{\mathcal{K}}_{\chi_i}, 1 \leq i \leq s$, is defined in [2]. Moreover if $C^{\mathcal{K}}_{\chi_i}$ is the part of the character table of $\overline{I_{\chi_i}} = I_{\chi_i}/V$ consisting of the columns corresponding to the classes of $\overline{I_{\chi_i}}$ which fuse to $\mathcal{K}$ in G, then the character table of H at the classes $\mathcal{K}_1, \ldots, \mathcal{K}_t$ is given by the matrix product $C^{\mathcal{K}}_{\chi_i}.F^{\mathcal{K}}_{\chi_i}, 1 \leq i \leq s$.

The group $GL_{n+1}(q)$ acts transitively on the non-zero vectors of the (n+1)-dimensional vector space $V_{n+1}(q)$ over the Galois field with q elements. For $v_0 = \begin{bmatrix} 1 \\ 0 \\ \vdots \\ 0 \end{bmatrix} \in V_{n+1}(q)$ the stabilizer of v_0 under $GL_{n+1}(q)$ is the group:

$$\left\{ \left[\begin{array}{c|c} 1 & a\,b\ldots c \\ \hline 0 & \\ \vdots & A \\ 0 & \end{array} \right] \;\middle|\; \begin{array}{l} A \text{ is an } n \times n \text{ non-singular matrix over } GF(q) \\ \text{and } v = (a\,b\ldots c) \in GF(q)^n \end{array} \right\}.$$

Since

$$\left[\begin{array}{c|c} 1 & v \\ \hline 0 & \\ \vdots & A \\ 0 & \end{array} \right] = \left[\begin{array}{c|c} 1 & vA^{-1} \\ \hline 0 & \\ \vdots & I \\ 0 & \end{array} \right] \left[\begin{array}{c|c} 1 & 0\ldots 0 \\ \hline 0 & \\ \vdots & A \\ 0 & \end{array} \right], \quad v = (ab\ldots c) \;\in GF(q)^n,$$

we get $H_n = V_n(q).GL_n(q)$ where

$$V = V_n(q) = \left\{ \left[\begin{array}{c|c} 1 & v \\ \hline 0 & \\ \vdots & I \\ 0 & \end{array} \right] \mid v = (a\,b\ldots c) \in GF(q)^n \right\} \text{ and}$$

$$G = GL_n(q) = \left\{ \left[\begin{array}{c|c} 1 & 0\ldots 0 \\ \hline 0 & \\ \vdots & A \\ 0 & \end{array} \right] \;\middle|\; \begin{array}{l} A \text{ is an } n \times n \text{ non-singular matrix} \\ \text{over } GF(q) \end{array} \right\}.$$

We set $H = V.G$ where V and G are given as above. The action of G on V is as follows:

$$\left[\begin{array}{c|c} 1 & 0\ldots 0 \\ \hline 0 & \\ \vdots & A \\ 0 & \end{array} \right]^{-1} \left[\begin{array}{c|c} 1 & v \\ \hline 0 & \\ \vdots & I \\ 0 & \end{array} \right] \left[\begin{array}{c|c} 1 & 0\ldots 0 \\ \hline 0 & \\ \vdots & A \\ 0 & \end{array} \right] = \left[\begin{array}{c|c} 1 & vA \\ \hline 0 & \\ \vdots & I \\ 0 & \end{array} \right].$$

V is an elementary abelian group of order q^n where q is a prime power and G acts on Irr(V) as follows: For $\chi \in \text{Irr}(V)$ and $g = \left[\begin{array}{c|c} 1 & 0\ldots 0 \\ \hline 0 & \\ \vdots & A \\ 0 & \end{array} \right] \in G$, we have:

$$\chi^g\left(\left[\begin{array}{c|c} 1 & v \\ \hline 0 & \\ \vdots & I \\ 0 & \end{array}\right]\right) = \chi\left(\left[\begin{array}{c|c} 1 & vA \\ \hline 0 & \\ \vdots & I \\ 0 & \end{array}\right]\right)$$

We see that the number of orbits of G on $\mathrm{Irr}(V)$ coincide with the number of orbits of G on V. Since G has two orbits on V, therefore G has two orbits on $\mathrm{Irr}(V)$. One orbit consists of the identity character χ_0 alone with $I_{\chi_0} = V_n(q).GL_n(q)$. The other orbit has size $q^n - 1$ and a representative of this orbit may be taken to be any non-identity character χ_1 with $I_{\chi_1} = V_n(q).(V_{n-1}(q).GL_{n-1}(q))$. Thus $\overline{I_{\chi_0}} = GL_n(q)$ and $\overline{I_{\chi_1}} = V_{n-1}(q).GL_{n-1}(q)$ and therefore we can find the Fischer matrices for the group $H_n = V_n(q).GL_n(q)$ inductively.

Let χ be an irreducible character of $V_n(q)$ and $\mathcal{K}$ be a conjugacy class of $GL_n(q)$. Let $x = \left[\begin{array}{c|c} 1 & 0\ldots 0 \\ \hline 0 & \\ \vdots & A \\ 0 & \end{array}\right] \in \mathcal{K}$ and $\mathcal{K}_1, \ldots, \mathcal{K}_t$ be all the conjugacy classes of H which map to $\mathcal{K}$. To explain $\mathcal{K}_1, \ldots, \mathcal{K}_t$, we must consider the action of $V_n(q)$ on the coset $V_n(q)x$. If vx is an element of this coset, then conjugation by $u \in V_n(q)$, gives: $u^{-1}vxu = xux^{-1}u^{-1}vx$. Now if we set $K_x = \{xux^{-1}u^{-1} | u \in V_n(q)\} = \{x^{-1}u^{-1}xu | u \in V_n(q)\}$, then

$$K_x = \left\{\left[\begin{array}{c|c} 1 & v(I-A) \\ \hline 0 & \\ \vdots & I \\ 0 & \end{array}\right] \mid v = (a\, b \ldots c) \in GF(q)^n \right\}.$$

Obviously $K_x \leq V_n(q)$ and $K_x \leq ker\chi$.

2. Fischer matrices for the group $\mathbf{H_2 = V_2(q).GL_2(q)}$

The group H_2 is of order $q^3(q-1)^2(q+1)$. Since the group H_2 is a split extension, we can obtain all the conjugacy classes of H_2 by analysing each coset $V_2(q).x$, where $x \in GL_2(q)$.

In [3], Steinberg determined all the conjugacy classes of $GL_2(q)$ and furthermore, obtained the character table of this group.

According to [3], the group $GL_2(q)$ has four types of conjugacy classes. After analysing each of these four types we obtain that the group H_2 has ten types of conjugacy classes. Note that the notation used for the conjugacy classes of $GL_2(q)$, $GL_3(q)$ and $GL_4(q)$ are as in [3]. In this section we give Fischer matrices at each of the conjugacy classes of the group $GL_2(q)$ because this will suffice to yield the whole character table of H_2 by matrix multiplication. In Sections 3 and 4 we will do the same.

Fischer matrices for $GL_n(q)$ have been determined by B.Fischer in [1]. In particular if x lies in the identity class of $GL_n(q)$, then $K_x = 1$ and therefore we have two conjugacy classes in $V_n(q).GL_n(q)$ which map ton the identity class. Therefore all the irreducible characters of $V_n(q)$ have K_x in their kernels. Now $GL_n(q)$ has two orbits on Irr($V_n(q)$) with representatives χ_0 and χ_1. Therefore $F^{id}_{\chi_0} = \begin{bmatrix} 1 & 1 \end{bmatrix}, F^{id}_{\chi_1} = \begin{bmatrix} q^n - 1 & -1 \end{bmatrix}$ and the Fischer matrix at the class identity is $F^{id} = \begin{bmatrix} 1 & 1 \\ q^n - 1 & -1 \end{bmatrix}$. Therefore in the rest of this paper we will give Fischer matrices at non-identity classes. We set $S_i = \begin{bmatrix} 1 & 1 \\ q^i - 1 & -1 \end{bmatrix}$ and $T_i = \begin{bmatrix} 1 & 1 & 1 \\ q^i - q & 0 & -q \\ q-1 & -1 & q-1 \end{bmatrix}$ where i is a natural number. So $F^{id} = S_n$ for the group H_n and we will show that the Fischer matrices for the groups $H_n, 2 \leq n \leq 4$, are of the forms S_i and T_i where i is a suitable natural number.

If x lies in the class $\mathcal{K}$ of the type A_1, A_2, A_3 and B_1 of $GL_2(q)$ with condition $\rho^a \neq 1, \rho^b \neq 1, a \neq mult(q+1)$, then $K_x = V_2(q)$ and therefore we have only one conjugacy class in H_2 which map to these types of conjugacy classes. Thus the Fischer matrix at each of these types of conjugacy classes is (1).

If x lies in the the class $\mathcal{K}$ of the type A_2 (with condition $\rho^a = 1$) and A_3 (with condition $\rho^a \neq 1, \rho^b = 1$) of $GL_2(q)$, then $|K_x| = q$ and $V_2(q)/K_x \cong V_1(q)$, therefore we have two conjugacy classes in H_2 which map to the class $\mathcal{K}$. Thus the Fischer matrix at each class $\mathcal{K}$ of this type is S_1.

We thus obtain the following irreducible characters for H_2 : $(q-1)$ irreducible characters of degree 1, $(q-1)$ irreducible characters of degree q, $\frac{1}{2}q(q-1)$ irreducible characters of degree $(q-1)$, $\frac{1}{2}(q-1)(q-2)$ irreducible characters of degree $(q+1)$, $(q-1)$ irreducible characters of degree q^2-1 and one irreducible character of degree $(q-1)^2(q+1)$.

3. Fischer matrices for the group $\mathbf{H_3 = V_3(q).GL_3(q)}$

The group H_3 is of order $q^6(q^3-1)(q^2-1)(q-1)$. As in the case of H_2, we can calculate the character table of this group by Fischer's method. For this purpose, we must first obtain all the conjugacy classes of the group H_3. Since H_3 is a split extension, therefore for each conjugacy class $\mathcal{K}$ of $GL_3(q)$ we analysed the coset $V_3(q).x, x \in \mathcal{K}$ and obtained the conjugacy classes of H_3. It turns out that H_3 has twenty seven types of conjugacy classes.

Now, we construct the Fischer matrices of H_3. If x lies in the class $\mathcal{K}$ of the type $A_1, A_2, A_3, A_4, A_5, A_6, B_1$ or C_1 of $GL_3(q)$ with conditions $\rho^a \neq$

$1, \rho^b \neq 1, \rho^c \neq 1, a \neq b \neq c \neq a, b \neq mult(q+1), a \neq mult(q^2+q+1)$, then $K_x = V_3(q)$ and therefore we have only one conjugacy class in $V_3(q).GL_3(q)$ which map to these types of conjugacy classes. Thus the Fischer matrix at each of these classes is (1).

If x lies in the class $\mathcal{K}$ of the type A_3(with condition $\rho^a = 1$), A_4(with conditions $\rho^a \neq 1, \rho^b = 1$), A_5(with conditions $\rho^a \neq 1, \rho^b = 1$), A_5(with conditions $\rho^a = 1, \rho^b \neq 1$), A_6(with conditions $\rho^a \neq 1, \rho^b \neq 1, \rho^c = 1$) or B_1(with condition $\rho^a = 1$) of $GL_3(q)$, then $|K_x| = q^2$ and $V_3(q)/K_x \cong V_1(q)$, therefore we have two conjugacy classes in H_3 which map to the class $\mathcal{K}$. Thus the Fischer matrix at each of the classes $\mathcal{K}$ of these types is S_1.

If x lies in the class $\mathcal{K}$ of the type A_4(with conditions $\rho^a = 1, \rho^b \neq 1$), of $GL_3(q)$, then $|K_x| = q$ and $V_3(q)/K_x \cong V_2(q)$, therefore we have two conjugacy classes in H_3 which map to the class $\mathcal{K}$. Thus the Fischer matrix at the class $\mathcal{K}$ is S_2.

If x lies in the class $\mathcal{K}$ of the type A_2(with condition $\rho^a = 1$) of $GL_3(q)$, then $|K_x| = q$ and there are three conjugacy classes in H_3 which map to the class $\mathcal{K}$. Thus the Fischer matrix at the class $\mathcal{K}$ is T_2.

We thus obtain $q^2 + q - 1$ irreducible characters for H_3 which are listed in Table I.

Degree	Frequency
1	$q-1$
q^2+q	$q-1$
q^3	$q-1$
q^2+q+1	(q-1)(q-2)
$q(q^2+q+1)$	(q-1)(q-2)
$(q+1)(q^2+q+1)$	$\frac{1}{6}(q-1)(q-2)(q-3)$
$(q-1)(q^2+q+1)$	$\frac{1}{2}q(q-1)^2$
$(q^2-1)(q-1)$	$\frac{1}{3}q(q-1)(q+1)$
q^3-1	$q-1$
$q(q^3-1)$	$q-1$
$(q+1)(q^3-1)$	$\frac{1}{2}(q-1)(q-2)$
$(q-1)(q^3-1)$	$\frac{1}{2}q(q-1)$
$(q^2-1)(q^3-1)$	$q-1$
$(q-1)(q^2-1)(q^3-1)$	1

Table I

4. Fischer matrices for the group $\mathbf{H_4 = V_4(q).GL_4(q)}$

The group H_4 is of order $q^{10}(q^4-1)(q^3-1)(q^2-1)(q-1)$. As is the case for H_2 and H_3 we can calculate the character table of this group by Fischer's method. For this purpose, first we must obtain all the conjugacy classes of the group H_4. Since H_4 is a split extension, therefore for each conjugacy class $\mathcal{K}$ of $GL_4(q)$ we analyse the coset $V_4(q).x, x \in \mathcal{K}$ and obtain all the conjugacy classes of H_4. It turns out that H_4 has seventy three types of conjugacy classes.

Now we construct the Fischer matrices of H_4. Since the calculation of these matrices is very long, we only give Fischer matrices for each conjugacy classes of $GL_4(q)$ in the following table:

type	condition	Fischer matrix
A_1	$\rho^a, \rho^b, \rho^c, \rho^d = 1$	S_4
$A_i, 1 \leq i \leq 14$	$\rho^a, \rho^b, \rho^c, \rho^d \neq 1$	(1)
$B_i, 1 \leq i \leq 3$	$\rho^a, \rho^b \neq 1,$	(1)
$C_i, 1 \leq i \leq 3$	none	(1)
D_1	$\rho^a \neq 1$	(1)
E_1	none	(1)
A_2	$\rho^a = 1$	T_3
A_3	$\rho^a = 1$	S_2
A_4	$\rho^a = 1$	T_2
A_5	$\rho^a = 1$	S_1
A_6	$\rho^a = 1, \rho^b \neq 1$	S_3
A_7	$\rho^a = 1, \rho^b \neq 1$	T_2
A_7	$\rho^b = 1, \rho^a \neq 1$	S_1
A_8	$\rho^a = 1, \rho^b \neq 1$	S_1
A_8	$\rho^b = 1, \rho^a \neq 1$	S_1
A_9	$\rho^a = 1, \rho^b \neq 1$	S_2
A_{10}	$\rho^a = 1, \rho^b \neq 1$	S_1
A_{10}	$\rho^b = 1, \rho^a \neq 1$	S_2
A_{11}	$\rho^a = 1, \rho^b \neq 1$	S_1
A_{12}	$\rho^a = 1, \rho^b, \rho^c \neq 1$	S_2
A_{12}	$\rho^c = 1, \rho^a, \rho^b \neq 1$	S_1
A_{13}	$\rho^a = 1, \rho^b, \rho^c \neq 1$	S_1
A_{13}	$\rho^c = 1, \rho^a, \rho^b \neq 1$	S_1
A_{14}	$\rho^a = 1, \rho^b, \rho^c, \rho^d \neq 1$	S_1
B_1	$\rho^a = 1$	S_2
B_2	$\rho^a = 1$	S_1
B_3	$\rho^a = 1, \rho^b \neq 1$	S_1
D_1	$\rho^a = 1$	S_1

Table II

Therefore we obtain all the Fischer matrices at each conjugacy class of $GL_4(q)$ and this enables us to compute the whole character table of the group H_4. The group H_4 has $q^4+q^3+q^2-q-1$ irreducible characters. The degrees of the irreducible characters of H_4 are listed in Table III

Degree	Frequency
1	$q-1$
$q(q^2+q+1)$	$q-1$
$q^2(q^2+1)$	$q-1$
$q^3(q^2+q+1)$	$q-1$
q^6	$q-1$
$(q+1)(q^2+1)$	$(q-1)(q-2)$
$q(q+1)^2(q^2+1)$	$(q-1)(q-2)$
$q^3(q+1)(q^2+1)$	$(q-1)(q-2)$
$(q+1)(q^2+1)(q^2+q+1)$	$\frac{1}{2}(q-1)(q-2)(q-3)$
$q(q+1)(q^2+1)(q^2+q+1)$	$\frac{1}{2}(q-1)(q-2)(q-3)$
$(q+1)^2(q^2+1)(q^2+q+1)$	$\frac{1}{24}(q-1)(q-2)(q-3)(q-4)$
$(q^2-1)(q^2+1)(q^2+q+1)$	$\frac{1}{4}q(q-1)^2(q-2)$
$(q^2-1)^2(q^2+1)$	$\frac{1}{3}q(q_1)^2(q+1)$
$(q^2+1)(q^2+q+1)$	$\frac{1}{2}(q-1)(q-2)$
$q^2(q^2+1)(q^2+q+1)$	$\frac{1}{2}(q-1)(q-2)$
$q(q^2+1)(q^2+q+1)$	$(q-1)(q-2)$
$(q-1)(q^2+1)(q^2+q+1)$	$\frac{1}{2}q(q-1)^2$
$q(q-1)(q^2+1)(q^2+q+1)$	$\frac{1}{2}q(q-1)^2$
$(q-1)^2(q^2+1)(q^2+q+1)$	$\frac{1}{8}q(q^2-1)(q-2)$
$(q-1)^3(q+1)(q^2+q+1)$	$\frac{1}{4}q^2(q-1)(q+1)$
$q^2(q-1)^2(q^2+q+1)$	$\frac{1}{2}q(q-1)$
$(q-1)^2(q^2+q+1)$	$\frac{1}{2}q(q-1)$
q^4-1	$q-1$
$(q^4-1)(q^2+q)$	$q-1$
$q^3(q^4-1)$	$q-1$
$(q^4-1)(q^2+q+1)$	$(q-1)(q-2)$
$q(q^4-1)(q^2+q+1)$	$(q-1)(q-2)$
$(q+1)(q^4-1)(q^2+q+1)$	$\frac{1}{6}(q-1)(q-2)(q-3)$
$(q-1)(q^4-1)(q^2+q+1)$	$\frac{1}{2}q(q-1)^2$
$(q+1)(q^2-1)(q^4-1)$	$\frac{1}{3}q(q-1)(q+1)$
$(q^3-1)(q^4-1)$	$q-1$
$q(q^3-1)(q^4-1)$	$q-1$
$(q+1)(q^3-1)(q^4-1)$	$\frac{1}{2}(q-1)(q-2)$
$(q-1)(q^3-1)(q^4-1)$	$\frac{1}{2}q(q-1)$
$(q^2-1)(q^3-1)(q^4-1)$	$q-1$
$(q-1)(q^2-1)(q^3-1)(q^4-1)$	1

Table III.

References

[1] B.Fischer, Clifford matrices, in *Representations of Finite Groups and Finite Dimensional Algebras* (G.O.Michler and C.M.Ringel (eds.), Birkhauser-Verlag, 1991), 1–16.

[2] R.J.List, On the characters of $2^{n-\epsilon}.S_n$, *Arch. Math.* **51**(1988), 118–124.

[3] R.Steinberg, The representations of $GL(3,q)$, $GL(4,q)$, $PGL(3,q)$ and $PGL(4,q)$, *Canad. J. Math.* **3**(1951), 225–235.

THE LATTICE OF COMPACT REPRESENTATIONS OF AN INFINITE GROUP

DIKRAN DIKRANJAN[1]

Dept. di Matematica e Informatica, Università di Udine, Via Zanon 6, I-33100 Udine, Italy

Dedicated to the memory of Ivan Prodanov

Abstract

We consider the lattice of compact representations of an infinite group G, its subset of faithful representations and the interrelation between their global poset structure and the algebraic structure of G.

AMS classification numbers: primary 20C99, 20K45, 22D10; secondary 20A15, 20E05, 20E26, 20F50.

1. Introduction

A compact representation of a group G is a dense homomorphism $h : G \to K$ (i.e. $\overline{h(G)} = K$), where K is a compact topological group ([2], [46], [55], [57]). For two representations $h_1 : G \to K_1$ and $h_2 : G \to K_2$ we set $h_1 \leq h_2$ if there exists a continuous homomorphism $f : K_1 \to K_2$ such that $h_2 = f \circ h_1$ and we declare h_1 and h_2 *equivalent* if $h_1 \leq h_2$ and $h_2 \leq h_1$. Then there is a *set* $\tilde{\mathcal{R}}(G)$ of non-equivalent compact representation of G, such that any other compact representation of G is equivalent to some $h \in \tilde{\mathcal{R}}(G)$. With the above order $\tilde{\mathcal{R}}(G)$ becomes a complete lattice. While the famous Gel'fand-Raĭkov theorem guarantees that every group G has sufficiently many irreducible unitary representations in Hilbert spaces, it may happen that $\tilde{\mathcal{R}}(G)$ is trivial even for matrix groups (see J. von Neumann and Wigner [53] for the group $SL(2, \mathbf{C})$).

In this survey we discuss the size and other poset invariants of the lattice $\tilde{\mathcal{R}}(G)$ and its subset $\mathcal{R}(G)$ of *faithful* representations of G, i.e. monomorphisms. This is based on an "approximation" of $\mathcal{R}(G)$ and some of its important subsets by means of appropriate large canonic posets and provides useful global qualitative and quantitative information about the poset $\mathcal{R}(G)$ (Theorems 2 and 4, Corollary 3). This combinatorial presentation of $\mathcal{R}(G)$ permits an easy way of *simultaneous* calculation of the poset-invariants of $\mathcal{R}(G)$ and its subsets, and shows that in many cases they do not depend on the algebraic structure of the group G but only on its cardinality $|G|$ (such

[1]This work has been supported by funds MURST.

calculations were usually carried out *case by case* on various occasions [6], [13], [16], [61] etc.). In this setting natural questions arise related to the poset structure of $\mathcal{R}(G)$ which cannot be answered in ZFC, the Zermelo-Fraenkel axioms of set theory (instances to this effect are given in §§4 and 5: Corollary 4, Theorems 5 and 6).

To make the paper reasonably self-contained we recall in Section 2 some known facts about $\mathcal{R}(G)$ and provide several equivalent descriptions - via the lattice $\mathcal{B}(G)$ of precompact topologies on G and via almost periodic functions. The topological language is adopted further in the paper. From this point of view we present a blend of results on compact-like topologizations of groups in the spirit of Halmos' question on the algebraic structure of compact abelian groups ([42]). In 2.3 we give a very brief review on atoms and complementation of $\mathcal{R}(G)$ to underline the substantial difference between the *local* and *global* aspects of the impact of poset properties of $\mathcal{R}(G)$ on the algebraic structure of the group G.

In §3 we discuss varieties of groups in which the free groups are residually finite. This gives a possibility of a "topological view" on Burnside's Problem. In §4 we consider topologies which are closer to the compact ones (pseudocompact and countably compact) and we show that in many cases the poset of pseudocompact topologies contains complete Boolean algebras of the largest possible size. Chains in $\mathcal{R}(G)$ are studied in §5. A lot of open questions are given throughout the text.

We denote by $\mathbf{c}$ the cardinality of the continuum, by $\mathbf{Z}$ the integers, by $\mathbf{Q}$ the rationals, by $\mathbf{R}$ the reals, by $r(G)$ the free-rank of an abelian group G and set $\log \tau = \min\{\sigma \geq \omega : 2^\sigma \geq \tau\}$ for an infinite cardinal τ.

2. Compact representations of a group G and precompact topologies on G

2.1. Representations, topologies and almost periodic functions

A Hausdorff group topology on a group G is *precompact* if its completion $\hat{G}$ w. r. t. the two-sided uniformity is compact (in such a case all uniformities on G coincide [74]). Every precompact group topology on G defines a faithful compact representation, namely the inclusion $G \to \hat{G}$. Conversely, a faithful compact representation $h : G \to K$ generates a precompact group topology on G, namely the initial topology on G w.r.t. h. Denote by $\mathcal{B}(G)$ the poset of all precompact group topologies on G equipped with the ordering by inclusion. The above correspondence defines an order isomorphism between $\mathcal{R}(G)$ and $\mathcal{B}(G)$ which can be extended to $\tilde{\mathcal{R}}(G)$ and the complete lattice $\tilde{\mathcal{B}}(G)$ of those group topologies T on G such that the (Hausdorff) quotient group G/N, where N is the T-closure of $\{1\}$, is precompact. We prefer to keep the term *precompact* also for topologies from $\tilde{\mathcal{B}}(G)$. In these terms the top element of $\tilde{\mathcal{B}}(G)$ is the finest precompact topology on G, while the top element $\rho_G : G \to bG$ of $\tilde{\mathcal{R}}(G)$ is the so called *Bohr compactification* of G. In the future we often identify these two lattices by means of this isomorphism.

Denote by $\Sigma(G)$ the subset of $\tilde{\mathcal{R}}(G)$ consisting of all compositions $h \circ \rho_G$ where h is a continuous irreducible finite-dimensional unitary representation of the compact group bG. By the celebrated Peter-Weyl theorem [75] the lattice $\tilde{\mathcal{R}}(G)$ is generated by $\Sigma(G)$, namely, each $h \in \tilde{\mathcal{R}}(G)$ is the join of the elements of $\Sigma(G)$ below h. An elementary proof (i. e. with no recourse to functional analysis) of this theorem in the abelian case can be found in [30], the author is not aware if such a proof exists in the non-abelian case.

When G is abelian $\Sigma(G)$ coincides with $Hom(G, \mathbf{R}/\mathbf{Z})$ and so has a structure of an (abelian) group, the group of *characters* of G. Now the characters $\chi \leq h$ are precisely those $\chi \in \Sigma(G)$ which are continuous w. r. t. the (precompact) topology associated to h; these characters form a subgroup of $\Sigma(G)$. Thus $\tilde{\mathcal{B}}(G)$ is isomorphic to the lattice of all subgroups of $\Sigma(G)$. Precompact topologies correspond to point-separating groups of characters of G ([21], [60], [57]). In the non-abelian case the lattice $\tilde{\mathcal{B}}(G)$ is anti-isomorphic to the lattice of all closed normal subgroups N of bG, the subset $\mathcal{B}(G)$ corresponds to the family of those N such that $\rho_G^{-1}(N) = \{1\}$ ([60], [57]).

For every cardinal σ denote by $\tilde{\mathcal{B}}_\sigma(G)$ the subset of $\tilde{\mathcal{B}}(G)$ consisting of topologies of *weight* σ and set $\mathcal{B}_\sigma(G) = \tilde{\mathcal{B}}_\sigma(G) \cap \mathcal{B}(G)$ (i. e. topologies $T \in \mathcal{B}(G)$ having a local base at 1 of cardinality σ). These topologies correspond to compact representations having σ elements of $\Sigma(G)$ below them.

A complex valued bounded function f on an (abstract) group G is *almost periodic* if the set of all (left) translates of f is relatively compact in the space of all bounded complex valued functions on G equipped with the uniform topology ([45], 18.1). These are precisely the functions which can be factorized through ρ_G. The **C**-algebra $A(G)$ of almost periodic functions of G provides another description of $\tilde{\mathcal{R}}(G)$ ([55]). There is a bijection between closed (with respect to the uniform topology) subalgebras of $A(G)$ and compact representations of G. Faithful compact representations (precompact topologies) correspond to subalgebras of $A(G)$ separating the points of G.

2.2. Maximally almost periodic groups

Following von Neumann's [52] terminology a group G is *maximally* (resp. *minimally*) *almost periodic* if ρ_G is injective (resp. trivial). Obviously, there are arbitrarily large minimally almost periodic groups (just take infinite simple groups of cardinality $> \mathbf{c}$). Note that $\ker \rho_G$ is always contained in the commutator subgroup G' of G, since the quotient G/G' is abelian, thus maximally almost periodic. In general one can prove that if for $x \in G$ the conjugacy class x^G contains powers of x^m for each $m \in \mathbf{N}$ then $x \in \ker \rho_G$ ([45],

22.22). There exists a nilpotent group G of class 2 such that $\ker \rho_G = G'$, two examples to this effect can be found in Warfield [73] (Example 5.10 is a p-group of exponent p, where p may be any odd prime, while Example 5.11 is torsion-free). They show that the subgroup $\ker \rho_G$ may be maximally almost periodic and non-trivial. In view of the obvious isomorphism between $\tilde{\mathcal{B}}(G)$ and $\tilde{\mathcal{B}}(G/\ker \rho_G)$ we concentrate on maximally almost periodic groups from now on.

For a maximally almost periodic group G consider the minimal weight of a precompact topology on G, defined obviously by $\gamma(G) = \min\{\kappa : \mathcal{B}_\kappa(G) = \emptyset\}$, and the weight $\Gamma(G)$ of the finest precompact topology on G (clearly $\Gamma(G) = w(bG)$). Then $\log|G| \leq \gamma(G) \leq \max\{|G'|, \log|G|\} \leq \Gamma(G) \leq 2^{|G|}$ and $\mathcal{B}_\sigma(G) \neq \emptyset$ precisely for $\gamma(G) \leq \sigma \leq \Gamma(G)$. ([28], Lemma 2.1). For abelian groups $\gamma(G) = \log|G|$, $\Gamma(G) = 2^{|G|}$ and $|\mathcal{B}_\sigma(G)| = 2^{\sigma|G|}$ for each σ with $\gamma(G) \leq \sigma \leq \Gamma(G)$ ([6], Theorem 5.3), the last two equalities remain true also for groups with $|G/G'| = |G|$ ([61], Theorem 2.13). It follows from Corollary 2 below that this and other cardinal invariants of $\mathcal{B}_\sigma(G)$ coincide with those of a canonical poset which *does not depend on G, but only on $|G|$* (in fact, only on $2^{|G|}$).

A topological space is *pseudocompact* if every real-valued continuous function defined on it is bounded ([44]). Pseudocompact groups are precompact ([22]), so that the poset $\mathcal{P}(G)$ of pseudocompact topologies on G is contained in $\mathcal{B}(G)$. In terms of almost periodic functions $\tau \in \mathcal{B}(G)$ is pseudocompact if and only if every continuous complex valued function of (G, τ) is almost periodic.

One can introduce compact representations and maximal almost periodicity for a *topological* group G requiring continuity wherever necessary ([57]). It is known that a locally compact connected maximally almost periodic group is isomorphic to $\mathbf{R}^n \times G_0$, where $n \in \mathbf{N}$ and G_0 is compact (see Freudenthal [39] in the metrizable case, Weil [74] in general). It would be important to find a satisfying counterpart of Freudenthal's characterization of maximally almost periodic groups in the discrete case. Hall [41] proved that the free groups are maximally almost periodic (actually, residually finite). In Proposition 1 below we describe the varieties in which the free groups are maximally almost periodic. It would be interesting to characterize the varieties $\mathcal{V}$ in which *all* groups are maximally almost periodic. Warfield's example suggests that such varieties may consist of abelian groups only.

2.3. Atoms and complementation in $\mathcal{B}(G)$

The atoms (i. e. minimal elements) in the larger poset $\mathcal{H}(G)$ of all *Hausdorff* group topologies on a group G, under the name *minimal topologies*, were introduced apparently by Stephenson [68] and studied extensively since then ([56], [57], [69], [61], [63], [64], see the monograph [30] and the survey [9] for more details and extended bibliography; for atoms in $\mathcal{P}(G)$ see [35] and [37]). Obviously compact topologies are minimal and the atoms in $\mathcal{B}(G)$ are minimal topologies. Conversely, it was proved by Prodanov and Stoyanov [59] that every minimal topology on an abelian group is *precompact* although this is not true in the non-abelian case. In fact, according to [40] the infinite symmetric group S_α admits a unique minimal topology (so that the poset $\mathcal{H}(S_\alpha)$ is a complete lattice), although $\mathcal{B}(S_\alpha) = \emptyset$ [23] (for another extremal property of this topology see [64]). The first examples of minimal non-precompact matrix groups were given in [63].

Even if $\mathcal{B}(G)$ is big for an abelian group G, there may exist no atoms in $\mathcal{B}(G)$: i) if $G = \mathbf{Q}$ ([56]; more generally, for a torsion-free abelian group G of finite rank $\mathcal{B}(G)$ has atoms if and only if G is reduced [26]); ii) if $G = \mathbf{Z}(p^\infty)$ (more generally, $\mathcal{B}(G)$ has atoms for a divisible torsion group G if and only if $G \cong (\mathbf{Q}/\mathbf{Z})^n$ for some $n \in \mathbf{N}$, cf. [29], where also a description of the torsion abelian groups G with this property is given); iii) if G is divisible and $\mathbf{c} \leq |G| < 2^{\log|G|}$ (see [24] for a description of *all* divisible groups with atomless $\mathcal{B}(G)$). In [25] necessary conditions were found for abelian groups with $r(G) < \mathbf{c}$ ensuring that $\mathcal{B}(G)$ has atoms, and it was shown that these conditions are also sufficient when G is splitting. Schinkel [62] (Satz 3.6) resolved the general problem of describing the abelian groups such that $\mathcal{B}(G)$ has atoms and showed that in the particular case when $r(G) < \mathbf{c}$ the condition given in [25] is not sufficient in general (Beispiel 5.10, see also Satz 5.9 for a complete solution in this case).

According to Steiner [67] the lattice of *all* topologies on a set X is complemented. In general $\tilde{\mathcal{B}}(G)$ is not a Boolean algebra for a group G. More precisely, it is a complemented lattice if and only if G is a bounded semisimple (as a $\mathbf{Z}$-module) abelian group, so that $\tilde{\mathcal{B}}(G)$ is a Boolean algebra if and only if G is a finite cyclic group of square-free order ([60]). In the lattice $\tilde{\mathcal{B}}(\mathbf{Z}(p^\infty))$ only the top and the bottom element have complements ([60]). It is easy to see that for a torsion-free abelian group G, $T \in \tilde{\mathcal{B}}(G)$ has a complement if and only if the completion (G, T) is torsion-free (by (24.23) of [45] a compact abelian group is torsion-free if and only if its (discrete) Pontryagin dual is divisible).

3. The global structure of $\mathcal{B}(G)$

3.1. Precompact varieties

A group topology having a local base at 1 consisting of open normal subgroups is called *linear*. The pro-finite topology of a residually finite group is obviously both linear and precompact Hausdorff. According to Mal'cev's theorem [49] a finitely generated matrix group over arbitrary field is residually finite. This means that every finitely generated relatively free group admitting precompact Hausdorff group topology is residually finite, thus admits a *linear* precompact group topology. Essentially this proves the following proposition (see [37] for a complete proof).

Proposition 1. *For a variety $\mathcal{V}$ of groups the following are equivalent:*

(i) each $\mathcal{V}$-free group admits a Hausdorff precompact linear group topology;

(ii) each finitely generated $\mathcal{V}$-free group is maximally almost periodic;

(iii) each $\mathcal{V}$-free group is residually finite;

(iv) $\mathcal{V}$ is generated by its finite groups.

(v) $\mathcal{V}$ is generated by its compact groups.

Following [33] a variety satisfying these conditions will be called *precompact.* Most of the known varieties of groups (such as, for example, the varieties of all groups, Abelian groups, nilpotent groups, polynilpotent groups, soluble groups, metaabelian groups etc.) are precompact ([51]). The only examples of non-precompact varieties known to the author are among Burnside varieties $\mathcal{B}_n$. In fact, let n be an exponent such that the Restricted Burnside Problem has positive solution : for prime exponents this was proved by Kostrikin [48], for odd exponents by Zel'manov [77], who resolved recently also the case of even exponent ([78]). Let $m > 1$ and $B(m,n)$ denote the free group of m generators of the Burnside variety $\mathcal{B}_n$ of exponent n. Then there exists a normal subgroup N of $B(m,n)$ of finite index which is contained in any other normal subgroup of finite index. According to Burnside's theorem a finitely generated torsion matrix group over arbitrary field is finite. Thus for every finite-dimensional unitary representation $h : B(m,n) \to U(k)$ the subgroup $\ker h$ contains N. In particular, this yields that $\ker \rho_G = N$. To see that the variety $\mathcal{B}_n$ is not precompact it suffices to take the exponent n in such a way that $B(m,n)$ is infinite, i.e. Burnside's Problem ([7]) has negative solution (by Adian's results [1] it suffices to take n to have an odd divisor ≥ 665, or according to more recent estimates given by S. Ivanov [47], it is enough to take any $n \geq 2^{48}$ regardless of parity of n). Therefore, $\mathcal{B}_n$ is precompact if and only if the group $B(2,n)$ (and hence any other $B(m,n)$) is finite, so $\mathcal{B}_2$, $\mathcal{B}_3$, $\mathcal{B}_4$ and $\mathcal{B}_6$ are precompact (see [1]).

Although $\mathcal{B}(B(m,n)) = \emptyset$ for infinite $B(m,n)$, the discrete topology is always available in general, so that $|\mathcal{H}(G)| \geq 1$ on each group G.

Question 1. Is $|\mathcal{H}(B(m,n))| = 1$ in the case when $B(m,n)$ is infinite ?

We remind that $B(m,n) = A(m,n)/Z$, where $A(m,n)$ is Adian's group and $Z \cong \mathbf{Z}$ its center ([1]). Ol'shanskii [54] showed that $|\mathcal{H}(A(m,n)/Z^n)| = 1$, resolving thus the longstanding problem of Markov ([50]): Does there exist a (countably) infinite group G with $|\mathcal{H}(G)| = 1$? Under the assumption of

CH Shelah [66] constructed Jonsson groups and so obtained as a by-product another solution of Markov's problem based on completely different ideas. Hesse [43] gave a proof without the assumption of *CH*.

3.2. The quasi-isomorphism

Two partially ordered sets X and Y are *quasi-isomorphic* if each one of them is isomorphic to a subset of the other. Quasi isomorphic posets obviously share a lot of common properties, such as monotone cardinal invariants, maximum size of well ordered subsets, anti-well ordered subsets, chains and antichains etc. For cardinals σ and κ we denote by $\mathbf{P}_\sigma(\kappa)$ the poset of all subsets of cardinality σ of κ. Now we describe the poset $\mathcal{B}_\sigma(G)$ up to quasi-isomorphism. According to the poset isomorphisms described in §2.1-2.2 there exists an embedding $\mathcal{B}_\sigma(G) \hookrightarrow \mathbf{P}_\sigma(\Gamma(G))$. The following theorem produces an embedding in the opposite direction.

Theorem 1. ([5],[28]) *Let G be an infinite maximally almost periodic group satisfying $\gamma(G) < 2^{|G/G'|}$. Then $\Gamma(G) \geq 2^{|G/G'|}$. For every infinite cardinal σ satisfying $\gamma(G) \leq \sigma \leq 2^{|G/G'|}$ there exists an embedding $\mathbf{P}_\sigma(2^{|G/G'|}) \hookrightarrow \mathcal{B}_\sigma(G)$.*

The proof of this theorem extends and develops ideas from [6] (Theorems 1.1 and 5.3) and [61] (Theorem 2.13).

Corollary 1. *Let G be an infinite maximally almost periodic group satisfying $\gamma(G) < \Gamma(G) = 2^{|G/G'|}$. Then*

$$\mathcal{B}_\sigma(G) \overset{q.i.}{\cong} \mathbf{P}_\sigma(2^{G/G'}) \tag{1}$$

for all σ with $\gamma(G) \leq \sigma \leq \Gamma(G)$. In particular, $\mathcal{B}_{\Gamma(G)}(G) \overset{q.i.}{\cong} \mathcal{B}(G) \overset{q.i.}{\cong} \mathbf{P}(2^{G/G'})$.

Corollary 2. *Let G be an infinite maximally almost periodic group with $2^{|G|} = 2^{|G/G'|}$. Then $\Gamma(G) = 2^{|G|}$ and*

$$\mathcal{B}(G) \overset{q.i.}{\cong} \mathbf{P}(2^{|G|}) \text{ and } \mathcal{B}_\sigma(G) \overset{q.i.}{\cong} \mathbf{P}_\sigma(2^{|G|}) \tag{2}$$

for each σ satisfying $\gamma(G) \leq \sigma \leq 2^{|G|}$.

It follows from the above corollary that for a group G satisfying $|G| = |G/G'|$ and an infinite cardinal σ, $\mathcal{B}_\sigma(G) \overset{q.i.}{\cong} \mathbf{P}_\sigma(2^{|G|})$ if and only if $\gamma(G) \leq \sigma \leq 2^{|G|}$. In particular, $|\mathcal{B}_\sigma(G)| = |\mathbf{P}_\sigma(2^{|G|})| = 2^{\sigma|G|}$ as mentioned in §2.2. We omit giving further details and applications. For the definition of other monotone poset invariants (such as width, height and depth) the interested reader may see [6] or [61] and carry out the easy calculation in $\mathbf{P}_\sigma(2^{|G|})$ (as in [5] or [28]) to obtain the results from [6] or [61] as a consequence of (2). The reader should note that all these invariants depend only on the exponential of $|G|$, thus abelian groups G and H with $2^{|G|} = 2^{|H|}$ cannot be distinguished by any monotone invariant of the poset of precompact group topologies ([5], [28]). A cardinal invariant which can distinguish (from this point of view) abelian groups of the same cardinality is proposed in [28].

By a classical result of van der Waerden [72], $|\mathcal{B}(G)| = |\mathcal{B}_\omega(G)| = 1$ for a compact connected semisimple Lie group G, i.e. such a group admits a unique precompact topology, namely the given compact metrizable group topology. This shows in particular that if $\gamma(G) < 2^{|G/G'|}$ fails for a maximally almost periodic group G then $\mathcal{B}_\sigma(G) = \emptyset$ may occur for each $\log|G| < \sigma \leq 2^{|G|}$, i. e. $\Gamma(G) = \log|G|$.

Denote by $\mathcal{LB}(G)$ the poset of linear topologies in $\mathcal{B}(G)$, and set $\mathcal{LB}_\sigma(G) = \mathcal{LB}(G) \cap \mathcal{B}_\sigma(G)$.

Corollary 3. *If the group G is abelian or relatively free and residually finite, then $\gamma(G) = \log|G|$, $\Gamma(G) = 2^{|G|}$ and (2) holds for each σ satisfying $\log|G| < \sigma \leq 2^{|G|}$. In case G is relatively free and residually finite both $\mathcal{LB}(G) \overset{q.i.}{\cong} \mathbf{P}(2^{|G|})$ and $\mathcal{LB}_\sigma(G)) \overset{q.i.}{\cong} \mathcal{P}_\sigma(2^{|G|}))$ hold.*

It suffices to apply Corollary 2 since obviously $|G| = |G/G'|$ holds in both cases. The first part of Corollary 3, namely $\gamma(G) = \log|G|$, cannot be extended to arbitrary maximally almost periodic groups with $|G/G'| = |G|$. An example to this effect, with $|G/G'| = |G| = \gamma(G)$, is given in [5]).

4. The poset of pseudocompact topologies

4.1. Large posets and Boolean algebras in $\mathcal{P}_\sigma(G)$

For a group G and a cardinal σ set $\mathcal{P}_\sigma(G) = \mathcal{P}(G) \cap \mathcal{B}_\sigma(G)$. Here we shall need the Singular Cardinals Hypothesis (SCH), equivalent to: if $\tau \geq 2^\omega$ is a cardinal and $cf(\tau) \neq \omega$, then $\tau^\omega = \tau$.

Following [33] call an infinite cardinal τ *admissible* if there exists G with $\mathcal{P}(G) \neq \emptyset$ and $|G| = \tau$. It was proved by van Douwen [38] that under SCH τ is admissible if and only if $\tau \geq \mathbf{c}$ and $cf(\tau) > \omega$ in case τ is a strong limit cardinal, i. e. $\log\tau = \tau$. Later Comfort and Robertson [19] showed that for infinite cardinals τ and σ the existence of a group G with $\mathcal{P}_\sigma(G) \neq \emptyset$ and $|G| = \tau$ is equivalent to the following set-theoretic condition $Ps(\tau, \sigma)$ introduced essentially by Cater, Erdös and Galvin [8]: $\{0,1\}^\sigma$ contains a subset of cardinality τ whose projection on every countable subproduct $\{0,1\}^A$ is a surjection. An admissible τ satisfies $\tau \geq \mathbf{c}$; for fixed σ the least cardinal $\tau = \delta(\sigma)$ with $Ps(\tau, \sigma)$ satisfies $\log\sigma \leq \delta(\sigma) \leq (\log\sigma)^\omega$ ([8]). This suggested the introduction in [19] of the following set-theoretic assumption, denoted there by (M): $\delta(\sigma) = (\log \sigma)^\omega$ for every infinite σ. It was pointed out in [16] that (M) is properly weaker that SCH. Obviously, under (M) $Ps(\tau, \sigma)$ is equivalent to $(\log \sigma)^\omega \leq \tau \leq 2^\sigma$.

It was proved in [10] that an abelian non-torsion group G with $\mathcal{P}(G) \neq \emptyset$ satisfies $r(G) \geq \mathbf{c}$. Groups with $\mathcal{P}(G) \neq \emptyset$ were studied further in [12]-[16], [31]-[36], [70]. The aim of the next theorem is to show that in such a case $\mathcal{P}(G)$ is actually quite big.

Theorem 2. ([27]) *Let $\sigma' \geq \sigma \geq \omega_1$ be cardinals and let G be an infinite group which is either relatively free or abelian with $r(R) < \mathbf{c}$ or $r(G) = |G|$. If $Ps(|G|, \sigma)$ and $Ps(|G|, \sigma')$ hold, then there exists an embedding $\mathbf{P}_\sigma(\sigma') \hookrightarrow \mathcal{P}_\sigma(G)$.*

By the above theorem the following are equivalent for such a group G: a) $\mathcal{P}_{2^{|G|}}(G) \neq \emptyset$; b) $\mathcal{P}_{2^{|G|}}(G) \overset{q.i.}{\cong} \mathbf{P}(2^{|G|})$; c) $\mathcal{P}_\sigma(G) \overset{q.i.}{\cong} \mathbf{P}_\sigma(2^{|G|})$ for each σ with $Ps(|G|, \sigma)$. For connected topologies we have

Theorem 3. ([27]) *Let $\sigma' \geq \sigma \geq \omega_1$ be cardinals and let G be an infinite group which is either free or abelian. If both $\mathcal{P}_\sigma(G)$ and $\mathcal{P}_{\sigma'}(G)$ contain connected topologies, then $\mathcal{P}_\sigma(G)$ contains a copy of $\mathbf{P}_\sigma(\sigma')$ consisting of connected and locally connected topologies.*

Again, if $\mathcal{P}_{2^{|G|}}(G)$ contains a connected topology, then quasi isomorphisms as above are available for the subsets of connected topologies in $\mathcal{P}_{2^{|G|}}(G)$ etc. If $\mathcal{P}(G)$ contains a connected topology for a relatively free group G, then G is either abelian or a free group [33].

Now we set $\sigma' = \sigma$ to obtain also a necessary condition.

Theorem 4. ([27]) *Let G be an infinite group and let $\sigma \geq \omega_1$ be a cardinal.*

a) If G is free (free abelian), then the following conditions are equivalent:

a_1) $Ps(|G|, \sigma)$ holds;

a_2) $\mathcal{P}_\sigma(G) \neq \emptyset$;

a_3) for each one of the following four properties P: zero-dimensional, connected and locally connected, disconnected and locally connected, connected and non-locally-connected, $\mathcal{P}_\sigma(G)$ contains a copy of the Boolean algebra $\mathbf{P}(\sigma)$ consisting of topologies with property P.

b) If G is abelian, then the following conditions are equivalent:

b_1) $Ps(r(G), \sigma)$ and $|G| \leq 2^\sigma$ hold;

b_2) G admits a pseudocompact connected group topology of weight σ;

b_3) the subset of pseudocompact connected and locally connected group topologies of $\mathcal{P}_\sigma(G)$ contains a copy of the Boolean algebra $\mathbf{P}(\sigma)$.

c) If G is a (necessarily bounded) torsion Abelian group, then the following conditions are equivalent:

c_1) $Ps(|G|,\sigma)$ holds and for all $m \in \mathbf{N}$, $|mG|$ is either finite or admissible;

c_2) G has a pseudocompact group topology and $Ps(|G|,\sigma)$ holds;

c_3) $\mathcal{P}_\sigma(G)$ contains a copy of the Boolean algebra $\mathbf{P}(\sigma)$.

The equivalence of the first two conditions in a), b) and c) was established in [33] (see also [36]; the equivalence of the first two conditions in b) was proved, under the assumption of (M), also in [16]; the equivalence of c_2) with $\mathcal{P}_\sigma(G) \neq \emptyset$ was proved in [14], Theorem 3.24). In a) G can be taken (more generally) relatively free and residually finite, but then connectedness and local connectedness are not available in a_3). The condition b_1) implies $Ps(|G|,\sigma)$, these conditions are equivalent if $|G| = r(G)$, in particular if G is torsion-free.

Clearly $\mathbf{P}(\sigma)$ is the largest complete Boolean algebra contained in $\mathbf{P}_\sigma(2^{|G|})$, thus in $\mathcal{P}_\sigma(G)$ as well. The above theorems yield lower bounds concerning the size of various poset-invariants of $\mathcal{P}_\sigma(G)$: length of chains, anti-chains, height, depth, etc. and so improve results from [12] and [16], Theorem 6.2 and Corollary 6.3. The existence of chains in $\mathcal{P}_\sigma(G)$ with the same length as in $\mathbf{P}(\sigma)$ whenever $\mathcal{P}_\sigma(G) \neq \emptyset$ was proved in [32] for free and free abelian groups G.

4.2. Divisible groups, test groups and more

For a divisible group G every topology in $\mathcal{P}(G)$ is connected ([76]) thus for abelian G Theorem 4 yields that $\mathcal{P}_\sigma(G) \neq \emptyset$ for some σ if and only if $Ps(r(G),\sigma)$ and $|G| \leq 2^\sigma$ hold ([37]). Since admissible cardinals are uncountable, $r(G) = |G/t(G)|$ holds. This gives

Corollary 4. ([33]) *Let G be a divisible abelian group. Then $\mathcal{P}(G) \neq \emptyset$ implies $\mathcal{P}(G/t(G)) \neq \emptyset$ and $\log\log|G| \leq r(G)$. The converse is true under GCH.*

It was shown in [33] that the sufficiency of this condition cannot be proved in ZFC even for rather simple divisible groups. The following question remains open in the general case ([34]):

Question 2. Is $\mathcal{P}(G/t(G)) \neq \emptyset$ true for an Abelian group G with $\mathcal{P}(G) \neq \emptyset$?

The answer is "Yes" if also $\mathcal{P}(t(G)) \neq \emptyset$ holds (according to Theorem 7.18 of [34] $\mathcal{P}(G/t(G)) \neq \emptyset$ and $\mathcal{P}(G) \neq \emptyset$ are equivalent provided $\mathcal{P}(t(G)) \neq \emptyset$).

The question of whether $\mathcal{P}(G) \neq \emptyset$ and $Ps(|G|, \sigma)$ yield $\mathcal{P}_\sigma(G) \neq \emptyset$ for each cardinal σ was raised recently by Comfort and Remus [14], Question 3.7. Following [14] call a group with this property a *test group.* According to Theorem 4 relatively free groups, torsion abelian groups and abelian groups G with $r(G) = |G|$ are test groups (the last case was proved also in [16], Theorem 4.13, under the assumption of (M)). The next theorem shows that a divisible abelian group G may be a test group or not depending on the following relations between $|G|$ and $r(G)$: (t_1) $|G| < 2^{r(G)}$; (t_2) $2^{r(G)} = 2^{|G|}$ (t_3) $2^{r(G)} = 2^{<|G|} = \sup\{2^\xi : \xi < |G|\}$.

We abbreviate also (t_0) $Ps(|G|, 2^{|G|})$; (t_0^r) $Ps(r(G), 2^{r(G)})$. Obviously $(t_2) \Rightarrow ((t_1) \wedge (t_3))$ and $(t_0^r) \wedge (t_2) \Rightarrow (t_0) \Rightarrow (t_0^r)$.

Theorem 5. ([27]) *Let G be a divisible abelian group and $\mathcal{P}(G) \neq \emptyset$.*

a) If G is a test group, then $|G| < 2^{r(G)}$, i. e. (t_1) holds.

b) If $cf(|G|) = \omega$, and G is a test group then (t_3) holds;

c) ("G is a test group"$\wedge(t_0)) \Leftrightarrow ((t_0^r) \wedge (t_2))$;

d) Assume SCH. If G is a test group then $(t_3) \wedge (t_0^r)$ holds.

e) If $|G|$ is a strong limit cardinal, then G is a test group.

f) Assume GCH. Then G is a test group $\Leftrightarrow (t_2)$ (i. e. $r(G) = |G|$) holds.

Items a)-c) follow from Theorem 4, item d) follows from b) and c), item e) follows from Corollary 4 (since $\mathcal{P}(G) \neq \emptyset$ if and only if $r(G) = |G|$ for such a group) and implies f). Examples show that none of the following implications holds even under SCH: 1. ("test"$\wedge(t_0^r) \wedge (t_3)) \Rightarrow ((t_0) \vee (t_2))$; 2. $((t_0) \wedge (t_2)) \Rightarrow$ ("test"$\vee(t_0^r))$; 3. $((t_0^r) \wedge (t_1) \wedge (t_3)) \Rightarrow$"test". Comparing 1. and 2. with item f) of Theorem 5 we see that neither the necessity nor the sufficiency of (t_2) can be established or disproved in ZFC. Finally, 3. shows that in c) (t_2) cannot be replaced by the weaker condition $(t_1) \wedge (t_3)$.

It is easy to see that every torsion-free reduced abelian group G is residually finite, so $\mathcal{LB}(G) \neq \emptyset$ (note that this condition implies that G is reduced), but the following remains still open ([34]):

Question 3. Does every torsion-free reduced abelian group G with $\mathcal{P}(G) \neq \emptyset$ admit a linear pseudocompact group topology ?

The following question was proved to be independent of ZFC in [19]: Does there exist a cardinal σ such that for some abelian group G with $|G| < 2^\sigma$ there exists $T \in \mathcal{P}_\sigma(G)$ for which all Hausdorff quotients of (G, T) are minimal?

A topological space X is *countably compact* if and only if every countable open cover of X has a finite subcover. Countably compact spaces are pseudocompact. The algebraic structure of countably compact groups is still unclear. Even the first question, the characterization of cardinalities of these groups, remains open.

Question 4. ([38]) Does a countably compact group G satisfy $|G|^\omega = |G|$?

It was shown by van Douwen [38] (Examples) that this condition is sufficient for Boolean groups, namely if $\log\lambda \leq \kappa = \kappa^\omega \leq 2^\lambda$, then $G = \mathbf{Z}(2)^{(\kappa)}$ admits a countably compact group topology of weight λ. It was observed in [16] that this can be extended to $G = \mathbf{Z}(p)^{(\kappa)}$ for any prime p (actually, Theorem 7.1 of [16] offers also large chains and anti-chains of such topologies). Let us note that the proofs are based on the fact that two Boolean groups (or, more generally, two abelian groups of exponent p) are isomorphic if and only if they have the same cardinality. The same argument can be used to prove that the group $G = \mathbf{Q}^{(\kappa)}$ also has the same property (now embeddings into powers of the compact Pontryagin dual of $\mathbf{Q}$ should be used). In all these case the condition $|G|^\omega = |G|$ was proved to be *sufficient* for the existence of a countably compact group topology on G (note that in Question 4 we ask if it is *necessary*). For other classes of groups this condition is not sufficient. Only the trivial subgroup of a Hausdorff free group may be countably compact ([33]), while Tkačenko [70] showed that under CH the free abelian group of rank $\mathbf{c}$ admits countably compact group topologies. The question of which free abelian groups admit countably compact group topologies is open [33] (compare with Theorem 4).

4.3. Refinements

Here we shall be interested in maximal elements of $\mathcal{P}(G)$, i.e. when a topology $T \in \mathcal{P}(G)$ can be properly "expanded" or "refined" to a pseudocompact group topology on G. Following (the spirit of) [20], Definition 5.1, let us call a pseudocompact group (G,T) *maximal* if there is no $T' \in \mathcal{P}(G)$ with $T < T'$, i. e. when T is a maximal element of $\mathcal{P}(G)$ (for maximal elements of $\mathcal{H}(G)$ and their relation to the minimal ones see [58]). According to Theorem 2.4 of [20] compact metrizable groups are maximal pseudocompact groups (i. e. every element of $\mathcal{P}_\omega(G)$ is a maximal element of $\mathcal{P}(G)$ since a metrizable pseudocompact topology is actually compact). On the other hand it is known that under suitable hypotheses (e. g. commutativity, [19], Remark 4.10(c)) a compact group of uncountable weight is not maximal (see also [18] and [15]). It is proved in Theorem 7.3 of [20] that 0-dimensional pseudocompact abelian groups of uncountable weight are not maximal. The following question still seems to be open:

Question 5. ([20], Question 2.5) Does there exist a maximal pseudocompact group of uncountable weight?

We give now the following

Corollary 5. *Let G be a torsion abelian group with $\mathcal{P}(G) \neq \emptyset$. Then G is bounded and the following are equivalent:*

a) $\mathcal{P}(G)$ has maximal elements;

b) G admits compact metrizable group topologies;

c) all Ulm invariants of G or either finite or **c**.

The first assertion is Lemma 7.4 of [20]. The equivalence of a) and b) follows from Corollary 7.5 [20] which says that no pseudocompact torsion abelian group of uncountable weight is maximal. The equivalence of b) and c) follows from the structure theory of compact abelian groups ([45], 25.22).

Recently, A. V. Arhangel'skiĭ [3] has shown that if a compact group (G, T) admits a group topology properly containing T which is countably compact, then $|G|$ is Ulam-measurable. Comfort and Remus [17] prove results in the converse direction: If (G, T) is a compact group with $|G|$ Ulam-measurable, and if G satisfies a suitable additional condition, then there is a countably compact group topology T' on G which properly contains T. Recently Uspenskij [71] eliminated all additional conditions in this direction.

5. Chains

Definition 1. ([11]) Let λ and σ be cardinals and let $C(\sigma, \lambda)$ denote the set-theoretic assumption "there is a chain of length λ in $\mathbf{P}(\sigma)$" .

It is possible to give equivalent forms of this condition in terms of trees or densities of totally ordered sets (see J. E. Baumgartner [4], Theorem 2.1).

The following cardinal function was introduced by Shelah [65] in the context of trees: $Ded(\sigma) = \min\{\lambda : C(\sigma, \lambda) \text{ does not hold}\}$. By [4] $Ded(\omega) = (2^\omega)^+$ and $\sigma^+ < Ded(\sigma) \leq (2^\sigma)^+$, so that $Ded(\sigma) = (2^\sigma)^+ = \sigma^{++}$ when $\sigma = \omega$ or $\sigma^+ = 2^\sigma$ (so under GCH). Mitchell's model shows that $Ded(\omega_1) = 2^{\omega_1}$ is consistent with ZFC ([4]).

For a poset $(P, \leq)$ define the following cardinal invariants

$$Ded(P) = \min\{\lambda : P \text{ has no chain of size } \lambda\}.$$

and, defining a chain in a poset to be *bounded* if it has a top element,

$$Ded_b(P) = \min\{\lambda : P \text{ has no bounded chain of length } \lambda\}.$$

According to Corollary 2 $Ded_b(\mathcal{B}_\sigma(G)) = Ded(\sigma)$. To calculate $Ded(\mathcal{B}_\sigma(G))$ it is enough to compute $Ded(\mathbf{P}_\sigma(\kappa))$ for cardinals $\sigma \leq \kappa$. According to [5] $Ded(\mathbf{P}_\sigma(\kappa)) = Ded(\mathbf{P}_\sigma(\sigma^+))$ in case $\sigma < \kappa$, moreover $Ded(\mathbf{P}_\sigma(\sigma^+)) \leq$

$Ded(\sigma)^+$ and equality occurs if and only if $cf(Ded(\sigma)) = \sigma^+$. Therefore $Ded(\mathbf{P}_\sigma(\sigma^+)) = Ded(\sigma)$ when $\sigma^+ = 2^\sigma$ (so under GCH), but in general this equality cannot be determined in ZFC even for $\sigma = \omega_1$ ([5]). Note that for $\sigma < \alpha$ $\mathbf{P}_\sigma(\alpha)$ has the same chain lengths as $\mathbf{P}_\sigma(\sigma^+)$, although these two posets need not be quasi isomorphic.

The next theorem shows that even the simplest equation $Ded_b(\mathcal{B}_{\omega_1}(\mathbf{Z})) = Ded(\mathcal{B}_{\omega_1}(\mathbf{Z}))$ cannot be proved in ZFC (compare items b) and c)), while

$$Ded_b(\mathcal{B}_\sigma(G)) \le Ded(\mathcal{B}_\sigma(G)) \le Ded_b(\mathcal{B}_\sigma(G))^+$$

is true (in ZFC) for many groups. The proof is based on the "approximation" of $\mathcal{B}_\sigma(G)$ given in Corollary 2, and properties of the function $Ded(\sigma)$.

Theorem 6. *Let G be a maximally almost periodic group with $2^{|G|} = 2^{|G/G'|}$. Then $Ded(\mathcal{B}_{2^{|G|}}(G)) = Ded(2^{|G|})$.*

a) For every cardinal σ satisfying $\gamma(G) \le \sigma < 2^{|G|}$

$$Ded(\mathcal{B}_\sigma(G)) = Ded((\mathbf{P}_\sigma(\sigma^+)) \le Ded(\sigma)^+.$$

b) For every cardinal σ satisfying $\gamma(G) < \sigma^+ = 2^\sigma \le 2^{|G|}$

$$Ded(\mathcal{B}_\sigma(G)) = Ded_b(\mathcal{B}_\sigma(G)) = Ded(\sigma).$$

c) $Ded_b(\mathcal{B}_{\omega_1}(\mathbf{Z}))^+ = Ded(\mathcal{B}_{\omega_1}(\mathbf{Z})) = Ded(\omega_1)^+$ is consistent with ZFC.

This theorem shows that both functions $Ded(\mathcal{B}_\sigma(G))$ and $Ded_b(\mathcal{B}_\sigma(G))$ depend only on σ and the cardinality of G, but not the algebraic structure of G. In particular, we obtain

Corollary 6. ([11], Th. 3.4) *Let G be an abelian group. Then $\mathcal{B}(G)$ admits a chain of length λ if and only if $C(2^{|G|}, \lambda)$ holds.*

Let G be a maximally almost periodic group with $2^{|G|} = 2^{|G/G'|}$. By Theorem 4 $\mathcal{P}_\sigma(G)$ and $\mathcal{B}_\sigma(G)$ have the same bounded chain lengths whenever $\mathcal{P}_\sigma(G) \ne \emptyset$, i. e. $Ded_b(\mathcal{P}_\sigma(G)) = Ded_b(\mathcal{B}_\sigma(G)) = Ded(\sigma)$. By Theorem 2 a sufficient condition for the optimal equality $Ded(\mathcal{P}_\sigma(G)) = Ded(\mathcal{B}_\sigma(G))$ is $Ps(\tau, \sigma^+)$ (this yields the obviously necessary $Ps(\tau, \sigma)$). We are not aware if the condition $Ps(\tau, \sigma^+)$ is also necessary.

Acknowledgments. The author is grateful to A. Berarducci, W. Comfort, M. Forti, D. Remus, D. Shakhmatov and S. Watson for helpful comments and conversations. The author's thanks go also to the referee for her/his helpful suggestions and constructive criticism concerning the original version of the paper.

References

[1] S. Adian, *The Burnside problem and Identities in Groups* (Ergebnisse der Mathematik und Ihrer Grenzgebiete **95**, Springer Verlag, Berlin-Heidelberg-New York, 1979).

[2] E.M. Alfsen and P. Holm, A note on compact representations and almost periodicity in topological groups, *Math. Scand.* **10**(1962), 127–136.

[3] A.V. Arhangel'skiĭ, On countably compact topologies on compact groups and dyadic compacta, submitted for publication.

[4] J.E. Baumgartner, Almost disjoint sets, the dense set problem and the partition calculus, *Ann. of Math. Logic* **10**(1976), 401–439.

[5] A. Berarducci, D. Dikranjan, M. Forti and S. Watson, On long chains of "small" sets, preprint.

[6] S. Berhanu, W.W. Comfort and J.D. Reid, Counting subgroups and topological group topologies, *Pacific J. Math.* **116**(1985), 217–241.

[7] W. Burnside, On an unsettled question in the theory of discontinuous groups, *Quart. J. Pure Appl. Math.* **33**(1902), 230–238.

[8] F.S. Cater, P. Erdös and F. Galvin, On the density of λ-box products, *Gen. Topol. Appl.* **9**(1978) 307–312.

[9] W.W. Comfort, K.H. Hofmann and D. Remus, Topological groups and semigroups, in *Recent Progress in General Topology* (M. Hušek and J. van Mill (eds.), Elsevier Science Publishers B. V., Amsterdam, 1992.), 59–144).

[10] W.W. Comfort and J. van Mill, Concerning connected, pseudocompact abelian groups, *Topology Appl.* **31**(1989), 21–45.

[11] W.W. Comfort and D. Remus, Long chains of Hausdorff topological group topologies, *J. Pure Appl. Algebra* **70**(1991), 53–72.

[12] W.W. Comfort and D. Remus, Pseudocompact topological group topologies, *Abstracts Amer. Math. Soc.* **12** (1991), 289 (abstract 91T-54-45).

[13] W.W. Comfort and D. Remus, Topologizing a group, Prague Symposium 1991, Abstracts.

[14] W.W. Comfort and D. Remus, Abelian Torsion Groups with a Pseudocompact Group Topology, *Forum Math.*, to appear.

[15] W.W. Comfort and D. Remus, Pseudocompact refinements of compact group topologies, *Math. Z.*, to appear.

[16] W.W. Comfort and D. Remus, Imposing pseudocompact group topologies on Abelian groups, *Fund. Math.*, to appear.

[17] W.W. Comfort and D. Remus, Compact Groups of Ulam-Measurable Cardinality: Partial Converses to Theorems of Arhangel'skiĭ and Varopoulos, submitted for publication.

[18] W.W. Comfort and L. C. Robertson, Proper pseudocompact extensions of compact Abelian group topologies, *Proc. Amer. Math. Soc.* **86**(1982), 173–178.

[19] W.W. Comfort and L. C. Robertson, Cardinality constraints for pseudocompact and for totally dense subgroups of compact topological groups, *Pacific J.*

Math. **119**(1985), 265-285.

[20] W. W. Comfort and L. C. Robertson, Extremal phenomena in certain classes of totally bounded groups, *Dissertationes Math.* **272**(1988).

[21] W.W. Comfort and K.A. Ross, Topologies induced by groups of characters, *Fund. Math.* **55**(1964), 283–291.

[22] W.W. Comfort and K.A. Ross, Pseudocompactness and uniform continuity in topological groups, *Pacific J. Math.* **16**(1966), 483–496.

[23] S. Dierolf and U. Schwanengel, Un example d'un groupe topologique q-minimal mais non précompact, *Bull. Sci. Math.* **101**(1977), 265–269.

[24] D. Dikranjan, Minimal topologies on divisible groups (Lecture Notes in Math. **1060**, Springer-Verlag, Berlin-Heidelberg-New York, 1984), 216–226.

[25] D. Dikranjan, On a class of finite-dimensional compact abelian groups, in *Topology, theory and applications (Eger, 1983)* (A. Czaszar (ed.), Coll. Math. Soc. János Bolyai **41**, North-Holland, Amsterdam-New York, 1985), 215–231.

[26] D. Dikranjan, On a conjecture of Prodanov, *C. R. Acad. Bulgare Sci.* **38**(1985), 1117-1120.

[27] D. Dikranjan, Chains of pseudocompact group topologies, preprint.

[28] D. Dikranjan, On the poset of precompact group topologies (International Colloquium on Topology, Szekszàrd 1993).

[29] D. Dikranjan and Iv. Prodanov, A class of compact abelian groups, *Annuaire Univ. Sofia, Fac. Math. Méc.* **70**(1975/76), 191–206.

[30] D. Dikranjan, Iv. Prodanov and L. Stoyanov, *Topological groups: characters, dualities and minimal group topologies* (Pure and Applied Mathematics, **130**, Marcel Dekker Inc., New York- Basel, 1989).

[31] D. Dikranjan and D. Shakhmatov, Pseudocompact topologizations of groups, *Zborinik radova Filozofskog fakulteta u Nišu, Ser. Mat.* **4**(1990), 83–93.

[32] D. Dikranjan and D. Shakhmatov, Long chains of pseudocompact topological group topologies, unpublished notes, Moscow, 1990.

[33] D. Dikranjan and D. Shakhmatov, Algebraic structure of the pseudocompact groupsn (Report **91-19**, York University (Toronto), Canada, 1991), 1–37.

[34] D. Dikranjan and D. Shakhmatov, Algebraic structure of the pseudocompact groups (Report **1/93**, Udine University (Udine), 1992).

[35] D. Dikranjan and D. Shakhmatov, Compact-like totally dense subgroups of compact groups, *Proc. Amer. Math. Soc.* **114**(1992), 1119–1129.

[36] D. Dikranjan and D. Shakhmatov, Pseudocompact topologies on groups, *Topology Proc.* **17**(1992), 335–342.

[37] D. Dikranjan and D. Shakhmatov, Pseudocompact and countably compact Abelian groups: Cartesian products and minimality. *Trans. Amer. Math. Soc.* **335**(1993), 775–790.

[38] E. van Douwen, The weight of a pseudocompact (homogeneous) space whose cardinality has countable cofinality, *Proc. Amer. Math. Soc.* **80**(1980), 678–682.

[39] H. Freudenthal, Topologische Gruppen mit genügend vielen fastperiodischen Funktionen, *Ann. Math.* (2)**37**(1936), 57–77.

[40] E. D. Gaughan, Topological group structures on the infinite symmetric group, *Proc. Nat. Acad. Sci. U.S.A.* **58**(1967), 907–910.

[41] M. Hall, A topology for the free group and related topics, *Ann. Math.* **52**(1950), 127–139.

[42] P. Halmos, Comments on the real line, *Bull. Amer. Math. Soc.* **50**(1944), 877–878.

[43] G. Hesse, *Zur Topologisierbarkeit von Gruppen*, Dissertation, Universität Hannover, Hannover (Germany), 1979.

[44] E. Hewitt, Rings of real-valued continuous functions I, *Trans. Amer. Math. Soc.* **64**(1948), 45–99.

[45] E. Hewitt and K. Ross, *Abstract hamonic analysis I, II* (Springer Verlag, Berlin-Heidelberg-New York, 1963, 1970).

[46] P. Holm, On the Bohr compactification, *Math. Ann.* **156**(1964), 34–36.

[47] S. Ivanov, On the Burnside problem in periodic groups, *Bull. Amer. Math. Soc.* **27**(1992), 257–260.

[48] A. I. Kostrikin, The Burnside Problem, *Izv. Akad. Nauk SSSR, Ser. Mat.* **23**(1959), 3-34; *Amer. Math. Soc. Transl. (2nd ser.)* **36**(1964), 63–100.

[49] A. I. Mal'cev, On the faithful representation of infinite groups by matrices, *Mat. sb.* **8**(1940), 405–422; *Amer. Math. Soc. Transl.* (2)**45**(1965), 1–18.

[50] A. A. Markov, On free topological groups, *Izv. Akad. Nauk SSSR* **9**(1945), 3–64 ; *Transl. Amer. Math. Soc.* (1)**8**(1962), 195–272.

[51] H. Neumann, *Varieties of groups* (Ergebnisse der Mathematik und ihrer Grenzgebiete **37**, Springer-Verlag, Berlin-Heidelberg-New York).

[52] J. von Neumann, Almost periodic functions in a group I, *Trans. Amer. Math. Soc.* **36**(1934), 445–492.

[53] J. von Neumann and E. Wigner, Minimally almost periodic groups, *Ann. Math. (Ser. 2)* **41**(1940), 746–750.

[54] A. Ol'shanskii, A note on countable non-topologizable groups, *Vestnik Mosk. Gos. Univ. Matem. Mekh.* **3** (1980), 103 (in Russian).

[55] Iv. Prodanov, Compact representations of continuous algebraic structures, in *General Topology and its Relation to Modern Analysis and Algebra II. Proc. of the Second Prague Topology Symposium 1966* (Academia, Prague, 1967), 290–294.

[56] Iv. Prodanov, Precompact minimal group topologies and p-adic numbers, *Annuaire Univ. Sofia, Fac. Math. Méc.* **66**(1971/72), 249–266.

[57] Iv. Prodanov, Minimal compact representations of algebras, *Annuaire Univ. Sofia, Fac. Math. Méc.* **67**(1972/73), 507–542.

[58] Iv. Prodanov, Maximal and minimal topologies on abelian groups, in *Topology* (Colloq. Math. Soc. Jànos Bolyai **23**, Budapest, 1978), 985–997.

[59] Iv. Prodanov and L. Stoyanov, Every minimal abelian group is precompact, *C. R. Acad. Bulgare Sci.* **37**(1984), 23–26.

[60] D. Remus, *Zur Struktur des Verbandes der Gruppentopologien.*, Dissertation, Univ. Hannover, Federal Repubic of Germany, 1983 (Engl. summary: *Resultate Math.* **6**(1983), 151–152).

[61] D. Remus, Minimal precompact topologies on free groups, *J. Pure Appl. Algebra* **70**(1991), 147-157.

[62] F. Schinkel, *Zur algebraischen struktur minimaler abelscher Gruppen*, Dissertation, Universität Hannover, Hannover (Germany), 1990.

[63] U. Schwanengel, *Minimale topologische Gruppen*, Dissertation, Universität München (Germany), 1978.

[64] D. Shakhmatov, Character and pseudocharacter of minimal topological groups, *Mat. Zametki* **38**(1985), 634–640 (in Russian); *Math. Notes* **38**(1985), 1003–1006).

[65] S. Shelah, *Classification Theory* (Studies in Logic and the Foundation of Mathematics, North Holland, Amsterdam-New York-Oxford, 1978).

[66] S. Shelah, On a problem of Kurosh, Jonsson groups and applications, in *Word Problems II* (S.I. Adian, W.W. Boone and G. Higman (eds.), North-Holland, Amsterdam, 1980), 373–394.

[67] A. K. Steiner, The lattice of topologies: structure and complementation, *Trans. Amer. Math. Soc.* **122**(1966), 379–398.

[68] R. M. Stephenson, Jr., Minimal topological groups, *Math. Ann.* **192**(1971), 193–195.

[69] L. Stoyanov, Total minimality of the unitary groups, *Math. Z.* **187**(1984), 273–283.

[70] M. Tkačenko, Countably compact and pseudocompact topologies on free Abelian groups, *Soviet Math.* **34**(1990), 79-86; Russian original: *Izv. Vyssh. Uchebn. Zaved. Mat.* **1** 336:5 (1990), 68–75.

[71] V. V. Uspenskij, Real-valued measurable cardinals and sequentially continuous homomorphisms, preprint.

[72] B. L. van der Waerden, Stetigkeitssätze für halbeinfache Liesche Gruppen, *Math. Z.* **36**(1933), 780–786.

[73] R. Warfield, Jr., *Nilpotent groups* (LNM **513**, Springer, 1976).

[74] A. Weil, *L'integration dans les groupes topologiques et ses applications* (Hermann, Paris, 1951).

[75] H. Weyl and F. Peter, Die Vollstandigkeit der primitiven Darstellungen einer geschlossenen kontinuierlichen Gruppe, *Math. Ann.* **97**(1927), 737–755.

[76] H. Wilcox, Dense subgroups of compact groups, *Proc. Amer. Math. Soc.* **28**(1971), 578–580.

[77] E. I. Zel'manov, Solution of the restricted Burnside problem of odd exponent, *Izv. Akad. Nauk SSSR Ser. Mat.* **54**(1990), 42–59 (in Russian).

[78] E. I. Zel'manov, Solution of the restricted Burnside problem for 2-groups, *Mat. Sb.* **182**(1991) (in Russian).

AUTOMORPHISMS OF NILPOTENT AND RELATED GROUPS

MARTYN R. DIXON

Mathematics Department, University of Alabama, Tuscaloosa, Al 35487-0350, U.S.A.
E-mail: mdixon@ualvm.bitnet

1. Introduction

If G is a group then as usual we denote the automorphism group of G by $\operatorname{Aut} G$. In this paper we give a survey of results concerning $\operatorname{Aut} G$ in the case when G is nilpotent or satisfies some generalized form of nilpotency. The first part of the paper will be concerned with the situation in which $\operatorname{Aut} G$ is as small as it can be. For each group G we let $Z(G)$ denote the centre of G, although when no ambiguity arises we shall often simply denote $Z(G)$ by Z. As usual we write G' to denote the derived subgroup of G. Of course, $G/Z(G)$ is isomorphic to $\operatorname{Inn} G$, the group of inner automorphisms of G. In Section 2, we discuss the situation in which $\operatorname{Inn} G = \operatorname{Aut} G$. We give rather an old example due to Zaleskii [50] showing that this situation can hold for nilpotent groups and then discuss some consequences.

As usual we denote the factor group $\operatorname{Aut} G/\operatorname{Inn} G$ by $\operatorname{Out} G$, the group of outer automorphisms of G. Thus Section 2 is concerned with the situation when $\operatorname{Out} G = 1$. By contrast, in Section 3 we discuss the situation which in a sense is opposite to that of Section 2. In Section 3 we are interested in showing that often $\operatorname{Out} G$ is rather large. The discussion is concerned mostly with nilpotent p-groups in Section 3. However we also discuss recent work of Puglisi and some simple consequences.

There are certain other canonical subgroups of $\operatorname{Aut} G$ which play an important role in the theory of automorphism groups. Every automorphism of G induces an automorphism of G/Z in a natural way. Those automorphisms which act trivially on G/Z are called central automorphisms, and we can form $\operatorname{Aut}_c G$, the subgroup of $\operatorname{Aut} G$ consisting of all central automorphisms. $\operatorname{Aut}_c G$ contains the subgroup $C_{\operatorname{Aut} G}(G/Z, Z) = C_{\operatorname{Aut} G}(G/Z) \cap C_{\operatorname{Aut} G}(Z)$ of automorphisms which also act trivially on $Z(G)$. It is well known that this subgroup is isomorphic with $\operatorname{Hom}(G/Z, Z)$, the group of homomorphisms from G/Z to Z, the isomorphism here being given by

$$\alpha \mapsto \theta_\alpha \text{ where } (gZ)\theta_\alpha = g^{-1}g^\alpha,$$

for each $\alpha \in C_{\operatorname{Aut} G}(G/Z, Z)$.

This abelian subgroup of $\operatorname{Aut} G$ is very useful since it enables us to use the theory of the functor Hom and also the theory of abelian groups. More generally, if $A \leq Z(G)$ then $\mathrm{Hom}(G/A, A) \cong C_{\mathrm{Aut}G}(G/A) \cap C_{\mathrm{Aut}G}(A)$. In the special case when G is a nilpotent group of class 2 we see immediately that $\mathrm{Inn}\,G \leq C_{\mathrm{Aut}G}(G/Z, Z)$. We refer the reader to [34] for our notation and terminology, which is standard where not explained.

2. Semicomplete nilpotent groups

If G is a group with trivial centre satisfying $\mathrm{Aut}\,G = \mathrm{Inn}\,G$ then we shall say that G is *complete*. There are numerous examples of complete groups, the simplest one being the symmetric group on three symbols. It is also an easy exercise to show that if G is a non-abelian simple group then $\mathrm{Aut}\,G$ is complete. On a more difficult note, Wielandt [48] has shown that every finite group with trivial centre can be subnormally embedded in a finite complete group. Hence finite complete groups can be rather complex. However, a fairly complete (forgive the pun) description of finite metabelian complete groups has been given by Gagen and Robinson [19]. Metanilpotent complete groups have been studied in [22]. The article [37] is a nice survey of these and related results.

For infinite examples of complete groups Dyer and Formanek have shown that for every non-cyclic free group F, $\mathrm{Aut}\,F$ is complete. In [36], Robinson shows that infinite supersoluble groups are never complete, but gives examples of polycyclic complete groups, both with and without torsion. Thomas [40] generalizes the result of Wielandt mentioned above, showing that every centreless group can be embedded as an ascendent subgroup of a complete group.

In [42], Thomas constructs examples of uncountable complete locally finite p-groups, under the assumption of Jensen's combinatorial principle $\diamond$. Further examples of uncountable locally finite complete groups are given in [24] and [41]. A recent theorem of Puglisi [32] shows that countable locally finite p-groups always have non-trivial outer autmorphisms. However, the following question still seems to be unanswered:

Question. Is there a countable locally finite complete group?

We shall call a group G *semicomplete* if $\mathrm{Aut}\,G = \mathrm{Inn}\,G$. We are here interested in the case when $Z(G)$ is non-trivial which leads one naturally to think of nilpotent groups. Of course, the cyclic group of order 2 is semicomplete, but this exhausts the non-trivial abelian examples. In 1955, Haimo and Shenkman asked if there were any non-abelian semicomplete nilpotent groups. For finite groups a complete answer to this question was provided by Gaschütz [20] in 1965. We state this well known result as our first theorem.

Theorem 2.1. *Every finite nilpotent group of order at least 3 has a non-trivial outer automorphism.*

Since we are not really concerned with finite groups here we merely note that Gaschütz's proof uses cohomological machinery and depends upon the fact that if G is a finite p-group and A is a (finite) G-module then $H^1(G, A) = 0$ implies $H^n(H, A) = 0$ for all $H \leq G$ and for all $n \geq 1$. In fact, Gaschütz showed further:

Theorem 2.2. *A finite p-group of order at least p^2 admits a non-inner automorphism of order a power of p.*

An elementary proof (i.e. using no cohomology) of this result is given in Webb [46], where she proves that for a non-abelian p-group G, where p divides $|C_{\mathrm{Out}G}(Z(G))|$. As far as I am aware the following problem is still unsolved:

Question. Does every finite non-abelian p-group have a non-inner automorphism of order p?

As far as infinite nilpotent groups are concerned, Zaleskii obtained the following result in [49].

Theorem 2.3. *Every infinite nilpotent p-group possesses an outer automorphism.*

Zaleskii's result is, in a sense, easier than that of Gaschütz's since cardinality arguments applied to Hom can be used to handle a number of the cases that arise. An important ingredient in the proof is the role played by the group $C_{\mathrm{Aut}G}(G/N, N)$, for various central subgroups N of the nilpotent p-group G. The results of Gaschütz and Zaleskii show that a periodic nilpotent group with at least three elements always has an outer automorphism. A similar result holds for finitely generated torsion-free nilpotent groups, a fact which was first proved by Ree [33] in 1956. We give a proof of this due to Gupta and Stonehewer [21]. First we require the following lemma (see Webb [44, p.4]).

Lemma 2.4. *Let G be a nilpotent group and let $z \in G$ normalize the abelian subgroup V of G. Suppose some power of z centralizes V. Then $[z, v]$ has finite order for all $v \in V$.*

PROOF. The proof is by induction on the nilpotency class of G, the case when G is abelian being obvious. So suppose G is not abelian and let bars denote images module $Z(G)$. By hypothesis there exists an integer j such that $[z^j, V] = 1$. Since the hypotheses of the lemma hold for $\bar{V}$ and $\bar{z}$ in $\bar{G}$ it follows that if $v \in V$ then $[z, v]^t \in Z(G)$ for some integer t. Since V is abelian, $[z, v]^t = [z, v^t]$. Hence

$$1 = [z^j, v^t] = [z, v^t]^j = [z, v]^{tj},$$

and $[z, v]$ has finite order. □

We shall in fact prove rather more than Ree's original theorem. Recall that the number of infinite cyclic factors occurring in a finite cyclic series of a polycyclic group G is an invariant of G called the *Hirsch length of* G, denoted by $h(G)$. Finitely generated nilpotent groups are, of course, polycyclic and the following result holds.

Theorem 2.5. *If G is a finitely generated nilpotent group and $h(G) > 1$ then G has an outer automorphism.*

PROOF. It is clear that G has a normal subgroup N with G/N infinite cyclic. Since $h(N) = h(G) - h(G/N)$, it follows that N is an infinite finitely generated nilpotent group, so that $Z(N)$ is infinite. Since G is nilpotent, there is an element y of infinite order such that $y \in Z(N) \setminus [G, Z(N)]$. Let x be an element of G such that $G = N\langle x \rangle$. Define a map $\theta : G \to G$ by

$$x\theta = xy, \; n\theta = n \text{ for all } n \in N.$$

Then θ is an automorphism of G. Suppose θ is inner, say conjugation by $x^i u$ for some integer i and $u \in N$. Then $[N, x^i u] = 1$ and $[x, x^i u] = y$, and hence

$$[Z(N), x^i] = 1 \quad (1)$$

$$[u, x^i] = 1 \quad (2)$$

$$[x, u] = y. \quad (3)$$

Let $V = \langle Z(N), u \rangle$ so that V is abelian. Also V is normalized by x by (3) and (1) and (2) show x^i centralizes V if $i \neq 0$. By 2.4 it follows that $[x, u] = y$ has finite order, contrary to the choice of y. Hence $i = 0$ and $u \in Z(N)$. Then by (3), $y \in [G, Z(N)]$, again contrary to the choice of y. It follows that θ is not inner. □

This extension of Ree's result was originally due to Baumslag [2], who actually showed that every finitely generated nilpotent group of Hirsch length at least two has an outer automorphism of infinite order. This leaves us with the following unsolved problem:

Question. If G is a finitely generated nilpotent group of Hirsch length 1 does G have an outer automorphism?

A number of partial solutions to this problem have been obtained. In [21] Gupta and Stonehewer show that a 2-generator nilpotent group of class at most 4 has an outer automorphism. The best results in this direction have been obtained by U.H.M. Webb [45]. Among other things she proves the following.

Theorem 2.6. *Let G be a non-cyclic finitely generated nilpotent group whose torsion subgroup T satisfies one of the following:*

(i) T has prime exponent.

(ii) T has cyclic derived group and is not a 2-group.

(iii) T has nilpotency class at most 2.

Then G has an outer automorphism.

Webb obtains her results by considering the group $H = C_{\mathrm{Aut}G}(G/T) \cap C_{\mathrm{Aut}G}(T)$, where G is a finitely generated nilpotent group of Hirsch length 1. She obtains conditions on T which are equivalent to all elements of H being inner. The centre of T turns out to be a cyclic module over the ring $\mathbf{Z}\langle\phi\rangle$, where ϕ is the automorphism of G induced by conjugation by a generator of G/T.

T. Fournelle [11] has generalized both 2.3 and 2.5 as follows.

Theorem 2.7. *Suppose G is nilpotent and that G/Z is infinite. If either*

(i) G/Z is periodic, or

(ii) G/Z is finitely generated

then G has an outer automorphism.

The first example of a semicomplete nilpotent group was constructed by Zaleskii [50]. We give here a sketch of his construction.

Theorem 2.8. *There exist $2^{\aleph_0}$ torsion free semicomplete, class 2 nilpotent groups.*

PROOF. Let L, M, N be infinite disjoint sets of primes with $\mathbf{P} = L \cup M \cup N$ and $2 \notin N$. We denote the 3×3 matrix $\begin{bmatrix} 1 & a & c \\ 0 & 1 & b \\ 0 & 0 & 1 \end{bmatrix}$ by (a, b, c) with $a, b, c \in \mathbf{Q}$.

Then define

$$\begin{aligned} a_p &= (1/p, 0, 1/p^2) && \text{with } p \in L \\ b_q &= (0, 1/q, 1/q^2) && \text{with } q \in M \\ c_r &= (1/r, 1/r, 0) && \text{with } r \in N \end{aligned}$$

and set $G = \langle a_p, b_q, c_r \mid p \in L, q \in M, r \in N\rangle$, so that G is a torsion-free nilpotent group of class 2. The proof proceeds in several (non-trivial) steps.

Let $Q_0 = \{m/n \in \mathbf{Q} \mid n \text{ is square-free}\}$.

(1) $G' = Z(G) = \{(0, 0, \gamma) \mid \gamma \in Q_0\} \cong Q_0$.

By computation, it is easily shown that $G' = \{(0, 0, \gamma) \mid \gamma \in Q_0\}$. On the other hand $Z(G) \leq \{(0, 0, \gamma) \mid \gamma \in \mathbf{Q}\}$. Let $z = (0, 0, \gamma) \in Z(G)$. Since $z \equiv \prod_i a_{p_i}^{k_i} \prod_j b_{q_j}^{l_j} \prod_k c_{r_k}^{m_k} \bmod G'$ we easily show that $p_i | k_i$, $q_j | l_j$ and $r_k | m_k$. If

$k_i = s_ip_i, l_j = t_jq_j, m_k = u_kr_k$ then $a_{p_i}^{k_i} = (s_i, 0, s_i/p_i), b_{q_j}^{l_j} = (0, t_j, t_j/q_j)$ and $c_{r_k}^{m_k} = (|u_k|, |u_k|, \binom{|u_k|}{2} + |u_k|(r_k - 1)/2r_k)$.

In any case, since $2 \notin N$ the last entry of each of these has square-free denominator. In particular $Z(G) \leq G'$, and (1) holds.

Let v_p denote the function which associates with each rational number the exponent of p in its factorization, and define $Q_A = \{\alpha \in Q_0 | v_p(\alpha) \geq 0$ for $p \notin A\}$, for $A = L, M, N$. Then define

$$Q(L) = \{(\alpha, 0) \in \mathbf{Q} \times \mathbf{Q} \mid \alpha \in Q_L\};$$
$$Q(M) = \{(0, \alpha) \in \mathbf{Q} \times \mathbf{Q} \mid \alpha \in Q_M\};$$
$$Q(N) = \{(\alpha, \alpha) \in \mathbf{Q} \times \mathbf{Q} \mid \alpha \in Q_N\}.$$

Then, by computation and the obvious natural map we have:

(2) $G/Z \cong \langle Q(L), Q(M), Q(N) \rangle$

$$= \{(\rho, \sigma) \in Q_0 \times Q_0 \mid v_p(\rho) > 0 (p \in M), v_p(\sigma) \geq 0 (p \in L) \text{ and } \min(0, v_p(\sigma)) = \min(0, v_p(\rho))(p \in N)\}.$$

(3) *Every automorphism of G/Z is either the identity or inversion*

Let φ be an automorphism such that $\varphi(1,0) = (\lambda, \mu)$ and $\varphi(0,1) = (\tau, \sigma)$, using the identification in (2).

If $p \in L$ then $\varphi(1/p, 0) = (\lambda/p, \mu/p)$, since extraction of roots is unique in a torsion-free abelian group. Since $(1/p, 0) \in G/Z$, it follows that $(0, \mu/p) \in G/Z$. Hence $v_p(\mu) \geq 1$. Since L is infinite, $\mu = 0$. In a similar fashion $\tau = 0$. Also for $p \in N$, $\varphi(1/p, 1/p) = (\lambda/p, \sigma/p)$ and hence $\varphi(1/p, 1/p) - \lambda(1/p, 1/p) = (0, (\sigma - \lambda)/p) \in G/Z$. Thus $v_p(\sigma - \lambda) \geq 1$ and $\sigma = \lambda$. Hence $\varphi(0,1) = (0, \lambda)$ and $\varphi(1,0) = (\lambda, 0)$, and $Q(M)$ is invariant under φ. The induced action of φ on the quotient of G/Z by $\varphi(M)$ (which is a subgroup of Q_0) is therefore given by $\varphi(x) = \lambda x$, for $x \in (G/Z)/Q(M)$. The only such automorphisms occur when $\lambda = \pm 1$ so that (3) follows.

(4) *Every automorphism is trivial on Z and G/Z*

Let φ be an automorphism of G. It follows from (1) and (3) that φ acts trivially on Z. Suppose φ acts by inversion on G/Z so that $\varphi(1,0,0) = (-1, 0, \xi)$ for some $\xi \in Q_0$. Since $a_p^p = (1, 0, 1/p)$ it follows that $\varphi(a_p)^p = (-1, 0, \xi + 1/p)$ and hence since G is torsion-free,

$$\varphi(a_p) = (-1/p, 0, \xi/p + 1/p^2).$$

Hence $\varphi(a_p)a_p = (0, 0, 2/p^2 + \xi/p)$. Since $\varphi(a_p)a_p \in Z$, by assumption, it follows that $2/p^2 + \xi/p \in Q_0$. However $p \neq 2$ so $v_p(\xi) = -1$ for all $p \in L$, which is impossible since L is infinite. Hence φ acts trivially on G/Z.

To complete the proof of 2.8, let $\varphi \in \operatorname{Aut} G$. Then φ corresponds to the homomorphism $\psi \in \operatorname{Hom}(G/Z, Z)$ where for $g \in G$, $\psi(gZ) = g^{-1}\varphi(g)$. Set

$x = (1,0,0), y = (0,1,0)$ and let $\psi(xZ) = (0,0,\sigma), \psi(yZ) = (0,0,-\rho)$. Since G/Z is torsion-free and $\{xZ, yZ\}$ is a maximal linearly independent subset, the action of ψ is completely determined. On the other hand if $h = (\rho, \sigma, \tau)$ then $[x,h] = (0,0,\sigma)$ and $[y,h] = (0,0,-\rho)$. It follows that for each $g \in G$,

$$\psi(gZ) = g^{-1}\varphi(g) = [g,h]$$

and hence $\varphi(g) = h^{-1}gh$ for all $g \in G$. Here h need not be in G. We must finally show that $(\rho, \sigma) \in G/Z$. Certainly $\rho, \sigma \in Q_0 \cong Z$. Now

$$\psi(a_p^p Z) = \psi(xZ) = \sigma \text{ so } \psi(a_p Z) = \sigma/p \in Q_0$$

and hence $v_p(\sigma) \geq 0$ for all $p \in L$. Similarly $v_q(\rho) \geq 0$ for all $q \in M$, and $v_r(\sigma - \rho) \geq 0$ for all $r \in N$. If $\sigma = \lambda/r$ and $v_r(\rho) \geq 0$ then $\sigma - \rho = (\lambda - r\rho)/r$ so $r|(\lambda - r\rho)$ which is impossible since $r \not| \lambda$. So $v_r(\sigma) = -1 \iff v_r(\rho) = -1$. Hence $(\rho, \sigma) \in G/Z$ by the characterization of G/Z before (3). This completes the proof. □

Further examples have been given in Fournelle [13]. These examples are constructed using the cohomology of class 2 nilpotent groups. Fournelle's examples are of particular interest here, so we briefly describe the groups obtained. If π is an infinite set of odd primes whose complement within the set of all primes is also infinite then Fournelle's groups G have the property that G is semicomplete and

$$T(Z) = \oplus_{p \in \pi} C_p, T(G/Z)_p = C_p \oplus C_p.$$

Here $T(Z)$ denotes the torsion subgroup of Z and C_p denotes the cyclic group of order p and if H is a nilpotent group we let H_p denote the unique maximal p-subgroup of H. Thus Fournelle's examples have infinite torsion subgroup but have finite Sylow p-subgroups for all primes p. As far as I am aware no one has yet constructed a semicomplete nilpotent group of class greater than 2. The obvious extensions of the examples of Zaleskii do not seem to work.

The study of the general structure of semicomplete groups was initiated in Fournelle [12] where the following result was proved.

Theorem 2.9. *Suppose G is a semicomplete nilpotent group and* Aut G *is torsion-free. Then either*

(i) G is torsion-free, or

(ii) $G = H \times C_2$ where H is torsion-free, semicomplete and each element of H has a square root.

Here, both cases can occur.

This led Fournelle to ask whether every semicomplete nilpotent group has finite Sylow p-subgroups, for all primes p. A partial solution to this question is given in [10], where we prove:

Theorem 2.10. *Suppose G is a semicomplete nilpotent group, suppose $Q = G/Z(G)$ and p is a prime. If one of the following conditions holds then the Sylow p-subgroup of G is finite.*

(i) *$Q/Q'Q^p$ is finite.*

(ii) *Q_p, the Sylow p-subgroup of Q, is countable.*

(iii) *Q_p has finite exponent.*

(iv) *G is nilpotent of class 2.*

This result should perhaps be contrasted with a further result in [10] where we show that G can never be semicomplete if $Q/Q'Q^p$ is uncountable. Like many of the results obtained in this area, the proof of 2.10 is heavily dependent on cardinality arguments, and the theory of abelian groups. Rather than give a proof of this result, we illustrate some of the techniques involved by giving a proof of the following special case of Puglisi's Theorem mentioned at the start of this section. The proof is due to M. Evans.

Theorem 2.11. *Suppose G is a countable, hypercentral p-group. Then G is never semicomplete.*

PROOF. Assume the contrary, so $Q = G/Z(G) \cong \operatorname{Inn} G = \operatorname{Aut} G$. If B is a basic subgroup of Q_{ab} then the epimorphism of Szele[18, p. 152] implies that there is an induced monomorphism $\operatorname{Hom}(B, Z) \hookrightarrow \operatorname{Hom}(Q_{ab}, Z)$, which is a subgroup of $\operatorname{Aut} G$, by the remarks in the introduction. Since B is a direct product of cyclic groups, B must be finite, otherwise $\operatorname{Hom}(B, Z)$ is uncountable. Hence $Q_{ab} = B \oplus D$, for some divisible group D (this argument is really [51, Lemma 1], but we include it simply to illustrate techniques).

We claim that if R is the finite residual of Q then R is a divisible abelian group and Q/R is finite. If $B = C/Q'$ then there exists a finite group H such that $C = HQ'$. Let N be a normal subgroup of Q of finite index. Then for some natural number $k, \gamma_k(Q) \leq N$ and since $Q/Q'N$ is finite it follows that $E \leq Q'N$, where E is the inverse image of D in Q. Then

$$Q = CE = HNQ' = HN\gamma_k(Q)(\text{by [34, Vol. 1, Lemma 2.22]}) = HN.$$

Hence $|Q/N| \leq |H|$ and the fact that Q/R is finite follows. Furthermore R must be radicable so R is abelian, by [34, Vol. 2, p 125].

Hence $Q \cong \operatorname{Aut} G$ is a periodic divisible abelian-by-finite group. But [7, Theorem A] shows that in this case $\operatorname{Aut} G$ is finite. Hence G is nilpotent and Zaleskii's Theorem shows G is not semicomplete. □

For locally nilpotent groups in general there seem to be few results concerning the size of $\operatorname{Out} G$. One exception is the Theorem of Stonehewer [39] that a locally nilpotent FC-group always has outer automorphisms. This result has been generalized by Robinson, Stonehewer and Wiegold in [38], a special case of their theorem being that a locally nilpotent FC-group G such that G/Z is not finite has $2^{|G/Z|}$ locally inner automorphisms. This result has also been generalized in another direction in [28] to include the class of CC-groups, where a group G is called CC if $G/C_G(x^G)$ is Černikov for all $x \in G$.

3. The case when Out$G \neq 1$

Since nilpotent p-groups always have outer automorphisms it is reasonable, in light of 2.2, to ask what the outer automorphisms look like in such a situation.

Buckley and Wiegold [3,4] appear to be the first authors to ask the question as to the size and nature of Out G, when G is an infinite, non-abelian, nilpotent p-group, (although several authors, [25, 26, 43] had examined this question for abelian p-groups somewhat earlier). Buckley and Wiegold showed that in this case $|\operatorname{Out} G| \geq 2^{\aleph_0}$.

A feature of [49] was the introduction of a generalization of the notion of a basic subgroup of an abelian group. If G is a nilpotent p-group then we call a subgroup B of G a *basic subgroup* if $G' \leq B$ and B/G' is a basic subgroup of G/G'. Many of the elementary properties of basic subgroups were elucidated in [49] and these subgroups were seen to have many properties similar to those of the basic subgroups of an abelian group. For example, it is easy to show that if B is a basic subgroup of the nilpotent p-group G then $B/B'B^{p^n} \cong G/G'G^{p^n}$ for all natural numbers n, and if B is infinite then $|B| = |G/G'G^p|$.

Buckley and Wiegold showed further that if G has non-trivial divisible subgroups then $|\operatorname{Out} G|$ is as large as it can be, namely $2^{|G|}$ whereas, in the case when a basic subgroup B is infinite, $|\operatorname{Out} G| = 2^{|B|}$. However, a glance at the proofs shows that for many of the cases, they actually construct automorphisms of infinite order and use the Generalized Continuum Hypothesis at a crucial stage.

This reliance on GCH was subsequently removed by Menegazzo and Stonehewer in [27]. They also showed that a nilpotent p-group always has, with obvious exceptions, an outer automorphism of order p thus generalizing the results of Gaschütz. Using properties of basic subgroups of nilpotent p-groups, Menegazzo and Stonehewer obtained the following. (An automorphism α of the nilpotent group G acts *nilpotently* on G if $G \rtimes \langle \alpha \rangle$ is nilpotent.)

Theorem 3.1. *Let G be a nilpotent p-group.*

(i) If G is neither cyclic of order p nor isomorphic to a direct product of fewer than $p-1$ Prüfer p-groups then G has an outer automorphism of order p.

(ii) *If G is neither cyclic of order p nor divisible abelian then G has a non-inner p-automorphism acting nilpotently.*

(iii) *If G is not a Černikov group then* $\operatorname{Out} G$ *has an infinite elementary abelian p-subgroup.*

(iv) *If G has an infinite basic subgroup B then* $\operatorname{Out} G$ *has an elementary abelian p-subgroup of cardinality* $2^{|B|}$.

This result is best possible; a cyclic group of order p certainly has no automorphisms of order p, whereas a divisible abelian p-group of rank n has automorphism group $GL_n(R_p)$, where R_p is the ring of p-adic integers. Since the polynomial $1 + x + \ldots + x^{p-1}$ is irreducible in $R_p[x]$, a divisible abelian p-group of rank n cannot have an automorphism of order p if $n < p - 1$. On the other hand the divisible subgroups of a periodic nilpotent group are central (see [34, Vol. 1, Lemma 3.13]) so the only way for an automorphism of finite order to act nilpotently on a divisible abelian group is for it to act trivially. If G is a Černikov group then a well-known theorem of Baer asserts that a periodic subgroup of $\operatorname{Aut} G$ is finite, so certainly $\operatorname{Out} G$ cannot contain infinite p-subgroups in this case.

There is one remaining small gap in 3.1 which is covered by the following lemma.

Lemma 3.2. *Suppose G is a nilpotent non-Černikov p-group and suppose a basic subgroup of G is finite. Then* $\operatorname{Out} G$ *contains an uncountable elementary abelian p-subgroup.*

PROOF. If B is a finite basic subgroup of G then $G = BZ$, by [49, XV]. By [49, XVI], Z contains a quasicyclic p-group. In fact, since $G/B \cong Z/B \cap Z$, the divisible part of Z has infinite rank. Hence there is a divisible subgroup $D = \oplus_{i=1}^{\infty} D_i$, where $D_i \cong C_{p^\infty}$, and a subgroup A such that $Z = D \oplus A$ and $B \cap Z \leq A$. Every automorphism α of Z acting trivially on A can be extended to an automorphism $\bar{\alpha}$ of G by

$$(bz)\bar{\alpha} = b(z\alpha), \quad \text{whenever } b \in B, z \in Z.$$

For each $r \geq 0$ let $I_r = \{rp+1, rp+2, \ldots, (r+1)p\}$. Then we can define an automorphism, α_r, of order p of $\operatorname{Dr}_{j \in I_r} D_j$ by

$$(x_{rp+1}, \ldots, x_{(r+1)p}) \mapsto (x_{(r+1)p}, x_{rp+1}, \ldots, x_{rp+p-1}).$$

If I is a subset of the natural numbers then we obtain an automorphism α_I of D by letting α_I act on $\operatorname{Dr}_{j \in I_r} D_j$ like α_r if $r \in I$ and acting trivially on $\operatorname{Dr}_{j \in I_r} D_j$ if $r \notin I$. If J is also a subset of the natural numbers then α_I and α_J commute. If we extend such α_I to automorphisms of G, as described above, we obtain the result. □

The recent results of Puglisi [32] show that for countable locally finite p-groups the group of outer automorphisms has cardinality $2^{\aleph_0}$. However, close examination of the proofs shows that we have:

Lemma 3.3. *Suppose G is an infinite countable locally finite p-group. If G is not divisible-by-finite then* $\operatorname{Out} G$ *contains an uncountable elementary abelian p-subgroup.*

PROOF. If every finite subgroup F of G has the property that $C_G(F) \neq Z(G)$ then as in [32, Lemma 1] we can construct a certain uncountable abelian subgroup A of Linn G, the group of locally inner automorphisms of G. It is easily seen that for a locally finite group $\pi(\operatorname{Linn} G) = \pi(\operatorname{Inn} G)$, so A is a p-group. However, an uncountable abelian p-group has uncountable rank so A contains an uncountable elementary abelian p-subgroup in this case.

So we may assume G has a finite subgroup F such that $C_G(F) = Z(G)$. If G is not abelian-by-finite then the result follows by the proof of [32, Theorem 2], where the automorphisms constructed all have order p and commute. So we may assume G is an infinite abelian-by-finite group, containing a finite subgroup F such that $C_G(F) = Z(G)$. It follows from [32, Corollary 1] that $Z(G)$ contains an infinite elementary abelian subgroup. Let $Q = G/Z$. If a basic subgroup of Q_{ab} is finite then as in the proof of 2.11, G is divisible-by-finite contrary to hypothesis. Hence Q_{ab} has an infinite basic subgroup and again, as in the proof of 2.11, $\operatorname{Aut} G$ contains an infinite elementary abelian p-subgroup in this case. □

It seems as though we ought to be able to eliminate all cases but those of Černikov p-groups in the above result, but I have been unable to do so.

It would be interesting to know to what extent 2.11 and 3.1 could be extended to the class of all hypercentral p-groups, but I know of no results. In particular:

Question. Is there a semicomplete, hypercentral p-group?

As mentioned above the automorphisms constructed by Buckley and Wiegold were, for the most part, of infinite order. This led Menegazzo and Stonehewer to ask:

Question. Which infinite nilpotent p-groups have automorphisms of infinite order?

The best answer at the moment appears to be: Not all of them.

In the papers of Heineken and Liebeck [23] and Webb [47] the authors describe a general method for constructing nilpotent p-groups, ($p \geq 3$) which we now describe. Let D be a digraph with vertex set $\Gamma = \{v_i : i \in I\}$.

Define $G = G(D)$ to be the group with generators $\{x_i : i \in I\}$ and relations $\{x_j^p = [x_i, y_i], [[x_i, x_j], x_k] = 1 \mid i, j, k \in I\}$ where $y_i = \prod\{x_j \mid v_j v_i$ is an edge of $\Gamma\}$, this product being taken in some fixed order. Clearly G is a nilpotent p-group of class 2 and exponent p^2, and it is shown in [23, 3.2] that under certain conditions on Γ, $\operatorname{Aut} G / \operatorname{Aut}_c G \cong \operatorname{Aut} \Gamma$. If we now consider the following directed graph

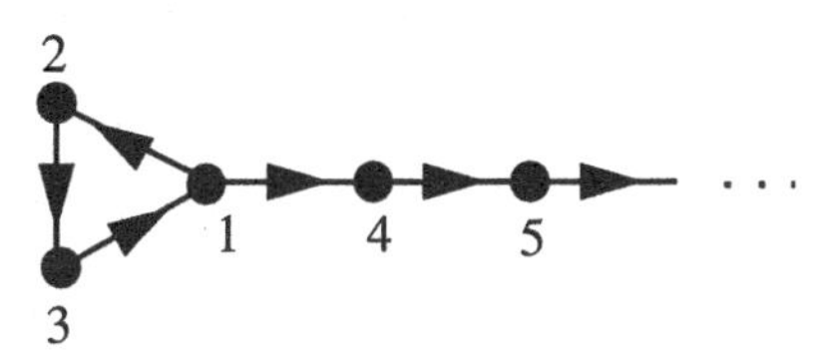

the corresponding p-group will be

$$G = \langle x_i : i \in \mathbf{N}, [x_i, x_j, x_k] = 1, x_1^p = [x_1, x_2 x_4], x_i^p = [x_i, x_{i+1}] \text{ for } i \geq 2, i \neq 3, x_3^p = [x_3, x_1], [x_i, x_j]^p = 1\rangle.$$

Clearly $G' = Z(G)$ is elementary abelian and since $\operatorname{Aut} \Gamma = 1$, $\operatorname{Aut} G = \operatorname{Aut}_c G \cong \operatorname{Hom}(G/Z, Z)$, since central automorphisms always fix G'. Thus $\operatorname{Aut} G$ is an uncountable elementary abelian p-group in this case, and in particular, not every nilpotent p-group has automorphisms of infinite order. (We remark that recently M. Pettet [31] has generalized the results of [23] and [47].)

The question of Menegazzo and Stonehewer seems to be rather difficult to answer. In [8] a characterization was given of those nilpotent p-groups with a central automorphism of infinite order. In this case, we obtain two rather tidy results. The first of these essentially follows from the work of Buckley and Wiegold [3].

Theorem 3.4. *If G is a nilpotent p-group of infinite exponent then $\operatorname{Aut}_c G$ contains an uncountable torsion free abelian subgroup.*

For groups of finite exponent the proof is rather different and more general.

Theorem 3.5. *Suppose G is a group and $Z(G)$ has finite exponent. The following are equivalent.*

(i) G has a central automorphism of infinite order.

(ii) G has an infinite abelian direct factor.

(iii) $\operatorname{Aut}_c G$ contains an uncountable torsion free abelian subgroup.

This latter result fails for locally finite groups in general if $Z(G)$ has infinite exponent, as can be observed by taking a direct product of groups defined by the digraph shown earlier, one for each prime p.

Central automorphisms of infinite groups have been the subject of several other recent papers. In [16, Lemma 2] Franciosi and de Giovanni extend certain results of Adney and Ti-Yen [1] to give the following result, which should be contrasted with 3.5.

Lemma 3.6. *Suppose G is a p-group with no non-trivial abelian direct factor. Then each central automorphism of G of finite order has p-power order.*

In contrast to 3.5, Pettet [29, Theorem 4.1] has shown that if G is a periodic group of finite exponent and if $\operatorname{Aut} G$ is nilpotent then $\operatorname{Aut} G$ is also periodic of finite exponent.

4. Final remarks

We conclude this survey with a discussion of results concerning the imposition of finiteness and related conditions on automorphisms of nilpotent groups, of which the result of Pettet mentioned above is just one example. Perhaps the fundamental result here is the following theorem of Robinson [35, Theorem B].

Theorem 4.1. *Suppose G is a group such that $\operatorname{Aut} G$ is a periodic nilpotent group. Then*

(i) $\operatorname{Aut} G$ *has finite exponent.*

(ii) The elements of G which have finite order form a subgroup of finite exponent.

(iii) If $\operatorname{Aut} G$ *is also countable then $G/Z(G)$ is finite and the torsion subgroup of G is finite.*

In particular, G is nilpotent and the imposition of these conditions on $\operatorname{Aut} G$ restricts the structure of G rather dramatically. Note also that this theorem implies that certain groups can never be the full automorphism group of a group, a theme which has been the subject of several recent papers (see [6], [7], [9], [17], [30], [35]). Pettet [30] has generalized this result of Robinson also.

Franciosi and de Giovanni [15-17] have also considered the effect of placing conditions on the automorphism group. In [16], they prove results concerning p-groups with locally nilpotent automorphism group, and show that the structure of such groups is rather restricted. In [15] and [17] Franciosi and de Giovanni study finite-by-nilpotent groups with restrictions on $\operatorname{Aut}_c G$. The main result of [17] shows that if G is a finite-by-nilpotent group such that $\operatorname{Aut}_c G$ is periodic then $G/Z(G)$ has finite exponent. In [15] they prove the following theorem.

Theorem 4.2. *Suppose G is a finite-by-nilpotent group. If* $\mathrm{Aut}_c G$ *is countable then the Sylow p-subgroups of G are finite, for all primes p.*

In particular, and this is where we came in, a semicomplete nilpotent group with $\mathrm{Aut}\, G$ countable has finite Sylow p-subgroups.

Finally we mention a paper of Curzio, Robinson, Smith and Wiegold [5], which considers central automorphisms of hypercentral groups. They show that if G is a nilpotent group and $\mathrm{Aut}_c G$ has finite exponent n then $G/Z(G)$ and $T(G)$, the torsion subgroup of G, both have finite exponent which can be bounded as functions of the nilpotency class of G and the number n. By contrast, no such result holds for hypercentral groups as the example $C_{2^\infty} \rtimes C_2$ shows.

References

[1] J.E. Adney and Ti Yen, Automorphisms of a p-group, *Illinois J. Math.* **9**(1965), 137–143.

[2] G. Baumslag, *Lecture Notes on Nilpotent Groups* (Amer. Math. Soc. Regional Conference Series 2, 1971).

[3] J. Buckley and J. Wiegold, On the number of outer automorphisms of an infinite nilpotent p-group, *Arch. Math.* **31**(1978), 321–328.

[4] J. Buckley and J. Wiegold, On the number of outer automorphisms of an infinite nilpotent p-group II, *Arch. Math.* **36**(1981), 1–5.

[5] M. Curzio, D.J.S. Robinson, H. Smith and J. Wiegold, Some remarks on central automorphisms of hypercentral groups, *Arch. Math.* **53**(1989), 327–331.

[6] M.R. Dixon and M.J. Evans, Divisible Automorphism Groups, *Quart. J. Math. Oxford (2)* **41**(1990), 179–188.

[7] M.R. Dixon and M.J. Evans, Periodic divisible-by-finite automorphism groups are finite, *J. Algebra* **137**(1991), 416–424.

[8] M.R. Dixon and M.J. Evans, On groups with a central automorphism of infinite order, *Proc. Amer. Math. Soc.* **114**(1992), 331–336.

[9] M.R. Dixon, Countable periodic CC-groups as automorphism groups, *Proc. Edinburgh Math. Soc.* **35**(1992), 295–299.

[10] M.R. Dixon and E.M. Rigsby, The Sylow p-subgroups of semicomplete nilpotent groups, *Proc. Amer. Math. Soc.*, **19**(1993), 341–349.

[11] T.A. Fournelle, Outer automorphisms of nilpotent groups, *Bull. London Math. Soc.* **13**(1981), 129–132.

[12] T.A. Fournelle, Nilpotent groups without periodic automorphisms, *Bull. London Math. Soc.* **15**(1983), 590-595.

[13] T.A. Fournelle, Torsion in semicomplete nilpotent groups, *Math. Proc. Cambridge Philos. Soc.* **94**(1983), 191–202.

[14] T.A. Fournelle, Pseudocomplete nilpotent groups, *Proc. Amer. Math. Soc.* **88**(1983), 1–7.

[15] S. Franciosi and F. de Giovanni, A note on groups with countable automorphism group, *Arch. Math.* **47**(1986), 12–16.

[16] S. Franciosi and F. de Giovanni, On torsion groups with nilpotent automorphism group, *Comm. Algebra* **14**(1986), 1909–1935.

[17] S. Franciosi and F. de Giovanni, On central automorphisms of finite-by nilpotent groups, *Proc. Edinburgh Math. Soc.* **33**(1990), 191–201.

[18] L. Fuchs, *Infinite Abelian Groups, Vol. I, II* (Academic Press, New York, London, 1970).

[19] T.M. Gagen and D.J.S. Robinson, Finite metabelian groups with no outer automorphisms, *Arch. Math.* **32**(1979), 417–423.

[20] W. Gaschütz, Nichtabelsche p-Gruppen besitzen äussere p-Automorphismen, *J. Algebra* **4**(1966), 1–2.

[21] N.D. Gupta and S.E. Stonehewer, Outer automorphisms of finitely generated nilpotent groups, *Arch. Math.* **31**(1978), 1–10.

[22] B. Hartley and D.J.S. Robinson, On finite complete groups, *Arch. Math.* **35**(1980), 67–74.

[23] H. Heinneken and H. Liebeck, The occurrence of finite groups in the automorphism group of nilpotent groups of class 2, *Arch. Math.* **25**(1974), 8–16.

[24] K. Hickin, Complete universal locally finite groups, *Trans. Amer. Math. Soc.* **239**(1978), 213–227.

[25] P. Hill, On the automorphism group of an infinite primary abelian group, *J. London Math. Soc.* **41**(1966), 731–732.

[26] S. Khabbaz, Abelian torsion groups having a minimal system of generators, *Trans. Amer. Math. Soc.* **98**(1961), 527–538.

[27] F. Menegazzo and S.E. Stonehewer, On the automorphism group of a nilpotent p-group, *J. London Math. Soc. (2)* **31**(1985), 272–276.

[28] J. Otal, J.M. Pena and M.J. Tomkinson, Locally inner automorphisms of CC-groups, *J. Algebra* **141**(1991), 382–398.

[29] M.R. Pettet, Central automorphisms of periodic groups, *Arch. Math.* **51**(1988), 20–33.

[30] M.R. Pettet, Almost-nilpotent periodic groups as automorphism groups, *Quart. J. Math. Oxford (2)* **41**(1990), 93–108.

[31] M.R. Pettet, Characterizing inner automorphisms of groups, *Arch. Math.* **55**(1990), 422–428.

[32] O. Puglisi, A note on the automorphism group of a locally finite p-group, *Bull. London Math. Soc.* **24**(1992), 437–441.

[33] R. Ree, The existence of outer automorphisms of some groups, *Proc. Amer. Math. Soc.* **7**(1956), 962–964.

[34] D.J.S. Robinson, *Finiteness Conditions and Generalized Soluble Groups, I, II* (Springer Verlag, Berlin-Heidelberg-New York, 1972).

[35] D.J.S. Robinson, Infinite torsion groups as automorphism groups, *Quart. J. Math. Oxford (2)* **30**(1979), 351–364.

[36] D.J.S. Robinson, Infinite soluble groups with no outer automorphisms, *Rend. Sem. Mat. Univ. Padova* **62**(1980), 281–294.

[37] D.J.S. Robinson, Recent results on finite complete groups, in *Algebra Carbondale 1980* (Springer Lecture Notes 848, Springer Verlag, Berlin-Heidelberg-New York, 1981), 178–185.

[38] D.J.S. Robinson, S.E. Stonehewer and J. Wiegold, Automorphism groups of FC-groups, *Arch. Math.* **40**(1983), 401–404.

[39] S.E. Stonehewer, Automorphisms of locally nilpotent FC-groups, *Math. Z.* **148**(1976), 85–88.

[40] S. Thomas, The automorphism tower problem, *Proc. Amer. Math. Soc.* **95**(1985), 166–168.

[41] S. Thomas, Complete existentially closed locally finite groups, *Arch. Math.* **44**(1985), 97–109.

[42] S. Thomas, Complete universal locally finite groups of large cardinality, in *Logic Colloquium '84, Manchester 1984* (Stud. Logic Foundations of Math. **120**, North Holland, Amsterdam, New York, 1986), 227–301.

[43] E.A. Walker, On the orders of the automorphism groups of infinite torsion abelian groups, *J. London Math. Soc.* **35**(1960), 385–388.

[44] U.H.M. Webb, Ph.D. Thesis, University of Warwick, 1978.

[45] U.H.M. Webb, Outer automorphisms of some finitely generated nilpotent groups I, *J. London Math. Soc. (2)* **21**(1980), 216–224.

[46] U.H.M. Webb, An elementary proof of Gaschütz' theorem, *Arch. Math.* **35**(1980), 23–26.

[47] U.H.M. Webb. The occurrence of groups as automorphisms of nilpotent p-groups, *Arch. Math.* **37**(1981), 481–498.

[48] H. Wielandt, Eine Verallgemeinerung der invarianten Untergruppen, *Math. Z.* **45**(1939), 209 244.

[49] A.E. Zaleskii, A nilpotent p-group possesses an outer automorphism, *Dokl. Akad. Nauk SSSR* **196**(1972), 751-754; *Soviet Math. Doklady* **12**(1972), 227–230.

[50] A.E. Zaleskii, An example of a torsion-free nilpotent group having no outer automorphisms, *Mat. Zametki* **11**(1972), 221-226; *Math. Notes* **11**(1972), 16–19.

[51] J. Zimmerman, Countable torsion FC-groups as automorphism groups, *Arch. Math.* **43**(1984), 108–116.

GENERATION OF ORTHOGONAL GROUPS OVER FINITE FIELDS

A.G. EARNEST

Southern Illinois University, Carbondale, Illinois 62901–4408, U.S.A.

Results on two-element generation have been established for many important families of finite groups, encompassing all finite simple groups. For example, theorems of this type were proved for the finite non-abelian simple groups of Lie type, the symplectic groups over finite fields, and the sporadic simple groups by Steinberg [13], Stanek [12], and Aschbacher and Guralnick [2], respectively. A comprehensive survey of the status of results on this and related generation problems, some historical remarks, and an extensive list of references to the original literature can be found in a recent paper by DiMartino and Tamburini [6]. The purpose of the present paper is to report on some recent results on the generation of the finite orthogonal groups, including a proof that all such groups are generated by two elements. This proof is contained in joint work of Ishibashi and the author [10], except for several low-dimensional cases which are resolved in [4].

1. Background

The development of a general theory of the classical groups over arbitrary fields and division rings came to fruition in Dieudonné's fundamental volume *La Géométrie des Groupes Classiques* [5]. The definitive book of Hahn and O'Meara [7] provides a thorough modern treatment of this theory in the broader context in which the underlying ring of scalars is kept as general as possible. In this paper we will consider the classical case of the orthogonal group $O(V)$ of a nonsingular finite-dimensional quadratic space (V, q) over a finite field F of characteristic p; that is, $O(V) = \{\sigma \in \mathrm{Aut}_F(V) : q(\sigma x) = q(x)$ for all $x \in V\}$. The fundamental properties of these orthogonal groups for the case of odd characteristic are described in [1]. Many excellent references detailing the structure of the underlying quadratic spaces are available; an integrated treatment including the characteristic 2 case appears, e.g., in [11].

For a given dimension n (with n even when $p = 2$) there are exactly two non-isometric nonsingular quadratic spaces over F, which are distinguished by their discriminants when p is odd and by their Arf invariants when $p = 2$. In the binary case $n = 2$, the two spaces can be explicitly described as follows. If V is isotropic (i.e., there exists $0 \neq v \in V$ with $q(v) = 0$), then V is a hyperbolic plane; that is, $V = Fu + Fv$ for vectors u, v with $q(u) = q(v) = 0$ and $b(u, v) = 1$ (we will say that such u, v form a hyperbolic pair). If V is anisotropic, then V is isometric to the quadratic space (K, N), where $K = F(\lambda)$ is the quadratic field extension obtained by adjoining to F a root λ of a suitable irreducible quadratic polynomial $f(X)$ and N is the norm from K to F. The polynomial $f(X)$ can be taken to be $X^2 - D$, D a nonsquare in $\dot{F}$, when $p \neq 2$, and to be $X^2 + X + D$, where $D \in \dot{F}$ is not of the form $\beta^2 + \beta$, when $p = 2$.

2. Two-element generation

Ishibashi [9] proved that the unitary group on a nonsingular Hermitian space with hyperbolic rank at least one over a finite field of characteristic not 2 is generated by three elements. In recent joint work of Ishibashi and the author [10], completed in [4], this result is refined and extended for the orthogonal groups as follows:

Theorem 1. *Let $O(V)$ be the orthogonal group of a nonsingular quadratic space over a finite field. Then $O(V)$ is generated by two elements.*

To describe the method of proving this theorem and the form of the generators obtained, we begin first with the binary case. Suppose first that V is a hyperbolic plane spanned by the hyperbolic pair u, v. In this case it is immediate from the definitions that the isometries $\sigma \in O(V)$ must have one of two forms: either (1) $\sigma(u) = \lambda u$ and $\sigma(v) = \lambda^{-1}v$, or (2) $\sigma(u) = \lambda v$ and $\sigma(v) = \lambda^{-1}u$, for some $\lambda \in \dot{F}$. It follows that $O(V)$ is generated by the elements Δ and ϕ defined by $\Delta(u) = v$, $\Delta(v) = u$ and $\phi(u) = \alpha u$, $\phi(v) = \alpha^{-1}v$, where α is a fixed generator for the multiplicative group $\dot{F}$. On the other hand, if V is anisotropic, then (V, q) is isometric to the space (K, N) with $K = F(\lambda)$ as described in the previous section. In this case, we consider the subgroup $O^+(V)$ of index 2 in $O(V)$ consisting of the isometries of determinant 1 when $p \neq 2$, and the isometries with Dickson invariant 0 when $p = 2$. Utilizing the nontrivial F-automorphism of K, it can be shown that the elements of $O^+(K)$ are precisely the multiplications by elements of norm 1. That is, $O^+(K)$ is isomorphic to the subgroup $K^1 = \{a \in K : N(a) = 1\}$ of $\dot{K}$, which is cyclic since K is a finite field. Taking a generator for $O^+(K)$ along with any element of order 2 which lies outside $O^+(K)$, we see that in this case, as in the preceding case, $O^+(V)$ is a dihedral group.

In the general case $n \geq 3$, V is always isotropic. For an isotropic vector $u \in V$ and any $x \in (Fu)^\perp$, the Eichler transformation $E_{u,x}$ in $O(V)$ is defined by the formula

$$E_{u,x}(z) = z + b(x, z)u - b(u, z)x - b(u, z)q(x)u,$$

where $b : V \times V \to F$ is the symmetric bilinear form on V defined by $b(v, w) = q(v + w) - q(v) - q(w)$. (These transformations masquerade under various aliases in the literature; for example, they are the Siegel transvections in [3].) For a set X of vectors in V, we denote $\{E_{u,x} : x \in X\}$ by $E(u, X)$. Now, as V is isotropic, V is split by a hyperbolic plane $\mathbb{H}$. Fix such a splitting $V = \mathbb{H} \perp L$, where $\mathbb{H}$ is spanned by a hyperbolic pair u, v. The following result, which is a direct consequence of Corollary 3.15, p. 71 of [3], plays a key role in the completion of the proof.

Lemma. *Let G be a subgroup of $O(V)$. If G contains ϕ, $E(u, L)$, $E(v, L)$ and some $\delta \notin O^+(V)$, then $G = O(V)$.*

The remainder of the proof now involves producing suitable generating pairs on a case-by-case basis, depending upon whether n and p are odd or even. Isolated low-dimensional cases over small fields need to be treated individually when the general constructions fail. All generating pairs produced in the various cases in [10] are of the uniform type $\{\tau\phi, \Delta E_{u,x}\rho\}$, where $\phi, \Delta \in O(\mathbb{H})$ are as described previously, x is some vector in L, and τ, ρ are elements of $O(L)$ (i.e., τ and ρ fix $\mathbb{H}$ pointwise).

The cases which escaped resolution in [10] are perhaps worthy of further comment. These occur when $n = 4$ and $F = \mathbb{F}_3$, giving rise to groups of order 1152 and 1440, depending upon whether the discriminant of V is a square or nonsquare in $\dot{F}$. The group structure in the case of a square discriminant is particularly interesting. Contrary to the general case for $n \geq 5$ in which the projective commutator subgroups $P\Omega(V)$ are simple groups, here we have $P\Omega(V) \cong PSL_2(\mathbb{F}_3) \times PSL_2(\mathbb{F}_3)$, and $PSL_2(\mathbb{F}_3)$ is also not simple. These remaining cases are completed in [4], where generating pairs of the form $\{\tau_1\phi, \tau_2 E_{u,x}\tau_3\}$ are produced with $\phi, E_{u,x}$ as before and τ_1, τ_2, τ_3 involutions. However, in contrast to the previous cases, the τ_i's required here do not fix $\mathbb{H}$.

3. Generation by symmetries

For any anisotropic vector $x \in V$, the formula $\tau_x(z) = z - b(x, z)q(x)^{-1}x$ defines an element τ_x of order 2 in $O(V)$ called the symmetry with respect to x. Let $S(V)$ denote the subgroup of $O(V)$ generated by all such symmetries. The fundamental Cartan-Dieudonné theorem asserts that $S(V) = O(V)$, except in the single case when $F = \mathbb{F}_2$ and V is an orthogonal sum of two hyperbolic planes (in which case $S(V)$ is a subgroup of index 2). Ishibashi [8] has shown that the orthogonal group of a nonsingular quadratic space of dimension n over a finite field of odd characteristic is generated by n symmetries, and that n is the minimal number. The following result from [4] gives a partial extension of this result to the characteristic 2 case.

Theorem 2. *Let $O(V)$ be the orthogonal group of a nonsingular quadratic space of dimension n over a finite field of characteristic 2 which is distinct* *from the field of two elements. Then $O(V)$ is generated by $n+1$ symmetries, two of which can be chosen to be the generators of $O(\mathbb{H})$, where $\mathbb{H}$ is any hyperbolic plane which splits V.*

4. Some related problems

In closing, I will mention some related generation problems which, as far as I know, remain open at this time.

First, I note that only in isolated instances do the generating pairs for $O(V)$ described in §2 contain an involution. However, in light of the positive results obtained to date for other families of 2-generated finite groups, the following questions may be reasonable.

Question 1. Is every finite orthogonal group generated by two elements, one of which is an involution?

Question 2. Is every finite orthogonal group generated by three involutions?

It may also be reasonable to hope to extend the result on 2-generation of the finite orthogonal groups to other finite unitary groups which are already known to be generated by three elements [9].

Question 3. Let $U(V)$ be the unitary group on a nonsingular Hermitian space with hyperbolic rank at least one over a finite field. Is $U(V)$ generated by two elements?

Finally, Theorem 2 above leaves several questions unresolved.

Question 4. Can the number of symmetries required in Theorem 2 be reduced from $n+1$ to n?

It is easy to see that n would be the minimal number possible. In fact, the whole question reduces to whether the orthogonal group for a space of dimension 4 is generated by 4 symmetries, as the proof for the general case for dimension ≥ 6 proceeds by induction from this case.

Question 5. Can Theorem 2 be extended to spaces of dimension at least 6 over the field $\mathbb{F}_2$?

Acknowledgements. I would like to take this opportunity to thank Professor Hiroyuki Ishibashi for introducing me to this subject and for generously sharing his expertise. My work on this paper was partially supported by a research grant from the National Security Agency.

References

[1] E. Artin, *Geometric Algebra* (Wiley Interscience, New York, 1957).

[2] M. Aschbacher and R. Guralnick, Some applications of the first cohomology group, *J. Algebra* **90**(1984), 446–460.

[3] R. Baeza, *Quadratic Forms over Semilocal Rings* (Lecture Notes in Mathematics **655**, Springer-Verlag, Berlin-New York, 1978).

[4] R.A. Catalpa, A.G. Earnest, G.T. Stewart and U.S. Schmidt, *Minimal generating sets for orthogonal groups over finite fields*, preprint.

[5] J. Dieudonné, *La Géométrie des Groupes Classiques*, 3rd ed. (Springer-Verlag, Berlin-New York, 1971).

[6] L. DiMartino and M.C. Tamburini, 2-generation of finite simple groups and some related topics, in *Generators and Relations in Groups and Geometries* (A. Barlotti, E.W. Ellers, P. Plaumann and K. Strambach (eds.), NATO ASI series, Kluwer, Dordrecht 1991), 195–233.

[7] A.J. Hahn and O.T. O'Meara, *The Classical Groups and K-Theory* (Grund. Math. Wiss. **291**, Springer-Verlag, Berlin-New York-Tokyo, 1989).

[8] H. Ishibashi, Generators of an orthogonal group over a finite field, *Czech. Math. J.* **28**(1978), 419–433.

[9] H. Ishibashi, Small systems of generators of isotropic unitary groups over finite fields of characteristic not two, *J. Algebra* **93**(1983), 324–331.

[10] H. Ishibashi and A.G. Earnest, Two-element generation of orthogonal groups over finite fields, *J. Algebra*, to appear.

[11] W. Scharlau, *Quadratic and Hermitian Forms* (Grund. Math. Wiss. **270**, Springer-Verlag, Berlin-Heidelberg-New York-Tokyo, 1985).

[12] P. Stanek, Two-element generation of the symplectic group, *Trans. Amer. Math. Soc.* **108**(1963), 429–436.

[13] R. Steinberg, Generators for simple groups, *Canad. J. Math.* **14**(1962), 277–283.

THE STRUCTURE OF CERTAIN COXETER GROUPS

V. FELSCH*, D.L. JOHNSON†, J. NEUBÜSER* and S.V. TSARANOV‡

*Lehrstuhl D für Mathematik, RWTH, Templergraben 64, 52062 Aachen, Germany
†Mathematics Department, University of Nottingham, University Park, Nottingham, NG7 2RD, U.K.
‡Institute for Systems Analysis, Academy of Sciences, 9 Prospekt 60–Let Oktyabrya, 117312 Moscow, Russia

Abstract

Recent work in graph theory has given rise to various classes of group presentations that are closely connected to those of Coxeter groups of a certain kind. We list here several examples of particular interest and investigate one such class in detail.

Introduction

Let $\Gamma = (V, E)$ be a finite graph with vertex set $V = \{x_1, \dots, x_n\}$ and adjacency matrix $E = (\varepsilon_{ij}) \in M_n(\{0,1\})$, so that ε_{ij} is 1 if x_i and x_j are joined by an edge of Γ and 0 otherwise. Then Γ determines a group $C(\Gamma)$ with generators V and defining relations

$$x_i^2 = 1, \;\; 1 \le i \le n, \quad (x_i x_j)^{2+\varepsilon_{ij}} = 1, \;\; 1 \le i < j \le n. \qquad (1)$$

$C(\Gamma)$ is called the *Coxeter group* of Γ, and Γ the *Dynkin diagram* of $C(\Gamma)$. For Coxeter groups, the classic text is [2], especially Chapter IV, and an up-to-date survey is [10]; [1] provides an excellent introduction.

Certain factor groups of such Coxeter groups have recently been introduced by Cameron, Seidel and Tsaranov [3, 4, 5, 15, 17]. These are defined in a natural way using Dynkin diagrams and are of two kinds. The first consists simply in adjoining new relators of a specific type to those in (1), while the second is slightly more complicated. Combinatorial considerations suggest that especial interest attaches to the case when the matrix E has multiple eigenvalues, and it is this that has motivated our choice of examples in what follows. Both the general construction and our specific examples are described in §1.

The method of attack is in outline very simple: hand calculations based on subgroup presentations resulting from machine calculations. It is important to emphasize however that, while our proofs can be made machine-independent with the benefit of hindsight, the use of the machine is essential in providing that hindsight. A description of our general strategy, and of which algorithms are involved, is given in §2.

Of the six classes of groups we have studied, we describe the results in detail in the case in which we have obtained the most detailed description. This occupies §§3-7, and §8 contains some commentary and speculation.

Notation and terminology are fairly standard throughout. Thus the symbol $\sim$ will always stand for commutation: $x \sim y$ means $xy = yx$. Further, we will say that a group G is a *retract* of a group S if there is a split epimorphism $\nu : S \twoheadrightarrow G$. This means that there is a homomorphism $\sigma : G \to S$ such that $\sigma\nu = 1_G$, whence $S \cong \operatorname{Ker}\nu \rtimes G$, the semi-direct product of $\operatorname{Ker}\nu$ and G.

Acknowledgements. It is a pleasure to record our thanks to a number of colleagues for useful discussion and valuable advice. Notable among these are P.J. Cameron, D.F. Holt, M. Geck, P.H. Kropholler, Π.M. Neumann, M.F. Newman, S.J. Pride, J.J. Seidel and R.M. Thomas. The second author is grateful to Lehrstuhl D für Mathematik, RWTH Aachen, for hospitality, and to the German Academic Exchange Service (DAAD) for support, while this work was in progress.

1. The groups

By a *face* of a graph Γ we mean a circuit with no chords, that is, a subgraph Δ with m vertices $y_1, \ldots, y_m$ and m edges $\{y_1, y_2\}, \ldots, \{y_{m-1}, y_m\}, \{y_m, y_1\}$ such that Δ is the full subgraph of Γ on $y_1, \ldots, y_m$. Then the element $(y_1 y_2 \cdots y_m y_{m-1} \cdots y_2)^2 \in C(\Gamma)$ is called a *Coxeter element* corresponding to the face Δ. Note that the relations (1) imply that

$$y_1 y_m y_{m-1} \cdots y_2 y_3 \cdots y_m = y_1 y_2 \cdots y_m y_{m-1} \cdots y_2 \text{ and}$$
$$y_2 \cdots y_m y_1 y_m y_{m-1} \cdots y_3 = (y_2 \cdots y_m)^{-1} \cdot (y_1 \cdots y_m y_{m-1} \cdots y_2) \cdot (y_2 \cdots y_m),$$

and hence all Coxeter elements corresponding to the same face are conjugates in $C(\Gamma)$. So, if we adjoin some Coxeter element to (1) as an extra relator, we will get a factor group which depends only on the corresponding face Δ, and not on the starting point y_1, say, nor on the sense in which it is traversed.

We are interested in the factor groups of $C(\Gamma)$ obtained by adjoining to (1) Coxeter elements corresponding to various combinations of faces of Γ. In particular, such factor groups of the Coxeter groups D_0, E_0, F_0, T_0 defined by the Dynkin diagrams in Fig. 1 have (up to symmetry) $12, 4, 4, 4$ such factor groups, respectively. Note that none of the groups D_0, E_0, F_0, T_0 is either finite or affine, and only one (E) is hyperbolic (see the list in [10]). A further class of considerable interest consists of such factor groups of the Coxeter group O_0, whose Dynkin diagram is the 1-skeleton of the octahedron.

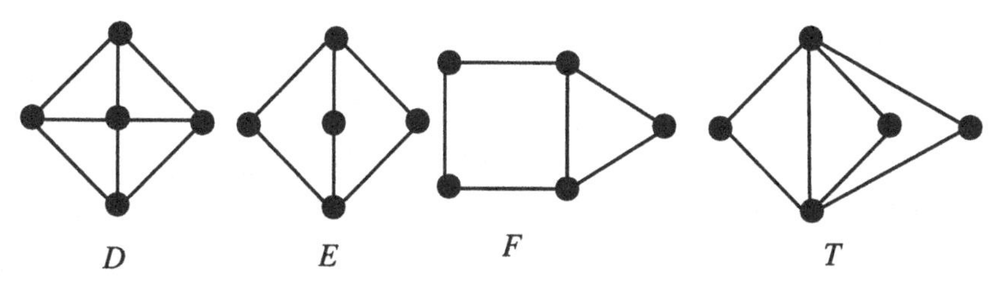

Figure 1.

This construction is slightly more general than that of *groups of signed graphs* introduced in [5] and defined as follows. Starting from any graph Γ, designate any set of edges as negative, and then adjoin to $C(\Gamma)$ as extra relators the Coxeter elements corresponding to all faces of Γ having an odd number of positive edges.

The second construction of Seidel and Tsaranov [15] depends only on the graph $\Gamma = (V, E)$. Taking $V = \{x_1, \ldots, x_n\}$ as above, define $\eta_{ij} = -1$ or $+1$ according as there is an edge between x_i and x_j or not, $1 \leq i, j \leq n$. Thus, if $E = (\varepsilon_{ij})$, $\eta_{ij} = 1 - 2\varepsilon_{ij}\ \forall i, j$. Then the *Tsaranov group* $Ts(\Gamma)$ of Γ has generators V and defining relations

$$x_i^3 = 1,\ \ 1 \leq i \leq n,\ (x_i x_j^{\eta_{ij}})^2 = 1,\ \ 1 \leq i < j \leq n. \tag{2}$$

Thus, for example, when the edge-set of Γ is empty, this is a presentation of the alternating group A_{n+2}. Of paramount interest here has been the group arising in this way from the hexagon, that is, from the case when Γ is a 6-cycle.

To see the connection between (2) and (1) we proceed as follows. First extend $Ts(\Gamma)$ by the involutory automorphism $y : x_i \mapsto x_i^{-1}$ $(1 \leq i \leq n)$, and then change to new generators $y_0 = y$, $y_i = yx_i$, $1 \leq i \leq n$. A simple calculation shows that the resulting group $Ts^*(\Gamma)$ is defined by the relations

$$\begin{array}{lll} y_i^2 = 1,\ \ 0 \leq i \leq n, & (y_0 y_i)^3 = 1,\ \ 1 \leq i \leq n, & \\ y_i \sim y_j\ \ if\ \ \eta_{ij} = -1, & y_i \sim y_0 y_j y_0\ \ if\ \ \eta_{ij} = +1, & i \leq i < j \leq n. \end{array}$$

Now replace the relation $y_i \sim y_0 y_j y_0$ in (2) by $(y_i y_j)^3 = 1$ whenever $\eta_{ij} = +1$. This is the Coxeter group $C(\Gamma')$, where Γ' is obtained from Γ by adjoining an isolated vertex x_0 and then taking the complement: $\Gamma' = \overline{\Gamma \cup \{x_0\}}$. It is an exercise to show that these new relations are consequences of (2), so that $Ts^*(\Gamma)$ is a natural homomorphic image of $C(\Gamma')$.

2. The method

Starting from a presentation of the group G in question, the machine computation proceeded in four steps, as follows:

1. finding all subgroups H of low index (say ≤ 5) in G;
2. determining the permutation representation ρ of G on the cosets of H in each case;
3. obtaining a presentation for the subgroups $K := \operatorname{Ker}\rho = \operatorname{core}(H)$;
4. calculating the rank and invariant factors of K/K' for each such K.

Originally (in 1991), steps 1 and 3 were handled by SPAS [16] using the Low Index Subgroups and Reduced Reidemeister-Schreier (or Modified Todd-Coxeter) algorithms, respectively. Step 2 was carried out using Cayley [6] and step 4 by using the Canberra Integer Matrix Diagonalization standalone [9]. The entire computation can now be performed using GAP 3.2 [14], which replaced all the old Aachen systems (CAS, SOGOS, and SPAS) and we have done this as a check. We emphasize at this point that machine output and hand calculations have been thoroughly checked against each other as far as possible.

3. *D*-groups

The 1-skeleton D of the square pyramid (see Fig. 2) is a graph with five vertices and five faces, four of degree three and one of degree four in each case. We study the Coxeter group $D_0 = C(D)$ and the factor groups of it obtained by adjoining to the relators the Coxeter elements corresponding to the various combinations of faces of D. There are 12 such groups all told, of which six correspond to signings of D. The possible combinations of triangular faces are shown in Fig. 2, and there are six more, denoted by a bar, obtained by adjoining the Coxeter element corresponding to the square face to each of these. These groups are related by natural homomorphisms as shown in Fig. 3, where all squares are pushouts.

We now attach the label x_5 to the vertex of degree four and x_1, x_2, x_3, x_4 to the others working counter-clockwise from the top, so that D_0 has the presentation with generators x_1, x_2, x_3, x_4, x_5 and relators

$$\begin{gathered} x_1^2,\ x_2^2,\ x_3^2,\ x_4^2,\ x_5^2,\ (x_1x_3)^2,\ (x_2x_4)^2,\ (x_1x_2)^3, \\ (x_2x_3)^3,\ (x_3x_4)^3,\ (x_4x_1)^3,\ (x_1x_5)^3,\ (x_2x_5)^3,\ (x_3x_5)^3,\ (x_4x_5)^3, \end{gathered} \tag{3}$$

and D_2, for example, is the factor group obtained by adjoining the extra relators

$$(x_1x_5x_1x_4)^2,\ (x_3x_5x_3x_2)^2$$

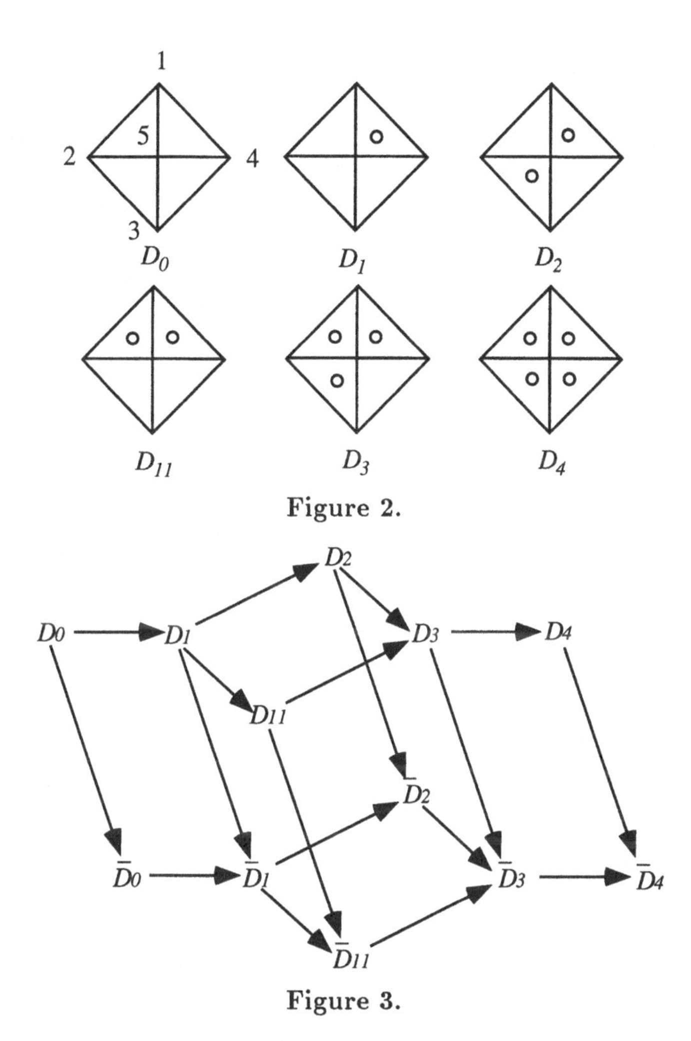

Figure 2.

Figure 3.

corresponding to the two faces marked in Fig. 2. Recall that the result of adjoining a Coxeter element depends only on the face and not on the initial vertex or the sense in which it is traversed.

Now consider the result of adjoining the relators $x_1x_5x_1x_2$, $x_3x_5x_3x_4$ to each of these groups. This yields in every case the Coxeter group of a tree with three vertices, that is, the symmetric group S_4. We denote the corresponding kernels by the symbol K, suitably suffixed and possibly barred as with the D's. In what follows, we shall describe the structure of most of these kernels in terms of possibly better-known objects, such as direct products of free groups.

Our use of the machine here is restricted to finding the presentations for these kernels that form the input for our hand calculations. Our proofs are thus machine-independent insofar as these presentations can be checked by hand either

(a) using the Reidemeister-Schreier rewriting process (index 24), or
(b) forming the corresponding extensions by S_4.

On the other hand, use of the machine was indispensable, for example in

(a) focussing attention on these kernels (using Low Index Subgroups, Reduced Reidemeister-Schreier, and Integer Matrix Diagonalization algorithms), and
(b) suggesting how they might be described (using the p-Quotient algorithm [11, 13]).

4. Some isomorphisms

Consider the effect on $\overline{D}_1$ of the change of generators

$$x_1 \mapsto x_4x_1x_4, \quad x_2 \mapsto x_2, \quad x_3 \mapsto x_5, \quad x_4 \mapsto x_4, \quad x_5 \mapsto x_3. \tag{4}$$

The relators (3) become

$$\begin{array}{c} x_4x_1^2x_4,\ x_2^2,\ x_5^2,\ x_4^2,\ x_3^2,\ x_4x_1x_4 \sim x_5,\ x_2 \sim x_4,\ (x_4x_1x_4x_2)^3, \\ (x_2x_5)^3,\ (x_5x_4)^3,\ (x_1x_4)^3,\ (x_4x_1x_4x_3)^3,\ (x_2x_3)^3,\ (x_5x_3)^3,\ (x_4x_3)^3, \end{array} \tag{5}$$

where $\sim$, both here and in what follows, denotes commutation, and the two extra relators

$$x_5 \sim x_4x_1x_4,\ x_2 \sim x_3x_4x_1x_4x_3$$

become

$$x_3 \sim x_1,\ x_2 \sim x_5x_1x_5. \tag{6}$$

Using the relations x_4^2, $x_2 \sim x_4$, $x_4x_1x_4 = x_1x_4x_1$, and $x_1 \sim x_3$, the relations (5), (6) are equivalent to (in the same order)

$$\begin{array}{c} x_1^2,\ x_2^2,\ x_5^2,\ x_4^2,\ x_3^2,\ x_4x_1x_4 \sim x_5,\ x_2 \sim x_4,\ (x_1x_2)^3,\ (x_2x_5)^3,\ (x_4x_5)^3, \\ (x_4x_1)^3,\ (x_4x_3)^3,\ (x_2x_3)^3,\ (x_3x_5)^3,\ (x_3x_4)^3,\ x_3 \sim x_1,\ x_2 \sim x_5x_1x_5. \end{array} \tag{7}$$

The presence of the relators

$$x_1^2,\ x_4^2,\ (x_1x_4)^3,\ (x_4x_5)^3,\ x_5 \sim x_4x_1x_4, \text{ and } x_4x_1x_4 = x_1x_4x_1$$

ensures that

$$\begin{aligned} 1 &= (x_1x_4)^3(x_4x_5)^3 = x_1x_4x_1x_4x_1x_5x_4x_5x_4x_5 \\ &= x_1x_4x_5x_1x_4x_1x_4x_5x_4x_5 \\ &= x_1x_4x_5x_1x_5x_4x_1x_4x_4x_5 \\ &= x_1x_4x_5x_1x_5x_1x_5x_1x_4x_1, \end{aligned}$$

so that the relation $(x_1x_5)^3 = 1$ is a consequence of the relations (7). Replacing $(x_4x_3)^3$ by $(x_1x_5)^3$ in (7) yields a presentation of D_{11}.

Since the relations

$$x_2 \sim x_5x_3x_5 \text{ and } x_4 \sim x_5x_3x_5$$

are preserved by the change of generators (4), we may adjoin either or both of them to the presentation of $\overline{D}_1$ to obtain three more isomorphisms between D-groups.

Proposition 1. $\overline{D}_1 \cong D_{11}$, $\overline{D}_2 \cong D_3$, $\overline{D}_{11} \cong D_3$, *and* $\overline{D}_3 \cong D_4$.

Two remarks are worth making here. Firstly, the kernels K of the natural maps onto S_4 are *not* preserved by these isomorphisms. Secondly, it has been shown by machine calculation that the 12 D-groups fall into *exactly* eight isomorphism classes: the six unbarred D's plus $\overline{D}_0$ and $\overline{D}_4$. In the following sections the structure of D_2 and its factor groups D_3, D_4, and $\overline{D}_4$ will be analysed in such detail that it is clear that they are pairwise non-isomorphic. The remaining groups can be distinguished by the following facts concerning the commutator factor groups of their normal subgroups with factor groups isomorphic to the symmetric group S_4.

group	commutator factor groups of normal subgroups with factors groups of type S_4
D_0	$Z_2^5 \times Z^5$ (4 times), $Z_2^{10} \times Z^2$, Z_3^5 (2 times), Z^8 (2 times)
$\overline{D}_0$	$Z_2^3 \times Z^5$ (6 times), $Z_2^{10} \times Z^2$, Z_3^5 (2 times)
D_1	$Z_2^5 \times Z^3$ (3 times), Z_2^{10} (3 times), Z^6
D_{11}	$Z_2^3 \times Z^3$ (3 times), Z_2^8 (3 times), Z_2^{10}
D_2	Z_2^8 (6 times), Z^6
D_3	$Z_2^3 \times Z^3$, Z_2^6 (4 times), Z_2^8 (2 times)
D_4	Z_2^6 (7 times)
$\overline{D}_4$	Z_2^6 (7 times)

Table 1.

5. The structure of D_2

The determination of the structure of D_2 started with the observation that the normal subgroup N of D_2 with $D_2/N \cong S_4$ and $N/N' \cong Z^6$ is in fact different from Z^6. This was obtained by calculating the p-central series of N for various primes using the Canberra p-Quotient program (pQ) and, when this produced equal ranks for various primes, also the lower central series of N using the new Nilpotent Quotient program (NQ) developed by W. Nickel in Canberra. In fact, as M.F. Newman then observed, the ranks of the factors of the lower central series after the first were exactly four times the corresponding ranks of the free group of rank 2, and this observation leads eventually to the conjecture which is proved in the rest of this section.

D_2 has the presentation with generators x_1, x_2, x_3, x_4, x_5 and relators

$$\begin{gathered} x_1^2,\ x_2^2,\ x_3^2,\ x_4^2,\ x_5^2,\ (x_1x_2)^3,\ (x_2x_3)^3, \\ (x_3x_4)^3,\ (x_4x_1)^3,\ (x_1x_5)^3,\ (x_2x_5)^3,\ (x_3x_5)^3,\ (x_4x_5)^3, \\ (x_1x_3)^2,\ (x_2x_4)^2,\ (x_4x_1x_5x_1)^2,\ (x_3x_2x_5x_2)^2. \end{gathered} \tag{8}$$

Note that the last four relators are equivalent to the commutators

$$x_1 \sim x_3,\quad x_2 \sim x_4,\quad x_3 \sim x_2x_5x_2,\quad x_4 \sim x_1x_5x_1.$$

Adjoining the new relators $x_1x_5x_1x_2$ and $x_3x_5x_3x_4$ yields the presentation

$$\langle x_1, x_5, x_3 \mid x_1^2,\ x_5^2,\ x_3^2,\ (x_1x_5)^3,\ (x_5x_3)^3,\ (x_1x_3)^2 \rangle \tag{9}$$

of the symmetric group S_4, which thus complements the normal closure K_2 of $x_1x_5x_1x_2$, $x_3x_5x_3x_4$ in D_2.

Machine calculations show that K_2 has a presentation with generators

$$\begin{aligned} y_1 &:= x_1x_5x_1x_2, & y_2 &:= x_2x_5x_2x_1, & y_3 &:= x_3x_1x_2x_1x_5x_3, \\ y_4 &:= x_3x_5x_3x_4, & y_5 &:= x_4x_5x_4x_3, & y_6 &:= x_1x_3x_4x_3x_5x_1, \end{aligned} \tag{10}$$

and defining relations

$$\begin{gathered} 1 \sim 2,\ 2 \sim 3,\ 3 \sim 1,\ 4 \sim 5,\ 5 \sim 6,\ 6 \sim 4, \\ 1 \sim 4,\ 2 \sim 5,\ 3 \sim 6,\ 12 \sim 56,\ 23 \sim 45, \end{gathered} \tag{11}$$

where for brevity we have written only the subscript i of each y_i.

To check this result by hand, we calculate the action (by conjugation in D_2) of x_1, x_3, x_5 on the elements (10) to obtain Table 2, where bars denote inverses. Thus, for example, the (4,1) entry contains the information $x_1y_4x_1 = y_6^{-1}y_5^{-1}$. Note that work is saved both here and in what follows by using the involutory automorphism of D_2 that sends x_1 to x_3, x_2 to x_4, x_5 to itself and thus acts as the permutation (14)(25)(36) on the generators of K_2.

	x_1	x_3	x_5
1	12	$\bar{3}\bar{2}$	$\bar{2}$
2	$\bar{2}$	2	$\bar{1}$
3	23	$\bar{2}\bar{1}$	123
4	$\bar{6}\bar{5}$	45	$\bar{5}$
5	5	$\bar{5}$	$\bar{4}$
6	$\bar{5}\bar{4}$	56	456

Table 2.

We now form the split extension of K_2 by S_4 using (9), (11) and Table 2. After changing generators to

$$x_1,\ x_3,\ x_5,\ x_2 = x_1x_5x_1y_1,\ x_4 = x_3x_5x_3y_4 \tag{12}$$

simple Tietze transformations produce the relators (8) of D_2.

Now consider the direct product

$$F = \mathop{\times}_{i=1}^{4} \langle a_i, b_i \mid \ \rangle$$

of four free groups of rank two, and the subgroup N generated by

$$z_1 = \overline{a_1}a_2,\ z_2 = \overline{a_2}a_3,\ z_3 = \overline{a_3}a_4,\ z_4 = \overline{b_3}b_4,\ z_5 = \overline{b_4}b_1,\ z_6 = \overline{b_1}b_2.$$

These elements will later be seen in the rôle of y_i, $1 \le i \le 6$, respectively.

Since N contains the derived group F' of F (in fact, $N' = F'$), it follows that N is normal in F. Moreover, N is complemented by the subgroup

$$A = \langle a_2, b_1 \mid a_2 \sim b_1 \rangle,$$

which is free abelian of rank 2.

Next, note that the z_i, abbreviated to subscripts, also satisfy the relations (11). Since there seems to be no simple reason why N need be finitely presented, although recent work [12] on the Bieri-Neumann-Strebel invariants gives finiteness conditions of the form $(\mathrm{FP})_m$ for such groups, it is so nevertheless, and we outline a proof in the next paragraph.

Returning to K_2, as defined by (11), we attempt to simulate the action of A on N by mapping the generators $y_1, \ldots, y_6$ of K_2 as shown in Table 3.

	a	b	$\overline{a}$	$\overline{b}$
1	1	$\overline{5}1 5$	1	$51\overline{5}$
2	2	2	2	2
3	3	3	3	3
4	4	4	4	4
5	5	5	5	5
6	$26\overline{2}$	6	$\overline{2}62$	6

Table 3.

To show that these are homomorphisms, observe that the relations in (11) involving generator 6 pass under a to

$$5 \sim 26\overline{2},\ \ 4 \sim 26\overline{2},\ \ 3 \sim 26\overline{2},\ \ 12 \sim 526\overline{2}.$$

The first and third of these are obvious, while the fourth follows from $12 \sim 56$ using $1 \sim 2$ and $2 \sim 5$. For the second, we deduce from (11) that

$$45 = (45)^{23} = 4^{23}5^{3} \ \Rightarrow\ 4^2 = (45)^{\overline{3}}\overline{5} \ \text{ and hence } \ 4^2 \sim 6,$$

as required. Thus, a defines an endomorphism of K_2. The relations in (11) involving 1 pass under b to

$$2 \sim \bar{5}15, \quad 3 \sim \bar{5}15, \quad 4 \sim \bar{5}15, \quad 56 \sim \bar{5}152,$$

which follow from (11) in like manner. The same goes for $\bar{a}$ and $\bar{b}$, which are then inverses of a and b, whence $a, b \in \mathrm{Aut}K_2$. Since their actions clearly commute, we obtain a homomorphism

$$\langle a, b \mid ab = ba \rangle \to \mathrm{Aut}K_2,$$

and the corresponding split extension is easily reduced to a presentation of F. The natural map $K_2 \to N$ is thus an isomorphism, and N is defined by the relations (11) as claimed.

We have thus succeeded in identifying the kernel K_2 as a fairly natural subgroup of the direct power F. To identify D_2 itself, consider the action of S_4 on F by permuting the subscripts of the a_i and b_i, $1 \leq i \leq 4$. To be more precise, take the presentation (9) with x_1, x_3, x_5 acting as the transpositions (23), (14), (13), respectively. It is then easy to check that the action of these on the generators z_i, $1 \leq i \leq 6$, of N is given by Table 2, and we can use (12) to identify D_2 as a subgroup of the split extension $F \rtimes S_4$ under this action.

Proposition 2. *With the notation and action just defined, the assignation*

$$x_1 \mapsto x_1, \quad x_2 \mapsto x_1x_5x_1a_1^{-1}a_2, \quad x_3 \mapsto x_3, \quad x_4 \mapsto x_3x_5x_3b_3^{-1}b_4, \quad x_5 \mapsto x_5 \qquad (13)$$

defines an embedding of D_2 into $F \rtimes S_4$.

6. The structure of D_3 and D_4

D_3 is obtained from D_2 by adjoining the relator $(x_1x_5x_1x_2)^2$, which is the element y_1^2 of K_2. It follows that K_3 has index 24 in D_3, has complement S_4 given by (9), and is defined on generators y_i, $1 \leq i \leq 6$, by relations (11) plus the squares of y_1, y_2, y_3 (whose normal closure in K_2 is equal to that of y_1^2 in D_2, by Table 2). The analysis now proceeds as in the previous section, but with F replaced by the group

$$F_1 = \mathop{\times}_{i=1}^{4} \langle a_i, b_i \mid a_i^2 \rangle.$$

Similarly, D_4 is obtained from D_2 by adjoining the relators $(x_1x_5x_1x_2)^2$ and $(x_3x_5x_3x_4)^2$, that is, y_1^2 and y_4^2. K_4 is then defined by (11) plus the squares of all six generators, and we replace F by

$$F_2 = \mathop{\times}_{i=1}^{4} \langle a_i, b_i \mid a_i^2, b_i^2 \rangle,$$

the fourth direct power of the infinite dihedral group.

Proposition 3. *With the notation and action defined above, the assignation (13) defines embeddings of D_3 and D_4 into $F_1 \rtimes S_4$ and $F_2 \rtimes S_4$, respectively.*

It follows from this that, while D_3 contains a copy of the free group of rank 2, D_4 is actually soluble (of derived length at most five). Since F_2 and the embedded image of K_4 have the same derived group, it follows that

$$K_4/K_4' \cong Z_2^6, \quad K_4' \cong Z^4.$$

Machine calculations using GAP show that the derived series of D_4 has length exactly five, with sections

$$Z_2, \ Z_3, \ Z_2^6, \ Z_2^5, \ Z^4.$$

7. The structure of $\overline{D}_4$

Because $x_3x_2x_1x_4x_1x_2 = y_3y_5^{-1}y_1y_2$, the group $\overline{D}_4$ is obtained from D_2 by adjoining as relators

$$y_1^2, \ y_4^2, \ (y_3y_5^{-1}y_1y_2)^2.$$

It follows that $\overline{K}_4$ again has a complement given by (9) and has relations (11), plus the squares of the generators together with the relator $(y_1y_2y_3y_5)^2$ and its images under the action in Table 2. These include

$$3 \sim 4, \ 1 \sim 6, \ 23 \sim 56,$$

which, together with (11) and the squares of the generators, define the group. Simple manipulation yields the presentation

$$\begin{aligned} \overline{K}_4 \ = \ & \langle\, y_1, y_2, y_3, y_4, y_5, y_6, z \,|\, y_1^2 = y_2^2 = y_3^2 = y_4^2 = y_5^2 = y_6^2 = z^2 = e, \\ & [y_1, y_5] = [y_2, y_6] = [y_2, y_4] = [y_3, y_5] = z, \\ & \text{all other pairs of generators commute} \rangle. \end{aligned}$$

Substituting the generators y_3 and y_6 by $z_3 := y_1y_3$ and $z_6 := y_4y_6$, respectively, we readily deduce the presentation

$$\begin{aligned} \overline{K}_4 \ = \ & \langle\, y_1, y_2, y_4, y_5, z, z_3, z_6 \,|\, y_1^2 = y_2^2 = y_4^2 = y_5^2 = z^2 = z_3^2 = z_6^2 = e, \\ & [y_1, y_5] = [y_2, y_4] = z, \ \text{all other pairs of generators commute} \rangle \end{aligned}$$

and hence the following description.

Proposition 4. *$\overline{D}_4$ is a split extension $\overline{K}_4 \rtimes S_4$ of order $3 \cdot 2^{10}$, where*

$$\overline{K}_4 \cong (D_8 \curlyvee D_8) \times Z_2 \times Z_2,$$

where the first factor on the right-hand side is the extra-special group equal to the central product of two dihedral groups of order 8.

8. Concluding remarks and open problems

1. Machine calculations show that the normal complement K_1 of $S_4 = \langle x_1, x_3, x_5\rangle$ in D_1 has generators

$$\begin{array}{lll} y_1 = x_4x_3x_4x_1x_2x_1, & y_2 = x_4x_3x_2x_3x_4x_1, & y_3 = x_4x_1x_3x_2x_3x_1, \\ y_4 = x_4x_3x_4x_5, & y_5 = x_1x_5x_3x_4x_1x_4x_3x_5, & y_6 = x_4x_5x_3x_5, \end{array}$$

and defining relations

$$1 \sim 2,\ \ 2 \sim 3,\ \ 3 \sim 1,\ \ 4 \sim 5,\ \ 5 \sim 6,\ \ 6 \sim 4,\ \ 1\bar{4} \sim 2\bar{5},\ \ 2\bar{5} \sim 3\bar{6},\ \ 3\bar{6} \sim 1\bar{4}.$$

Note that, in the presence of the first six relations, the last three are equivalent to

$$[1,6] = [3,4],\ \ [2,6] = [3,5],\ \ [1,5] = [2,4].$$

It would be interesting to have a realisation of K_1 similar to that obtained for K_2 in §5, but we have so far been unable to find one. However, machine calculations show easily that K_1/K_1' is free abelian of rank 6, i. e. isomorphic to K_2/K_2'. In contrast, for the corresponding kernel K_0 in the group D_0 one obtains that the commutator factor group K_0/K_0' is free abelian of rank 8.

2. The graph E of Fig. 1 has just three faces (all squares), so there are four groups E_i obtained by adjoining i Coxeter elements to the relators defining $C(E)$, $0 \leq i \leq 3$. Labelling the vertices as in the case of D, we obtain a presentation with generators $x_1, \ldots, x_5$ and relators

$$\begin{array}{c} x_1^2,\ x_2^2,\ x_3^2,\ x_4^2,\ x_5^2,\ (x_1x_3)^2,\ (x_2x_5)^2,\ (x_5x_4)^2,\ (x_4x_2)^2, \\ (x_1x_2)^3,\ (x_1x_5)^3,\ (x_1x_4)^3,\ (x_3x_2)^3,\ (x_3x_5)^3,\ (x_3x_4)^3. \end{array}$$

Since the result of identifying vertices x_2, x_5, x_4 in E is just the Dynkin diagram A_3 of S_4, it follows that E_0 has the group $\langle x_1, x_3, x_5\rangle \cong S_4$ as a retract. Thus, E_0 is a semi-direct product $B_0 \rtimes S_4$, where B_0 is the normal closure in E_0 of $\{x_5x_2^{-1}, x_5x_4^{-1}\}$. Since the adjoined Coxeter elements all lie in B_0, a similar conclusion holds for B_i, $i = 1, 2, 3$. It thus suffices to identify the groups B_i, $0 \leq i \leq 3$, and we state the following without proof.

Proposition 5.

(i) *For $0 \leq i \leq 3$, E_i is a split extension $B_i \rtimes S_4$.*

(ii) *B_0, B_1 have minimal presentations in which all relators are squares, and $B_0/B_0' \cong Z_2^{10}$, $B_0'/B_0'' \cong Z_2^{91} \times Z^{27}$, $B_1/B_1' \cong Z_2^8$, $B_1'/B_1'' \cong Z_2^9 \times Z^3$.*

(iii) *B_2 has derived factor group Z_2^6 and derived group $B_2' \cong Z^4 \rtimes Z_2$, where the involution acts by inversion.*

(iv) *(cf. Proposition 4) $B_3 \cong (D_8 \curlyvee D_8) \times Z_2 \times Z_2$.*

3. Since Coxeter groups are residually finite and have only finitely many conjugacy classes of elements of finite order, it follows that they all have a torsion-free subgroup of finite index. We conclude with two questions about the maximal elements of the sublattice of such subgroups. Firstly, to what extent are they unique? Secondly, what connection (if any) exists between such subgroups and graph groups [7] or graph products of groups [10]?

References

[1] C.T. Benson and L.C. Grove, *Finite reflection groups* (Bogden and Quigley, Tarrytown 1971).

[2] N. Bourbaki, *Groupes et algèbres de Lie.* Chs. iv-vi (Hermann, Paris, 1968).

[3] F.C. Bussemaker, P.J. Cameron, J.J. Seidel and S.V. Tsaranov, *Tables of signed graphs* (EUT Report 91-WSK-01, Eindhoven Inst. of Tech., 1991).

[4] P.J. Cameron, Two-graphs and trees, in *Proc. 2nd Japan Conf. Graph Theory and Combinatorics*, Hakone, 1990), to appear.

[5] P.J. Cameron, J.J. Seidel and S.V. Tsaranov, Signed graphs, root lattices, and Coxeter groups, *J. Algebra*, to appear.

[6] J.J. Cannon, An introduction to the group theory language, Cayley, in *Computational group theory* (M. D. Atkinson (ed.), Academic Press, London, 1984), 145–183.

[7] C. Droms, Isomorphisms of graph groups, *Acta Math. Acad. Sci. Hungar.* **38**(1981), 19–28.

[8] E.R. Green, *Graph products of groups*, PhD thesis, University of Leeds, 1990.

[9] G. Havas and L.S. Sterling, Integer matrices and abelian groups, *Symbolic and algebraic computation* (E.W.Ng (ed.), Lecture Notes in Computer Science, **72**, Springer, Berlin, 1979), 431–451.

[10] J.E. Humphreys, *Reflection groups and Coxeter groups* (Cambridge University Press, Cambridge, 1990).

[11] I.D. Macdonald, A computer application to finite p-groups, *J. Austral. Math. Soc.* **17**(1974), 102–112.

[12] H. Meinert, The higher geometric invariants of direct products of virtually free groups, preprint, Universität Frankfurt, 1992.

[13] M.F. Newman, Calculating presentations for certain kinds of quotient groups, in *SYMSAC '76, Proceedings of the 1976 ACM symposium on symbolic and algebraic computation* (R.D. Jenks (ed.), Assoc. Comput. Mach., New York, 1976), 2–8.

[14] M. Schönert et al., *GAP – groups, algorithms, and programming*, 2nd ed. (Lehrstuhl D für Mathematik, RWTH Aachen, 1993).

[15] J.J. Seidel and S.V. Tsaranov, Two-graphs, related groups, and root systems, *Bull. Soc. Math. Belgique* **42**(1990), 695–711.

[16] *SPAS – subgroup presentation algorithms system, version 2.5. User's reference manual* (Lehrstuhl D für Mathematik, RWTH Aachen, 1989).

[17] S.V. Tsaranov, On a generalization of Coxeter groups, *Algebras, Groups and Geometries* **6**(1989), 281–318.

N-FREE GROUPS AND QUESTIONS ABOUT UNIVERSALLY FREE GROUPS

BENJAMIN FINE*, ANTHONY M. GAGLIONE†1,
GERHARD ROSENBERGER‡ and DENNIS SPELLMAN*

*Fairfield University, Fairfield, CT, U.S.A.
†United States Naval Academy, Annapolis, MD, U.S.A.
‡Universität Dortmund, Dortmund, Germany
*Philadelphia, PA, U.S.A.

Abstract

This paper produces examples of groups which provide negative answers to earlier questions of Gaglione and Spellman. One such example is a consequence of some new results on constructing n-free groups which should also be of independent interest. Specifically, the primary focus is on answering: (1) Does there exist an integer $r \geq 3$ such that every finitely-generated, non-abelian r-free group is a model of the elementary theory of the non-abelian free groups? In particular, does $r = 3$ satisfy this condition? and (2) Let G be a non-abelian group. Is it the case that G satisfies precisely the same universal sentences as the non- abelian free groups if and only if there is an ordered abelian group Λ and a Λ-tree T such that G acts via Λ-isometries on T freely and without inversions?

1. Introduction and preliminaries

For a positive integer n a group G is n-free if every subgroup generated by n or fewer distinct elements is free (necessarily if rank $\leq n$).

G is *locally free* if it is n-free for all positive integers n; moreover, following Graham Higman [15], G is *countably free* if every countable subgroup is free. From straightforward topological considerations an orientable surface group of genus g is $(2g-1)$-free. This was generalized by B. Baumslag [2] to show that certain cyclically pinched one relator groups are 2-free and extended by Rosenberger [19] to show that these groups are also 3-free (redone in a different manner by G. Baumslag and P. Shalen [4]). More recently, Fine, Rohl and Rosenberger [8], [9] showed that certain HNN extensions of free groups are 2-free and 3-free.

In a different direction, Gaglione and Spellman in a series of papers [10], [11] and [12] studied groups which have in common various first-order properties of

[1]The research of this author was partially supported by the Naval Research Laboratory. Identification Systems Branch, Radar Division.

the non-abelian free groups. Let L_o be the first-order language with equality containing a binary operation symbol $\cdot$, a unary operation symbol $^{-1}$ and a constant symbol 1. A *universal sentence* (*existential sentence*) of L_o is one of the form $\forall\overline{x}\phi(\overline{x})$ ($\exists\overline{x}\phi(\overline{x})$) where $\overline{x}$ is a tuple of distinct variables, $\phi(\overline{x})$ is a formula of L_o containing no quantifiers and containing at most the variables in $\overline{x}$. Since the negation of a universal sentence is logically equivalent to an existential sentence and vice-versa, two L_o-structures A and B satisfy precisely the same universal sentences of L_o if and only if they satisfy precisely the same existential sentences of L_o. We say that A and B have the *same universal theory* in that case. It is well known that any two non-abelian free groups have the same universal theory. Furthermore, we define for an L_o-structure A the *(first-order) theory* of A to be the set $Th(A)$ of all sentences of L_o true in A. Vaught has shown that $Th(A) = Th(B)$ whenever A and B are free groups each of infinite rank and Tarski has conjectured that $Th(A) = Th(B)$ whenever A and B are any two non-abelian free groups whatsoever.

Let $\forall$ be the set of universal sentences of L_o and $\exists$ be the set of existential sentences of L_o. For each cardinal r let F_r be a free group of rank r. Let ω be the first limit ordinal so that $\omega = \aleph_0$. To conform to the notation of earlier papers of Gaglione and Spellman [10] and [11], we put $\Phi = Th(F_2) \cap (\forall \cup \exists)$ and $\Sigma = \bigcap_{2 \leq r \leq \omega} Th(F_r)$.

Thus, a group has the same universal theory as some non-abelian free group (hence any non-abelian free group) if and only if it is a model of Φ. We call models of Φ *universally free groups* while Remeslennikov [18] calls them $\exists$-free groups. (For a complete treatment of models see Chang and Keisler [6].) Moreover, Σ is the set of all sentences of L_o true in every non-abelian free group. In particular, $\Phi \subseteq \Sigma$. Writing $\mathbf{M}(S)$ for the class of all models of a set S of sentences of L_o, we conclude that $\mathbf{M}(\Sigma) \subseteq \mathbf{M}(\phi)$.

It is an easy consequence of Theorem 5.2.6., p. 308 and Exercise 5.2.9., p.319 of Chang and Keisler [6] that every non-abelian, locally free group is universally free. Furthermore, it follows easily from Theorem 3.1.6., p.138 of Chang and Keisler [6] (a version of the Löwenheim-Skolem Theorem) that every non-abelian, countably free group is a model of Σ. (A somewhat more delicate argument with the conclusion that every non-abelian, non-free, countably free group is a model of the set of sentences of L_o true in every infinite rank free group appears in Gaglione and Spellman [11]. Pending verification of the Tarski conjecture this conclusion is apparently stronger than being a model of Σ.) In this paper we provide a negative answer to the following question of Gaglione and Spellman [10]: "Does there exist an integer $r \geq 3$ such that every finitely-generated, non-abelian r-free group is a model of Σ. ? In particular, does $r = 3$ satisfy this condition?" We also answer negatively the following question of Gaglione and Spellman [13]: "Let G be a non-abelian group. Is it the case that G is a model of Σ if and only if there is an ordered abelian group Λ and a Λ- tree T such that G acts freely on T without inversions?" (Bass [1] calls such groups *tree-free.*) A different counterexample,

$$G =< x, y, t; tx^2y^2t^{-1} = xyx^{-1}y^{-1} >,$$

of a non-abelian, tree-free group which is not universally free is given by Remeslennikov [18]. (We note that every universally free group is tree-free, see [14].)

Motivated by these examples we proved the following general result:

Theorem 4.1. *Let G be a one-relator group with presentation*

$$G =< B_1, \ldots, B_n; \prod_{i=1}^{n} W_i = 1 >$$

where $B_1, \ldots, B_n$ are nonempty pairwise disjoint subsets of generators each of size ≥ 2 and for $i = 1, \ldots, n$ we have that $W_i = W_i(B_i) \neq 1$ are nontrivial words neither proper powers nor primitive elements in the free group on B_i. Then G is n-free.

In this paper $< \ldots; \ldots >$ indicates a presentation of a group in terms of generators and defining relations. We reserve $< a_1, \ldots, a_n >$ for the subgroup of a group generated by the elements $a_1, \ldots, a_n$. For groups A and B, $A * B$ indicates their free product. Finally, if G is any group, we write the *commutator* of $x, y \in G$ as $[x, y] = xyx^{-1}y^{-1}$. We now make explicit the constructions of 3-free groups to which we previously alluded. Suppose p and q are integers with $\min\{p, q\} \geq 2$. Let $F =< a_1, \ldots, a_p; >$ be free of rank p and $\overline{F} =< b_1, \ldots, b_q; >$ be free of rank q. Suppose $1 \neq u = u(a_1, ..., a_p)$ is not a proper power in F and $1 \neq v = v(b_1, \ldots, b_q)$ is not a proper power in $\overline{F}$. Rosenberger [19] has shown:

Theorem 1.1. (Rosenberger) *Under the above hypotheses, the free product with amalgamation $K =< F * \overline{F}; u = v >$ is 3-free.*

Fine, Rohl and Rosenberger [9] call a maximal two-generator subgroup $N =< u, v >$ of a group G a *strongly maximal two-generator subgroup* provided for each $x \in G$ there is $y \in G$ such that $< u, xvx^{-1} > \subseteq < u, yvy^{-1} >$ and $< u, yvy^{-1} >$ is a maximal two-generator subgroup of G. They proved

Theorem 1.2. (Fine, Rohl and Rosenberger) *Let F be a free group, u and v be non-trivial elements of F such that $< u, v >$ is a strongly maximal two-generator subgroup of F and let K be the HNN extension $< F, t; t^{-1}ut = v >$. Then K is 3-free.*

Related to Theorem 1.2, we have the following result which appeared in [8]:

Theorem 1.3. (Fine and Rosenberger) *Let F be a free group and let u and v be non-trivial elements of F which generate a maximal two-generator subgroup of F. Also let K be the HNN extension $< F, t; t^{-1}ut = v >$. Then K is 2-free.*

We end our preliminaries with a theorem which we shall need that appeared in [11]:

Theorem 1.4. (Gaglione and Spellman) *Every finitely presented universally free group is residually free.*

We note that Remeslennikov [18] has proven the stronger result that every finitely-generated universally free group is residually free.

2. Two more questions

In [11] Gaglione and Spellman posed a pair of questions such that a positive answer to either would imply the existence of a finitely presented, non-abelian 3-free group which is not universally free. We shall show that both questions have positive answers.

Question 2.1. (Gaglione and Spellman [11]) Does there exist a finitely presented, non-abelian 3-free group all of whose free homomorphic images are abelian?

Suppose G were such a group with presentation

$$< a_1, \ldots, a_m; R_1(a_1, \ldots, a_m) = \ldots = R_n(a_1, \ldots, a_m) = 1 > \tag{2.1}$$

Then the universal sentence

$$\forall x_1 \ldots \forall x_m (\bigwedge_{i=1}^{n} (R_i(x_1, \ldots, x_m) = 1 \rightarrow \bigwedge_{i<j} (x_i x_j = x_j x_i)) \tag{2.2}$$

would be true in every free group but false in G. To see that (2.2) would be true in any free group, let F be a free group. Let $(x_1, \ldots, x_m)$ be a fixed ordered m-tuple of elements of F. Either

(a) $(\bigwedge_{i=1}^{n}(R_i(x_1, \ldots, x_m) = 1)$ is true in F, or

(b) $(\bigwedge_{i=1}^{n}(R_i(x_1, \ldots, x_m) = 1)$ is false in F.

If case (b) holds, then the conditional

$$\bigwedge_{i=1}^{n}(R_i(x_1,\ldots,x_m)=1) \to \bigwedge_{i<j}(x_ix_j=x_jx_i)$$

is vacuously true in F. If, on the other hand, (a) holds, then the mapping $a_i \mapsto x_i (1 \leq i \leq m)$ determines an epimorphism from G onto the subgroup $\overline{F} = < x_1,\ldots,x_m >$ of F. By the Nielsen-Schreier subgroup theorem, $\overline{F}$ is free. Therefore, by the hypotheses G is constrained to satisfy, $\overline{F}$ is abelian. Hence, the conditional

$$\bigwedge_{i=1}^{n}(R_i(x_1,\ldots,x_m))=1 \to \bigwedge_{i<j}(x_ix_j=x_jx_i)$$

is verified in case (a) also. Since $(x_1,\ldots,x_m)$ was an arbitrary m-tuple of elements of F we must have the conditional verified in all instances. Therefore (2.2) holds in F.

Thus, we have proven:

Proposition 2.1. *A positive answer to Question 2.1 implies the existence of a finitely presented, non-abelian 3-free group which is not universally free, i.e., the group described in Question 2.1 is not universally free.*

Definition 2.1. Let F be a free group. Let u and v be elements of F such that

(1) u and v are not conjugate in F;
(2) u and v do not commute;
(3) for every endomorphism $h : F \to F$ such that $h(u)$ and $h(v)$ do not commute it is the case that $h(u)$ and $h(v)$ are not conjugate in F.

Then we say that u and v are F*-strongly-non-conjugate.*

Question 2.2. (Gaglione and Spellman [11]) Does there exist an integer $r \geq 3$ and a free group $F = < a_1,\ldots,a_r; >$ containing elements $u = u(a_1,\ldots,a_r)$ and $v = v(a_1,\ldots,a_r)$ such that

(1) u and v generate a strongly maximal 2-generator subgroup of F and
(2) u and v are F-strongly-non-conjugate?

If the above Question 2.2 has an affirmative answer, then the universal sentence (depending on r, u and v)

$$\begin{aligned}&\forall y \forall x_1 \ldots \forall x_r((v(x_1,\ldots,x_r)=y^{-1}u(x_1,\ldots,x_r)y)\\ &\to (u(x_1,\ldots,x_r)v(x_1,\ldots,x_r)=v(x_1,\ldots,x_r)u(x_1,\ldots,x_r)))\end{aligned} \tag{2.3}$$

would be true in every non-abelian free group but false in

$$K = < F, t; t^{-l}ut = v > . \tag{2.4}$$

To see that (2.3) would be true in any non-abelian free group, it suffices to show that it is true in $F = < a_1, \ldots, a_r; >$ (recall $r \geq 3$) since (2.3) is a universal sentence and the non-abelian free groups have the same universal theory. To that end, let $(x_1, \ldots, x_r)$ be a fixed ordered r-tuple of elements of F and y a fixed element of F. Let the endomorphism $h : F \to F$ be determined by $a_i \mapsto x_i (1 \leq i \leq r)$. Writing $\bar{u}$ for the image $u(x_1, \ldots, x_r)$ of $u = u(a_1, \ldots, a_r)$ and $\bar{v}$ for the image $v(x_1, \ldots, x_r)$ of $v = v(a_1, \ldots, a_r)$, we see that either

(a) $\overline{u}\,\overline{v} \neq \overline{v}\,\overline{u}$ or

(b) $\overline{u}\,\overline{v} = \overline{vu}$.

In case (b), the conditional

$$\bar{u}\bar{v} \neq \bar{v}\bar{u} \to (\overline{v} \neq y^{-1}\overline{u}y)$$

is vacuously true in F. In case (a) since u and v are F-strongly-non-conjugate, the conditional

$$\bar{u}\bar{v} \neq \bar{v}\bar{u} \to (\overline{v} \neq y^{-1}\overline{u}y)$$

is also verified. Thus, for arbitrary elements $y, x_1, \ldots, x_r \in F$, the conditional $\bar{u}\bar{v} \neq \bar{v}\bar{u} \to (\overline{v} \neq y^{-1}\overline{u}y)$ and its contrapositive $(\overline{v} = y^{-1}\overline{u}y) \to \bar{u}\bar{v} = \bar{v}\bar{u}$ hold in F. Hence (2.3) holds in every non-abelian free group. We also note that K given in (2.4) is 3-free by Theorem 1.2. Thus we have proven:

Proposition 2.2. *A positive answer to Question 2.2 implies the existence of a finitely presented, non-abelian 3-free group which is not universally free, i.e., the group K constructed from the elements u and v and the free group F of Question 2.2 is not universally free.*

3. The counterexamples

Let (g, n) be an ordered pair of integers with $min\{g, n\} \geq 2$. Let $G(g, n)$ be the group with presentation

$$G(g, n) = < a_1, \ldots, a_{2g-2}, b_1, b_2; \prod_{i=1}^{g-1} [a_{2i-1}, a_{2i}] = [b_1, b_2] b_1^n >$$

By Theorem 1.1, $G(g, n)$ is 3-free. Furthermore in [3] B. Baumslag, F. Levin and G. Rosenberger showed that $G(2, n)$ is not residually free whenever $n \geq 4$.

However, Bass [1] has shown that every cyclically pinched product of free groups $G = < F * \overline{F}; u = v >$ with $u \in F$ not a proper power in F and $v \in \overline{F}$

not a proper power in $\bar{F}$ acts freely without inversions on a $(\mathbf{Z} \times \mathbf{Z})$-tree (lexicographic order on $\mathbf{Z} \times \mathbf{Z}$), i.e., such a group is tree-free.

Thus, not only is $G(2,4)$ a finitely presented, non-abelian 3-free group which is not universally free (according to Theorem 1.4), it is also a non-abelian tree free group which is not universally free. This gives a negative answer to the Question posed in Section I. We shall return to the groups $G(g, 2g)$ after constructing examples yielding positive answers to the Questions posed in Section II.

Let $F = < a, b, c; >$ be free on a, b, c. Let $u = b$ and $v = [a, c](b^2a^2c^2)^5b$. Certainly $< u, v >$ is a maximal two-generator subgroup of F because $u = b$ is primitive in F. If $< u, xvx^{-1} >$ with $x \in F$ is contained in the maximal two-generator subgroup $< g, h >, g, h \in F$, then $\{g, h\}$ is Nielsen equivalent to a pair $\{b, k\}$ because $u = b$ is primitive in F. We may assume that k as a freely reduced word in a, b, c neither starts nor ends with a non-trivial power of b, that is, that $\{b, k\}$ is Nielsen- reduced. Now let $\overline{F} = < \overline{a}, \overline{c}; >$ be free on a, c. Let $\rho : F \to \overline{F}, f \overset{\rho}{\to} \overline{f}$ be the epimorphism determined by

$$\begin{cases} a \mapsto \overline{a} \\ b \mapsto 1 \\ c \mapsto \overline{c} \end{cases}$$

Let N be the kernel of ρ. Under ρ the subgroups $< b, xvx^{-1} > = < u, xvx^{-1} > \subseteq < g, h > = < b, k >$ are mapped onto $< \bar{x}\bar{v}\bar{x}^{-1} > \subseteq < \overline{k} >$. But $\overline{v} = [\overline{a}, \overline{c}](\overline{a}^2\overline{c}^2)^5 \neq 1$ is not a proper power in F; so, no conjugate $xv^\epsilon x^{-1}$ in $\overline{F}$ can be either. Hence, $< \bar{x}\bar{v}\bar{x}^{-1} > \subsetneq < \overline{k} >$ is impossible. Therefore, $k \equiv xv^\epsilon x^{-1} (mod N)$ where $\epsilon = \pm 1$. Writing now xvx^{-1} as a freely reduced word in b and k gives $< b, k > = < b, yvy^{-1} >$ for a suitable $y \in F$. It follows that $< u, v >$ is a strongly maximal two-generator subgroup of F. Therefore, by Theorem 1.2, $K = < F, t; t^{-1}ut = v >$ is 3-free.

Now let F_n be free of rank n and let $h : K \to F_n, k_1 \overset{h}{\mapsto} k$ be an epimorphism. We wish to show that $n \leq 1$. F_n could not be free on $\bar{a}, \bar{b}, \bar{c}$ and $\bar{t}$ since these satisfy the relation $\overline{t}^{-1}\overline{u}\overline{t} = \overline{v}$ in F_n. Moreover, $n = rank(F_n)$ is the minimum cardinality of a set of generators for the finitely-generated free group F_n. It follows that, a priori, $0 \leq n \leq 3$. The equation $\overline{t}^{-1}\overline{u}\overline{t} = \overline{v}$ is written out as $\overline{t}^{-1}\overline{b}\overline{t} = [\overline{a}, \overline{c}](\overline{b}^2\overline{a}^2\overline{c}^2)^5\overline{b}$, which may be rearranged to yield $[\overline{c}, \overline{a}][\overline{t}^{-1}, \overline{b}] = (\overline{b}^2\overline{a}^2\overline{c}^2)^5$.

But J. Comerford, L. P. Comerford and C. C. Edmunds [5] have shown that if $m > 0$ is an integer and $[x_1, y_1][x_2, y_2] = A^m \neq 1$ in a free group, then $m \leq 3$. Thus, $\overline{b}^2\overline{a}^2\overline{c}^2 = 1$ in F_n. Now by a result of Lyndon and Schutzenberger [17], if $M, N, P \geq 2$ are integers and $x^N y^M z^P = 1$ in a free group, then x, y and z are constrained to lie in a common cyclic subgroup. Hence, $< \overline{b}, \overline{a}, \overline{c} >$ is cyclic, from which we infer $[\overline{a}, \overline{c}] = 1$. Therefore, $[\overline{t}^{-1}, \overline{b}] = 1$ and $F_n = < \overline{a}, \overline{b}, \overline{c}, \overline{v} >$ has no choice other than to be abelian. That is, F_n is cyclic and $0 \leq n \leq 1$.

In conclusion, $K =< F, t; t^{-1}ut = v >$ with $F =< a, b, c; >, u = b$ and $v = [a, c](b^2a^2c^2)^5b$ is a finitely presented, non-abelian 3-free group whose only free homomorphic images are abelian. Such a group K cannot be universally free according to Proposition 2.1.

We now tackle the second of the two questions posed in Section II. This time let F $= < a, b, c, d; >$ be free on a, b, c and d. Let $u = [a, b]$ and $v = [c, d]^2a^7$. We shall first establish that $< u, v >$ is a maximal two-generator subgroup of F. To that end let $x, y \in F$ and assume $< u, v >$ is contained in $< x, y >$. Write $F = F_1 * F_2$ with $F_1 =< a, b; >$ and $F_2 =< c, d; >$. Apply Nielsen's cancellation method with respect to the syllable length L in $F_1 * F_2$ and a suitable ordering. Recalling that $[a, b]$ cannot be a proper power in F by a result of Schutzenberger [24], we deduce from $u \in F_1, L(v) = 2$ that $\{x, y\}$ is Nielsen equivalent to a Nielsen reduced pair $\{x', y'\}$ with $x' = u$ and $L(y') = 2$. Representing v in $x' = u$ and y' gives that $\{x', y'\}$ is Nielsen equivalent to $\{u, v\}$. Thus, $< u, v >$ is a maximal two-generator subgroup of F as claimed. In fact $< u, v >$ is a strongly maximal subgroup of F. If $< u, zvz^{-1} >, z \in F$, is contained in $< g, h >$ with $g, h \in F$ then $\{g, h\}$ is Nielsen equivalent to a Nielsen reduced pair $\{u, w\}, w \in F$, because $L(u) = 1$. Replacing u by a conjugate $kuk^{-1}, k \in F_1$, and z by $k^{-1}z$ if necessary, we may assume that $z = 1$ or $z \neq 1$ and, as a reduced word in the free product $F_1 * F_2, z$ does not start with a non-trivial element of F_1. We now replace $\{u, w\}$ by $\{z^{-1}uz, z^{-1}wz\}$. Recall that we get $v \in< z^{-1}uz, z^{-1}wz >$. We may apply the Nielson reduction method so that at each step $z^{-1}uz$ remains unchanged or is replaced by a conjugate to get from $\{z^{-1}uz, z^{-1}wz\}$ a Nielsen reduced pair $\{z_1uz_1^{-1}, w_1\}, z_1, w_1 \in F$. Representing v in $z_1uz_1^{-1}$ and w_1 gives that $\{z_1^{-1}uz_1^{-1}, w_1\}$ is Nielsen equivalent to the pair $\{z_1uz_1^{-1}, v\}$ by the arguments given in the proof of Theorem (2.1) from [20]. Hence, $\{u, v\}$ is Nielsen equivalent to a pair $\{u, z_2vz_2^{-1}\}, z_2 \in F$, and $< u, v >$ is a strongly maximal two-generator subgroup of F. We wish to show that u and v are F- strongly-non-conjugate. To that end we shall demonstrate the apparently stronger condition that if $h : F \to F, x \overset{h}{\mapsto} \overline{x}$ is any endomorphism of F such that $\overline{u}$ and $\overline{v}$ do not commute, then no conjugate of $\overline{u}$ commutes with $\overline{v}$. $[\overline{u}, \overline{v}] \neq 1$ implies $\overline{u} \neq 1$ and $\overline{v} \neq 1$. But $\overline{u} = [\overline{a}, \overline{b}] \neq 1$ so $\overline{a} \neq 1$ also. Assume there is an element $x \in F$ with $[x\overline{u}x^{-1}, \overline{v}] = 1$. Then $\overline{v}$ is a power of $x\overline{u}x^{-1}$ because $1 \neq x\overline{u}x^{-1} = [x\overline{a}x^{-1}, x\overline{b}x^{-1}]$, so $x\overline{u}x^{-1}$ cannot be a proper power in F. That is, $[\overline{c}, \overline{d}]^2a^7 = \overline{v} = [x\overline{a}x^{-1}, x\overline{b}x^{-1}]^n$ for some $n \in \mathbf{Z}$. If $|n| \geq 2$, then by a result of Lyndon and Schutzenberger [17], $\overline{a}$ must be a power of $[x\overline{a}x^{-1}, x\overline{b}x^{-1}]$, which we claim is impossible. To see that $\overline{a}$ must be a power of $[x\overline{a}x^{-1}, x\overline{b}x^{-1}]$ we note that $\overline{a}^7[\overline{d}, \overline{c}]^2[x\overline{a}x^{-1}, x\overline{b}x^{-1}] = 1$ which (if $|n| > 1$) implies that $\overline{a}, [\overline{d}, \overline{c}]$ and $[x\overline{a}x^{-1}, x\overline{b}x^{-1}]$ lie in a cyclic subgroup by the result of Lyndon and Schutzenberger previously stated. But $[x\overline{a}x^{-1}, x\overline{b}x^{-1}]$, being a non-trivial commutator, cannot be a proper power. Thus, it generates the cyclic subgroup, whence $\overline{a}$ is a power of $[x\overline{a}x^{-1}, x\overline{b}x^{-1}]$. To see that this is impossible we note that if $\overline{a} \in \gamma_w(F) - \gamma_{w+1}(F)$ has weight w, then $[x\overline{a}x^{-1}, x\overline{b}x^{-1}] \neq 1$ can have weight no smaller than $w+1$. Hence, we have $n = \epsilon = \pm 1$ and $\overline{a}^7 = [\overline{d}, \overline{c}]^2[x\overline{a}x^{-1}, x\overline{b}x^{-1}]^\epsilon$.

However, Duncan and Howie [7] have generalized the result of Comerford, Comerford and Edmunds to show that if $m > 0$ is an integer and $[x_1, y_1] \ldots [x_g, y_g] = A^m \neq 1$ in a free group, then $m \leq 2g - 1$. Applying this result with $g = 3$ and observing that $\overline{a}^7 \neq 1$ is the product of three commutators, we deduce a contradiction. Thus, we conclude that $[x\overline{u}x^{-1}, \overline{v}] \neq 1$ for all $x \in F$. We have succeeded in showing that if $F =< a, b, c, d; >, u = [a, b]$ and $v = [c, d]^2 a^7$, then u and v are a pair of F-strongly-non-conjugate elements of F which generate a strongly maximal two-generator subgroup of F. It follows that if $K =< F, t; t^{-1}ut = v >$ where F, u and v are as above, then K is a finitely presented, non- abelian 3-free group by Theorem 1.2, which is not universally free according to Proposition 2.2.

4. n-free groups - another counterexample

An orientable surface group T_g of genus g has a one relator presentation

$$T_g =< a_1, \ldots, a_{2g}; \prod_{i=1}^{g} [a_{2i-1}, a_{2i}] = 1 >$$

This has rank $2g$. If H is a subgroup of T_g of infinite index, then it is the fundamental group of an infinite cover of a surface of genus g, and therefore H is a free group. If H is a subgroup of T_g of finite index, then H is the fundamental group of a finite cover of a surface of genus g, and therefore it is a surface group of genus $g_1 \geq g$ - and hence of rank no less than that of T_g. Therefore, if $rank(H) \leq 2g - 1$, then H cannot have finite index and H must be a free group. Thus, T_g is $(2g - 1)$ - free. A non-orientable surface group of genus g has a one relator presentation

$$S_g =< a_1, \ldots, a_g; \prod_{i=1}^{g} a_i^2 = 1 >$$

and an identical argument shows that this must be $(g-1)$-free. Generalizing these results algebraically, we obtain

Theorem 4.1. *Let G be a one-relator group with presentation*

$$G =< B_1, \ldots, B_n; \prod_{i=1}^{n} W_i = 1 >$$

where $B_1, \ldots, B_n$ are non-empty pairwise disjoint subsets of generators each of size ≥ 2 and for $i = 1, \ldots, n$ we have that $W_i = W_i(B_i) \neq 1$ are non-trivial *words neither proper powers nor primitive elements in the free group on B_i. Then G is n-free.*

Theorem 4.2. *Let G be a one-relator group with presentation*

$$G =< B_1, \ldots, B_n; \prod_{i=1}^{n} V_i^{t_i} = 1 >$$

where $B_1, \ldots, B_n$ are non-empty pairwise disjoint subsets of generators each of size ≥ 2 and for $i = 1, \ldots, n$ we have that $V_i = V_i(B_i) \neq 1$ are non-trivial words neither proper powers nor primitive elements in the free group on B_i and $t_i \geq 1$. Then G is (n-1)-free.

Before proving these two theorems we observe the following consequence of Theorem 4.1.

Corollary 4.3. *Let $g \geq 2$ be an integer. The group $G(g, 2g)$ generated by $a_1, \ldots, a_{2g-2}, b_1, b_2$ subject to the single defining relation $\prod_{i=1}^{g-1}[a_{2i-1}, a_{2i}] = [b_1, b_2]b_1^{2g}$ is g-free.*

PROOF. Put

$$\begin{aligned} B_i &= \{a_{2i-1}, a_{2i}\}, i = 1, \ldots, g-1 \\ B_g &= \{b_1, b_2\} \\ W_i &= [a_{2i-1}, a_{2i}], i = 1, \ldots, g-1 \\ W_g &= b_1^{-2g}[b_2, b_1] \end{aligned}$$

in Theorem 4.1. □

$G(g, 2g)$ however is not universally free since it violates the theorem of Duncan and Howie [7]

$$\forall x_1 \forall y_1 \ldots \forall x_g \forall y_g \forall z((\prod_{i=1}^{g}[x_i, y_i] = z^{2g}) \rightarrow (z = 1))$$

valid in every free group. Hence, although every non-abelian, locally free group is universally free, for every integer $n > 0$, however large, there is a finitely presented, non-abelian n-free group that fails to be universally free - let alone a model of $\Sigma = \bigcap_{2 \leq r \leq \omega} Th(F_r)$.

PROOF OF THEOREM 4.1: Our proof will be based on Nielsen reduction in free products as introduced by Zieschang [25]. We need the following two technical lemmas.

The first lemma is a direct analogue of Theorem 2.12 using Theorem 2.11 in [21].

Lemma 1. *Let F be a free group with free generating system B. Let $n \geq 1$ be an integer and suppose that $B_1, \ldots, B_n$ are non-empty, pairwise disjoint subsets of B. For each integer $1 \leq j \leq n$ let F_j be the subgroup of F generated by B_j. Let $W_j \in F_j - \{1\}$ be neither a proper power nor a primitive element in F_j. Let $W = W_1 \ldots W_n$. For any integer $1 \leq m \leq 2n$ let $\{x_1, \ldots, x_m\}$ be any system in F and put $X = < x_1, \ldots, x_m >$.*

If $W^\alpha \in X$ for some non-zero integer α, then one of the following cases occurs,

(a) *$\{x_1, \ldots, x_m\}$ is Nielsen equivalent to a system $\{y_1, \ldots, y_m\}$ with $y_1 = W^\beta$ and β is the smallest positive integer for which a relation $W^\beta \in X$ occurs.*

(b) *We have $rank(X) = m = 2n$ and $\{x_1, \ldots, x_{2n}\}$ is Nielsen equivalent to a system $\{y_1, \ldots, y_{2n}\}$ with $y_{2i-1}, y_{2i} \in F_i$ and $W_i \in < y_{2i-1}, y_{2i} >$ for $i = 1, \ldots, n$.*

The proof of the lemma is the same as the proof of Theorem 2.12 in [21] in conjunction with Theorem 2.11.

Lemma 2. (Zieschang [25], Rosenberger [23]) *Let $G = G_1 *_A G_2$ and assume that a length L and order are introduced in G as in [25] and [23]. If $\{x_1, \ldots, x_m\}$ is a finite system of elements in G, then there is a Nielsen transformation from $\{x_1, \ldots, x_m\}$ to $\{y_1, \ldots, y_m\}$ for which one of the following cases holds.*

(i) *Each $w \in < y_1, \ldots, y_m >$ can be written as*

$$\prod_{i=1}^{q} y_{\nu_i}^{\epsilon_i},$$

$\epsilon = \pm 1$, $\epsilon_i = \epsilon_{i+1}$ if $\nu_i = \nu_{i+1}$ with $L(y_{\nu_i}) \leq L(w)$ for $i = 1, \ldots, q$.

(ii) *There is a product*

$$a = \prod_{i=1}^{q} y_{\nu_i}^{\epsilon_i}, a \neq 1$$

with $y_{\nu_i} \in A (i = 1, \ldots, q)$, and in one of the factors G_j there is an element $x \in G_j$ with $x^{-1}ax \in A$.

(iii) *Of the y_i there are $p \geq 1$ contained in a subgroup of G conjugate to G_1 or G_2 and a certain product of them is conjugate to a non-trivial element of A.*

(iv) *There is a $g \in G$ such that for some $i \in \{1, \ldots, m\}$ we have $y_i \notin gAg^{-1}$ but for a suitable positive integer k we have $y_i^k \in gAg^{-1}$.*

The Nielsen transformation can be chosen so that $\{y_1, \ldots, y_m\}$ is smaller than $\{x_1, \ldots, x_m\}$ or the lengths of the elements of $\{x_1, \ldots, x_m\}$ are preserved.

Further if $\{x_1, \ldots, x_m\}$ is a generating system for G, then we find that $p \geq 2$ because then the conjugations determine a Nielsen transformation. If we are interested in the combinatorial description of $< x_1, \ldots, x_m >$ in terms of generators and relations in case (iii) we again find that $p \geq 2$ possibly after suitable conjugations.

Now suppose G is as in the statement of the theorem and suppose $X = < x_1, \ldots, x_m >$ is an m-generator subgroup of G. If $n = 1$ that G is 1-free (torsion-free) follows from the Freiheitssatz. If $n = 2$ then the fact that G is 2-free follows directly from Lemma 2. If $n = 3$, then the fact that G is 3-free follows from Rosenberger [19]. Now consider $n \geq 4$. Consider the free product with amalgamation decomposition for G

$$G = G_1^{(1)} *_{A(1)} G_2^{(1)}$$

where $G_1^{(1)} =< B_1, \ldots, Bn_1 >$ is the free group on $B_1 \cup \ldots \cup B_{n-1}$, $G_2^{(1)} = < B_n >$ is the free group with B_n as free basis and $A(1) =< W_1, \ldots, W_{n-1} >= < W_n >$.

Assume first that $m \leq 2n - 2$. If case (i) of Lemma 2 holds, then X is free – so assume that this case doesn't hold. Then from Lemma 1 and Lemma 2 we may assume that one of the following holds.

(a) X is contained in a conjugate of $G_1^{(1)}$.

(b) $x_1 = y(W_1 \ldots W_{n-1})^\alpha y^{-1}$ for some y in G and some $\alpha \neq 0$.

(c) Two of $\{x_1, \ldots, x_m\}$ are in a conjugate of $G_2^{(1)}$.

Replacing $W_1 \ldots W_{n-1}$ by W_n we can say that X is free or we may assume that at least one of $x_1, \ldots, x_m$ is contained in a conjugate of $G_2^{(1)}$. Therefore assume that $x_1 = y_1 z_1 y_1^{-1}$ with z_1 in $G_2^{(1)}$ and y_1 in G. Now change the factorization. Let $G = G_1^{(2)} *_{A(2)} G_2^{(2)}$ where $G_1^{(2)} =< B_1, \ldots, B_{n-2} >$ is the free group on $B_1 \cup \ldots \cup B_{n-2}$, $G_2^{(2)} =< B_{n-1}, B_n >$ is the free group on $B_{n-1} \cup B_n$ and

$$A(2) =< W_1 \ldots W_{n-2} >=< W_{n-1} W_n >$$

Now apply the Nielsen reduction method to this factorization so that if $x_1 = y_1 z_1 y_1^{-1}$ at each step x_1 remains unchanged or is replaced by a conjugate (that this can be done this way is in [16]).

Since $x_1 = y_1 z_1 y_1^{-1}$ is in a conjugate of $G_2^{(1)}$ it is not conjugate to an element of $A(2)$. Assume that now $m \leq 2n - 3$. As before we get that X is free or without loss of generality we may assume that $x_1 = y_1 z_1 y_1^{-1}$ with z_1 in $G_2^{(1)}$ and y_1 in G and $x_2 = y_2 z_2 y_2^{-1}$ with z_2 in $G_2^{(2)}$ and y_2 in G.

Continue in this way by changing factorizations until we have $m \leq n$ and the final factorization $G = G_1^1 *_{A^1} G_2^1$ where $G_1^1 =< B_1 >$ is the free group with B_1 as free basis, $G_2^1 =< B_2, \ldots, B_n >$ is freely-generated by $B_2 \cup \ldots \cup B_n$

and $A^1 = < W_1 > = < W_2 \dots W_n >$. From the constructions in the previous factorizations $\{x_1, \dots, x_m\}$ is free or, without loss of generality, $x_i = y_i z_i y_i^{-1}$ with z_i in $< B_3, \dots, B_n >$ and y_i in G for $i = 1, \dots, n-2$. Applying the Nielsen method to this final factorization we have that X is free or we have the situation where x_{n-1}, x_n are in a conjugate of $< B_1 >$ and all the others are in conjugates of $< B_3, \dots, B_n >$. Since the relator contains elements of $< B_2 >$ from standard cancellation arguments in this final situation X must also be free. □

Remark. The proof, in fact, shows that G is $(n+1)$-free whenever $n \geq 2$.

We conjecture that in general if G is as in Theorem 4.1, then G is $(2n-1)$-free. However there are difficulties extending the proof above to beyond $n+1$.

If the W_i's are proper powers, then Theorem 2.12 in [21] takes the place of Lemma 1.

PROOF OF THEOREM 4.2. If $n = 1$ it follows as before from the Freiheitssatz. If $t_i = 1$ and V_i is primitive in the free group on B_i for some i, then the group is free – so the result is clear.

Assume then that $t_i \geq 2$ if V_i is primitive. The proof then goes through as the proof of Theorem 4.1 using Theorem 2.12 of [21] instead of Lemma 1. □

We note that this result is best possible because, for example, the non-orientable surface group of genus n is not n-free.

References

[1] H. Bass, Group actions on non-Archimedean trees, in *Arboreal Group Theory* (MSRI Pubs., Springer, NY, 1991), 69–131.

[2] B. Baumslag, Generalized free products whose two generator subgroups are free, *J. London Math Soc.* **43**(1968), 601–606.

[3] B. Baumslag, F. Levin, and G. Rosenberger, A cyclically pinched product of free groups which is not residually free, *Math. Z.* **212**(1993), 533–534.

[4] G. Baumslag and P. Shalen, Groups whose three generator subgroups are free, *Bull. Austral. Math. Soc.* **40**(1989), 163–174.

[5] J. Comerford, L.P. Comerford, and C. C. Edmunds, Powers as products of commutators, *Comm. Algebra* **19**(1991), 675–84.

[6] C.C. Chang and H.J. Keisler, *Model Theory*, 3rd edition (North Holland, The Netherlands, 1978).

[7] A. Duncan and J. Howie, The genus problem for one relator products of locally indicable groups, *Math. Z.* **208**(1991), 225–237.

[8] B. Fine and G. Rosenberger, On restricted Gromov groups, *Comm. Algebra,* **20**(1992), 2171–2181.

[9] B. Fine, F. Röhl and G. Rosenberger, On HNN groups whose three generator subgroups are free, in *Proc. of AMS Conf. on Infinite Groups and Group Rings (Tuscaloosa, AL, 1992)* (World Scientific, 1993, N.J.), 13–36.

[10] A.M. Gaglione and D. Spellman, Some model theory of free groups and free algebras, *Houston J. Math.* **19**(1993), 327–356.

[11] A.M. Gaglione and D. Spellman, More model theory of free groups, *Houston J. Math.*, to appear.

[12] A.M. Gaglione and D. Spellman, Even more model theory of free groups, in *Proc. of AMS Conf. on Infinite Groups and Group Rings* (Tuscaloosa, AL, 1992) (World Scientific, N.J., 1993), 37–40.

[13] A.M. Gaglione and D. Spellman, Does Lyndon's length function imply the universal theory of free groups?, *Contemp. Math.*, to appear.

[14] A.M. Gaglione and D. Spellman, Every 'universally free' group is tree-free, in *Proc. Ohio State-Denison Conf. for H. Zassenhaus, 1992* (World Scientific, N.J., 1993), 149–154.

[15] G. Higman, Almost free groups, *Proc. London Math. Soc.*, **3**(1951), 284–290.

[16] R.N. Kalia and G. Rosenberger, Über Untergruppen ebener diskontinuierlicher Gruppen, *Contemp. Math.* **33**(1984), 308–327.

[1] R.C. Lyndon and M.P. Schützenberger, The equation $a^M = b^N c^P$ in a free group, *Michigan Math. J.* **9**(1962), 289–298.

[18] V.N. Remeslennikov, $\exists$ -free groups and groups with a length function, preprint.

[19] G. Rosenberger, On one relator groups that are free products of two free groups with cyclic amalgamation, in *Proc. of Groups St. Andrews* (Cambridge U. Press, 1981), 328–344.

[20] G. Rosenberger, Über Darstellungen von Elementen und Untergruppen in freien Produkten in *Proc. of Groups Korea 1983* (Lecture Notes, 1098, Springer, 1984), 142–160.

[21] G. Rosenberger, Minimal generating systems for plane discontinuous groups and an equation in free groups, in *Proc. of Groups Korea 1988* (Lectures Notes, 1398, Springer, 1989), 170–186.

[22] G. Rosenberger, Applications of Nielsen's reduction method to the solution of combinatorial problems in group theory (London Math. Soc. Lecture Notes **36**, 1979), 339–358.

[23] G. Rosenberger, *Zum Rang und Isomorphieproblem für freie Producte mit Amalgam* (Habilitationschift, Hamburg, 1974).

[24] M. P. Schützenberger, Sur l'equation $a^{2+n} = b^{2+m}c^{2+p}$ dans un groupe libre, *C. R. Acad. Sci. Paris* **248**(1959), 2435–2436.

[25] H. Zieschang, Über die Nielsensche Kürzungsmethode in freien Produkten mit Amalgam, *Invent. Math.* **10**(1970), 4–37.

CLASSIFICATION OF ALL GENERATING PAIRS OF TWO GENERATOR FUCHSIAN GROUPS

BENJAMIN FINE* and GERHARD ROSENBERGER†

*Department of Mathematics, Fairfield University, Fairfield, Connecticut 06430, U.S.A.

† Fachbereich Mathematik Universität, Dortmund, Postfach 50 05 00, 44221 Dortmund, Germany

1. Introduction

A *Fuchsian group* is a discrete subgroup F of $PSL_2(\mathbb{R})$ or a conjugate of such a group in $PSL_2(\mathbb{C})$. A discrete subgroup G of $PSL_2(\mathbb{C})$ is *elementary* if any two elements of infinite order (regarded as linear fractional transformations) have at least one common fixed point. This is equivalent to the fact that the commutator of any two elements of infinite order has trace 2. The structure of elementary subgroups of $PSL_2(\mathbb{R})$ is well-known {see [6]} so for this paper we concentrate on non-elementary groups and use the term *Fuchsian group* to refer to a non-elementary discrete subgroup F of $PSL_2(\mathbb{R})$ or a conjugate of such a group in $PSL_2(\mathbb{C})$.

The purpose of this note is to present a complete classification, in one location, of the possibilities for generating pairs for two-generator Fuchsian groups. Specifically we prove the four main theorems listed below. These results have appeared in many different locations {see [26] for a discussion} but it would be convenient and important to have the proofs in just one place. The techniques we employ are straightforward and depend only on the properties of linear fractional transformations and their traces. Before stating the theorems we need some basic ideas from both Fuchsian group theory and abstract combinatorial group theory.

If a Fuchsian group F is finitely generated then F has a standard presentation called its *Poincare presentation* of the following form:

$$F = < e_1, \ldots, e_p, h_1, \ldots, h_t, a_1, b_1, \ldots, a_g, b_g; e_i^{m_i} = 1, i = 1, \ldots, p, R = 1 >$$

where $R = e_1 \ldots e_p h_1 \ldots h_t [a_1, b_1] \ldots [a_g, b_g]$ and $p \geq 0, t \geq 0, g \geq 0, p+t+g > 0$, and $m_i \geq 2$ for $i = 1, \ldots, p$. The e_i are elliptic generators, the h_i are parabolic or hyperbolic generators and the a_i, b_i are hyperbolic generators {see Section 2}. We say then that F has *signature* $(g; m_1, \ldots, m_p; t)$. From this presentation one can determine up to isomorphism all possibilities for presentations for two-generator Fuchsian groups. This has been done in several places [20], [22], [26]. Our techniques provide an independent proof of this result {Theorem A below}.

Suppose $G = < a, b >$ is a two-generator group. We write $(a, b) \overset{N}{\sim} (u, v)$ if there exists a Nielsen transformation from (a, b) to (u, v) {see [13] or [18] for necessary terminology on Nielsen transformations}. If the element a has finite order n then a transformation $(a, b) \to (a^m, b)$ with $1 \leq m < n$ and $\gcd(m, n) = 1$ is called an *E-transformation*. An *extended Nielsen transformation* is a finite sequence of Nielsen transformations and E-transformations. We write $(a, b) \overset{eN}{\sim} (u, v)$ if there exists an extended Nielsen transformation from (a, b) to (u, v). If $(a, b) \overset{eN}{\sim} (u, v)$ then $< a, b > = < u, v >$, that is the pairs (a, b) and (u, v) generate the same group.

We now state our main theorems. As remarked above Theorem A has appeared elsewhere however we state it with our main results since we need the classification of presentations.

Theorem A. *A two-generator Fuchsian group has one and only one of the following descriptions in terms of generators and relations:*

(1.1) $G = < A, B; >$ *G is a free group on A,B;*

(1.2) $G = < A, B; A^p = 1 >$ *for* $2 \leq p$;

(1.3) $G = < A, B; A^p = B^q = 1 >$ *for* $2 \leq p \leq q$ *and* $p + q \geq 5$;

(1.4) $G = < A, B; A^p = B^q = (AB)^r = 1 >$ *for* $2 \leq p \leq q \leq r$ *and* $(1/p) + (1/q) + (1/r) < 1$;

(1.5) $G = < A, B; [A, B]^p = 1 >$ *for* $2 \leq p$;

(1.6) $G = < A, B, C; A^2 = B^2 = C^2 = (ABC)^p = 1 >$ *for* $p = 2k + 1 \geq 3$.

Theorem B. *Let* $G = < U, V >$ *be a two-generator Fuchsian group.*

(2.1) If G is of type (1.1) or (1.5) then $(U, V) \overset{N}{\sim} (A, B)$.

(2.2) If G is of type (1.6) then $(U, V) \overset{N}{\sim} (AC, CA)$.

(2.3) If G is of type (1.2) or (1.3) then $(U, V) \overset{eN}{\sim} (A, B)$.

(2.4) If G is of type (1.4) then one and only one of the following cases holds

(a) $(U, V) \overset{eN}{\sim} (A, B)$;

(b) G is a $(2, 3, r)$ *triangle group with* $\gcd(r, 6) = 1$ *and* $(U, V) \overset{N}{\sim} (ABAB^2, B^2ABA)$;

(c) G is a $(2, 4, r)$ *triangle group with* $\gcd(r, 2) = 1$ *and* $(U, V) \overset{N}{\sim} (AB^2, B^3AB^3)$;

(d) G is a $(3, 3, r)$ *triangle group with* $\gcd(r, 3) = 1$ *and* $(U, V) \overset{N}{\sim} (AB^2, B^2A)$;

(e) G is a $(2, 3, 7)$ *triangle group and*

$$(U, V) \overset{N}{\sim} (AB^2ABAB^2AB^2AB, B^2ABAB^2ABABA).$$

The next two theorems classify the generating pairs and the corresponding types of Fuchsian groups by the trace of the commutator $[U, V]$.

Theorem C. *Suppose $U, V \in PSL_2(\mathbb{R})$ with $tr[U, V] > 2$ and suppose $G = < U, V >$ is non-elementary. Then G is discrete if and only if there is an extended Nielsen transformation from (U, V) to a pair (R, S) which satisfies $\{$after a suitable choice of signs$\}$*

(1) $0 \le trR \le trS \le |trRS|$;

(2) $trR = 2\cos(\pi/p)$ *or* $trR \ge 2$;

(3) $trS = 2\cos(\pi/q)$ *or* $trS \ge 2$ *and*

(4) $trRS = -2\cos(\pi/r)$ *or* $trRS \le -2$ *where* $p, q, r \in \mathbb{N} \setminus \{1\}$.

Moreover, if G is discrete then G is of type (1.1),(1.2),(1.3) or (1.4).

Theorem D. *Suppose $U, V \in PSL_2(\mathbb{R})$ with $0 \le trU, 0 \le trV$ and $tr[U, V] < 2$ and suppose $G = < U, V >$. Then G is discrete if and only if one of the following cases holds:*

(3.1) $tr[U, V] \le -2$;

(3.2) $tr[U, V] = -2\cos(\pi/p), p \in \mathbb{N} \setminus \{1\}$;

(3.3) $tr[U, V] = -2\cos(2\pi/p), p \in \mathbb{N} \setminus \{1\}$ *and* $\gcd(p, 2) = 1$;

(3.4) $tr[U, V] = -2\cos(6\pi/r), r \in \mathbb{N}, r \ge 5$ *with* $\gcd(r, 2) = 1$ *and* (U, V) *is Nielsen equivalent in a trace minimizing manner to a pair* (R, S) *which satisfies* $trR = trS = trRS$;

(3.5) $tr[U, V] = -2\cos(4\pi/r), r \in \mathbb{N}, r \ge 7$ *with* $\gcd(r, 6) = 1$ *and* (U, V) *is Nielsen equivalent in a trace minimizing manner to a pair* (R, S) *which satisfies* $trR = trS$ *and* $trRS = \frac{1}{2}(trR)^2$;

(3.6) $tr[U, V] = -2\cos(3\pi/r), r \in \mathbb{N}, r \ge 4$ *with* $\gcd(r, 3) = 1$ *and* (U, V) *is Nielsen equivalent in a trace minimizing manner to a pair* (R, S) *which satisfies* $trR = trS = trRS$;

(3.7) $tr[U, V] = -2\cos(4\pi/7)$ *and* (U, V) *is Nielsen equivalent in a trace minimizing manner to a pair* (R, S) *which satisfies* $trS = trRS = trR + 1$.

Moreover, if G is discrete then G is of type (1.1) in case (3.1), of type (1.5) in case (3.2), of type (1.6) in case (3.3), a $(2, 3, r)$ triangle group in case (3.4), a $(2, 4, r)$ triangle group in case (3.5), a $(3, 3, r)$ triangle group in case (3.6), and a $(2, 3, 7)$ triangle group in case (3.7).

Corollary. *Suppose $G = < U, V > \subset PSL_2(\mathbb{R})$ is a Fuchsian group. Then* $|tr[U, V] - 2| \ge 2 - 2\cos(\pi/7)$.

2. Preliminaries and notation

Our classification depends in large part on the traces of the generators so we first review the basic trace classification and trace identities. Recall that an element $A \in PSL_2(\mathbb{C})$ is a linear fractional transformation $z' = (az+b)/cz+d)$ with $ad-bc=1$. This can also be considered as the pair $\{\overline{A}, -\overline{A}\}$ with $\overline{A} = \begin{pmatrix} a & b \\ c & d \end{pmatrix} \in SL_2(\mathbb{C})$. We then use trA in an appropriate manner. The classification by trace is as follows.

If $A \in PSL_2(\mathbb{C})$ with $A \neq \pm I$ then

(1) A is *hyperbolic* if $trA \in \mathbb{R}$ and $|trA| > 2$;

(2) A is *parabolic* if $trA \in \mathbb{R}$ and $|trA| = 2$;

(3) A is *elliptic* if $trA \in \mathbb{R}$ and $|trA| < 2$;

(4) A is *loxodromic* if $trA \notin \mathbb{R}$.

Further A has finite order $p \geq 2$ if and only if $trA = \pm 2\cos(q\pi/p)$ with $1 \leq q < p$ and $\gcd(p,q) = 1$.

Now we define the *Chebyshev polynomials* $S_n(x)$ recursively by $S_0(x) = 0, S_1(x) = 1, S_n(x) = xS_{n-1}(x) - S_{n-2}(x)$ for $n \geq 2$ and $S_n(x) = -S_{-n}(x)$ if $n < 0$. These polynomials satisfy the following identities:

(1) $S_{n+m}(x) = S_n(x)S_{m+1}(x) - S_m(x)S_{n-1}(x)$;

(2) $S_n^2(x) - S_{n+1}(x)S_{n-1}(x) = 1$;

(3) $S_{mn}(x) = S_m((S_{n+1}(x) - S_{n-1}(x))S_n(x)$;

(4) $S_n(x + (1/x)) = (1-x^{2n})/(1-x^2)x^{n-1}$ for $n, m \in \mathbb{N} \cup \{0\}$;

(5) $S_{gcd(n,m)}(x) = gcd(S_n(x), S_n(x))$ for $m, n \in \mathbb{N}$.

Now let $A, B \in PSL_2(\mathbb{R})$ and let $x = trA$, $y = trB$ and $z = trAB$. We then have the following identities.

(1) $trAB^{-1} = xy - z$.

(2) $tr[A,B] = x^2 + y^2 + z^2 - xyz - 2$.

(3) $tr[A^n, B^m] - 2 = S_n^2(x)S_m^2(y)(tr[A,B] - 2)$ for $m, n \in \mathbb{N} \cup \{0\}$.

(4) $A^n = S_n(x)A - S_{n-1}(x)I$ for $I = \begin{pmatrix} 1 & 0 \\ 0 & 1 \end{pmatrix}$ and $\in \mathbb{N} \cup \{0\}$.

If $0 \leq x < 2$ then there is a $\theta \in \mathbb{R}, 0 < \theta \leq \pi/2$ with $x = 2\cos\theta$ and we have $S_n(x) = \frac{\sin(n\theta)}{\sin\theta}, n \in \mathbb{N} \cup \{0\}$.

If especially $x = \lambda_p = 2\cos(\pi/p), p \in \mathbb{N}, p \geq 2$ then $S_n(x) = \frac{\sin(n\pi/p)}{\sin(\pi/p)}$ for $1 \leq n < p$ and $S_n(x) > 1$ if $p \geq 4$ and $1 < n < p-1$.

If $x = 2$ then $S_n(x) = n$ for $n \in \mathbb{N} \cup \{0\}$. If $x > 2$ then there is a $\theta \in \mathbb{R}$ with $x = 2\cosh\theta$ and we have

$$S_n(x) = \frac{\sinh(n\theta)}{\sinh\theta} = \frac{\alpha^n - \alpha^{-n}}{\sqrt{x^2-4}}$$

for $n \in \mathbb{N} \cup \{0\}$ where $\alpha = (1/2)(x + \sqrt{x^2 - 4})$. Hence if $x > 2$ then $\lim_{n\to\infty} \frac{S_{n+1}(x)}{S_n(x)} = \alpha$.

Lemma 1. *Let $A, B \in PSL_2(\mathbb{R})$ with $|trA| \leq 2$. Then $tr[A, B] \geq 2$.*

PROOF. We may assume that $A = \begin{pmatrix} 1 & \lambda \\ 0 & 1 \end{pmatrix}$ or $A = \begin{pmatrix} t & s \\ -s & t \end{pmatrix}$ where $t = \cos\theta, s = \sin\theta$. Let $B = \begin{pmatrix} a & b \\ c & d \end{pmatrix}$. Note that $a^2 + b^2 + c^2 + d^2 \geq 2$. Then depending on the case $tr[A, B] = 2(ad - bc) + \lambda^2 c^2 \geq 2$ or $tr[A, B] = 2t^2(ad - bc) + s^2(a^2 + b^2 + c^2 + d^2) \geq 2(t^2 + s^2) = 2$. □

Note that from Lemma 1 if $A, B \in PSL_2(\mathbb{R})$ with $0 \leq trA, trB$ and $tr[A, B] < 2$ then automatically both $trA > 2$ and $trB > 2$. Moreover, $tr[A, B] < 2$ is equivalent to the fact that the fixed points of A and B are all distinct and separate each other.

If $A, B \in PSL_2(\mathbb{R})$ and $G = < A, B >$ we define the sets $E_G, L_G, M_G, E_G^\star$ and $M_G^\star$ as follows:

(1) $E_G = \{(U, V); U, V \in PSL_2(\mathbb{R}) \text{ and } (U, V) \overset{N}{\sim} (A, B)\}$;
(2) $L_G = \{(trU, trV, trUV); (U, V) \in E_G\}$;
(3) $M_G = \{trU; (U, V) \in E_G \text{ for some } V \in G\}$;
(4) $E_G^\star = \{(U, V); U, V \in PSL_2(\mathbb{R}) \text{ and } (U, V) \overset{eN}{\sim} (A, B)\}$;
(5) $M_G^\star = \{trU; (U, V) \in E_G^\star \text{ for some } V \in G\}$.

If $(U, V) \overset{N}{\sim} (A, B)$ then $tr[U, V] = tr[A, B]$ and hence the ternary form $f(x, y, z) = x^2 + y^2 + z^2 - xyz$ is invariant under automorphisms of the free group of rank two since

$$tr[A, B] = (trA)^2 = (trB)^2 + (trAB)^2 - trA.trB.trAB - 2.$$

Starting from the triple $(trA, trB, trAB) \in L_G$ we are able to obtain all triples $(u, v, w) \in L_G$ by repeated application of the following transformations:

$$O_1 : (u, v, w) \to (v, u, w)$$
$$O_2 : (u, v, w) \to (w, u, v)$$
$$O_3 : (u, v, w) \to (u, v, uv - w).$$

Indeed, each automorphism of the free group of rank 2 induces in a natural way such a transformation since $trA.trB - trAB = trAB^{-1}$. Let $H = < O_1, O_2, O_3 >$ the group generated by O_1, O_2, O_3. We have $< O_1, O_2 > \simeq S_3$ the symmetric group on three letters. H operates on L_G and it operates discontinuously if and only if M_G is discrete. It should be remarked that $H \simeq PGL_2(\mathbb{Z})$ if $tr[A, B] < 2$. We call the triples $(vw - u, v, w), (u, uw - v)$ and $(u, v, uv - w)$ the *neighbors* of $(u, v, w) \in L_G$ {see [29]}.

Lemma 2. *Suppose* $A, B \in PSL_2(\mathbb{R})$ *with* $0 \leq trA, trB$ *and* $tr[A,B] < 2$. *Let* $G = < A, B >$. *Then* M_G *is discrete and there is a pair* $(R,S) \in E_G$ *with* $2 < trR \leq trS \leq trRS \leq (1/2)(trR)(trS)$. *The triple* $(trR, trS, trRS)$ *is uniquely determined within* L_G *by these properties.*

Moreover, if $-2 < tr[A,B] < 2$ *then* $2 < trR < 3$ *and if* $-2 \leq tr[A,B] < 2$ *then* $2 < trR \leq 3$.

PROOF. Let $trA = x$, $trB = y$ and $trAB = z$. Since $tr[A,B] = tr[A,AB] < 2$ we must have from Lemma 1, $2 < x, y, z$. We may assume then that $2 < x \leq y \leq z$.

Assume there is no $(u,v,w) \in L_G$ with $2 < u \leq v \leq w \leq (1/2)uv$. Then $z > (1/2)xy$ Recall that also from Lemma 1, $2 < u, v, w$ for all $(u,v,w) \in L_G$ and $tr[A,B] = tr[U,V]$ if $(U,V) \in E_G$. To complete the proof of Lemma 2 we need the following propositions.

Proposition 1. *x,y,z as above. Then we must have* $yz - x > z \geq x, xz - y > z \geq y$ *and* $xy - z < z$.

PROOF. Assume $xz - y \leq z$. Then $z < z(x-1) \leq y$ and this contradicts $z \geq y$. Similarly $yz - x > z \geq x$. Lastly $xy - z < z$ since $z > (1/2)xy$. □

Proposition 2. *x,y,z as above. Then we must have* $xy - z < y$.

PROOF. Assume $y \leq xy - z$. Then $(x, y, xy - z) \in L_G$ where $x \leq y \leq xy - z$. By Proposition 1, $xy - z > xy - (xy - z) = z$, that is, $z < (1/2)xy$, and this is a contradiction. Therefore $xy - z < y$. □

We are now equipped to complete the proof of Lemma 2.

For $xy - z \leq x$ we consider the triples $(x,y,z), 2 < x \leq y \leq z$ and $(xy - z, x, y), 2 < xy - z \leq x \leq y$.

For $xy - z > x$ we consider the triples $(x,y,z), 2 < x \leq y \leq z$ and $(x, xy - z, y), 2 < x \leq xy - z \leq y$.

In either case we have $x + y + z > xy - z + x + y = x + xy - z + y$. Therefore we obtain sequences $(x_n), (y_n), (z_n)$ with $(x_n, y_n, z_n) \in L_G$ and $2 \leq x_n \leq y_n \leq z_n, z_n > (1/2)x_n y_n$(by assumption)$, y_n z_n - x_n > z_n, x_n z_n - y_n > z_n$ (from Proposition 1)$, z_{n+1} = y_n, y_{n+1} \leq y_n, x_{n+1} \leq x_n, x_n + y_n + z_n > x_{n+1} + y_{n+1} + z_{n+1}$ and either

(i) $x_{n+1} = x_n y_n - z_n, y_{n+1} = z_n$ (if $x_n y_n - z_n \leq x_n$) or

(ii) $x_{n+1} = x_n, y_{n+1} = x_n y_n - z_n$ (if $x_n y_n - z_n > x_n$).

Then each of the sequences $(x_n), (y_n), (z_n)$ converges.

Let $x_0 = \lim x_n, y_0 = \lim y_n, z_0 = \lim z_n$. We have $x_0 \leq x_n, y_0 \leq y_n, z_o \leq z_n$ for all $n \in \mathbb{N}$. Moreover we have

(i) $2 \leq x_0 \leq y_0 \leq z_0$;

(ii) $x_0^2+y_0^2+z_0^2-x_0y_0z_0 < 4$ in view of $tr[A,B] = x^2+y^2+z^2-xyz-2 < 2$;

(iii) $x_0y_0 - z_0 \leq y_0 \leq z_0$.

$x_0y_0 - z_0 < y_0$ cannot occur, for otherwise there would exist some y_n with $y_n < y_0$. Therefore $x_0y_0 - z_0 = y_0$. Similarly the case $x_0y_0 - z_0 < z_0$ cannot occur. Therefore $x_0y_0 - z_0 = z_0 = y_0$ and hence $x_0y_0 = 2y_0$, that is $x_0 = 2$. This gives $4 > x_0^2+y_0^2+z_0^2-x_0y_0z_0 = 4+2y_0^2-2y_0^2 = 4$ and this is inadmissible.

Therefore there exists a triple $(u,v,w) \in L_G$ with $2 < u \leq v \leq w \leq (1/2)uv$. For this triple $u+v+w \leq trX + trY + trXY$ for all $(X,Y) \in E_G$.

We now show the uniqueness. Assume that there are two different triples $(x_1,y_1,z_1),(x_2,y_2,z_2) \in L_G$ with $2 < x_i \leq y_i \leq z_i \leq (1/2)x_iy_i, i = 1,2$. Then there exists a finite sequence (u_m,v_m,w_m) with $m = 0$ to $m = p$ in L_G with $(u_0,v_0,w_0) = (x_1,y_1,z_1)$ and $(u_p,v_p,w_p) = \phi(x_2,y_2,z_2)$ for some $\phi \in\, < O_1,O_2 > \cong S_3$, such that $(u_{m+1},v_{m+1},w_{m+1})$ is a neighbor of (u_m,v_m,w_m), $0 \leq m \leq p-1$ and all members of the sequence are pairwise different. Further $2 < u_m,v_m,w_m$ as before.

Let $\max(u_m+v_m+w_m) = u_s+v_s+w_s$. Then we have $0 < s < p$ since the members of the sequence are pairwise different. The triples $(u_{s-1},v_{s-1},w_{s-1})$ and $(u_{s+1},v_{s+1},w_{s+1})$ are both neighbors of (u_s,v_s,w_s) and we have $u_{s-1}+v_{s-1}+w_{s-1} < u_s+v_s+w_s$ and $u_{s+1}+v_{s+1}+w_{s+1} < u_s+v_s+w_s$. Without loss of generality let $u_{s-1} = v_sw_s - u_s, v_{s-1} = v_s, w_{s-1} = w_s$ and $u_{s+1} = u_s, v_{s+1} = u_sw_s - v_s, w_{s+1} = w_s$. From $u_{s-1}+v_{s-1}+w_{s-1} < u_s+v_s+w_s$ we get that $u_{s+1} = v_sw_s - u_s < u_s$. Analagously , $v_{s+1} = u_sw_s - v_s < v_s$. From these we obtain $u_s^2, v_s^2 > (1/2)u_sv_sw_s$ and hence $u_s^2+v_s^2 > u_sv_sw_s = u_s^2+v_s^2+w_s^2 - tr[A,B]-2$. Therefore $w_s^2 < 4$ so $w_s < 2$ since $tr[A,B] < 2$. This contradicts $u_m,v_m,w_m > 2$ for all m. This therefore establishes the uniqueness of the triple $(trR, trS, trRS)$ with the required properties and also shows that M_G is discrete.

Now suppose that in addition $-2 < tr[A,B] < 2$. Then $(trR)^2+(trS)^2+(trRS)^2 - trR\cdot trS\cdot trRS = 2+tr[R,S] = 2+tr[A,B] > 0$. Since $trR \leq trS \leq trRS$ we have that $(1/3)trR\cdot trRS \leq trRS \leq (1/2)trR\cdot trS$. Thus $(1/2)trR\cdot trS - trRS \leq (1/6)trR\cdot trS$. From this we see that $0 < (trR)^2+(trS)^2+[(1/2)trR\cdot trS - trRS]^2 - [(1/2)trR\cdot trS]^2$ implies that $0 < 2(trS)^2[1-(1/9)(trR)^2]$ and therefore $trR < 3$. Analagously if $-2 \leq tr[A,B] < 2$ then $2 < trR \leq 3$. This completes the proof of Lemma 2. □

Lemma 3. *Suppose $A,B \in PSL_2(\mathbb{R})$ with $0 \leq trA, trB$ and $tr[A,B] > 2$. Let $G =< A,B >$. Then there is a pair $(R,S) \in E_G$ with $0 \leq trR \leq trS$ and $trRS < 0$.*

PROOF. Let $x = trA$, $y = trB$ and $z = trAB$ as in Lemma 2. If $z < 0$ then there is nothing to prove. Now let $z \geq 0$. We may assume that $0 \leq x \leq y \leq z$. To complete the proof of Lemma 3 we need the following propositions.

Proposition 3. *x,y,z as above. Then $xy - z < z$.*

PROOF. Assume $z \leq xy - z$. Then $x \geq 2$ since $y \leq z$ and $y \leq z = (1/2)xy - \sqrt{(1/4)x^2y^2 - x^2 - y^2 + c}$ where $c = tr[A,B] + 2 > 4$. This imples that $y^2(x-2) \leq x^2 - c < x^2 - 4$; because of $c > 4$ we get $x > 2$ and $x^2 \leq y^2 \leq x + 2$ which gives a contradiction. Hence $xy - z < z$. □

Proposition 4. *x,y,z as above. Then $xy - z < y$.*

PROOF. Assume $y \leq xy - z$. Then $0 \leq x \leq y \leq xy - z$ and by Proposition 3, $z = xy - (xy - z) < xy - z$ which gives a contradiction. Hence $xy - z < y$.□

Now we finish the proof of Lemma 3. If $xy - z < 0$ there is nothing to prove so assume that $xy - z \geq 0$. If $xy - z \leq x$ then we regard the triple $(xy - z, x, y), 0 \leq xy - z \leq x \leq y$. If $xy - z > x$ then we regard the triple $(x, xy - z, y), 0 \leq x \leq xy - z \leq y$. We now obtain a (finite or infinite) sequence $(x_n, y_n, z_n), 0 \leq x_n \leq y_n \leq z_n \in L_G$ where $(x_1, y_1, z_1) = (x, y, z)$ and if $n \geq 1$

$$(x_{n+1}, y_{n+1}, z_{n+1}) = (x_ny_n - z_n, x_n, y_n) \text{ if } 0 \leq x_ny_n - z_n \leq x_n \text{ or}$$
$$(x_{n+1}, y_{n+1}, z_{n+1}) = (x_n, x_ny_n - z_n, y_n) \text{ if } x_ny_n - z_n > x_n.$$

In any case, we have $x_ny_n - z_n \leq z_n, x_{n+1} \leq x_n, y_{n+1} \leq y_n, z_{n+1} \leq z_n$. Assume that this sequence (x_n, y_n, z_n) is infinite. Then the three sequences $(x_n), (y_n), (z_n)$ converge. Let $x_0 = \lim x_n, y_0 = \lim y_n, z_0 = \lim z_n$, and as in the proof of Lemma 2 we must have $x_0 = 2$ and $x_0^2 + y_0^2 + z_0^2 - x_0y_0z_0 = c$. From this it follows that $4 < c = 4 + 2z_0^2 - 2z_0^2 = 4$ which gives a contradiction. Therefore the (x_n, y_n, z_n) is finite, that is $x_ny_n - z_n < 0$ for some (x_n, y_n, z_n) completing the proof of Lemma 3. □

Note that if $0 \leq trA < 2$ then Lemma 3 is trivial. In this case just choose an $m \geq 1$ with $trA^{m-1}B \geq 0$ and $trA^mB < 0$.

Under the hypotheses of Lemma 3, M_G is in general not discrete. If there is a $U \in G =< A, B >$ with $trU \in M_G$ and $|trU| < 2$ which has infinite order then the set $\{trU^nV; n \in \mathbb{N}\}$ has an accumulation point for each V with $(U,V) \in E_G$. Therefore we get the following lemma.

Lemma 4. *Suppose $A, B \in PSL_2(\mathbb{R})$ with $tr[A,B] < 2$. Let $G =< A, B >$. Then the set M_G is discrete only if each $U \in G$ with $trU \in M_G$ and $|trU| < 2$ has finite order.*

The algorithm described in the proofs of Lemma 2 and 3 has been used [8], [24], [25] to obtain some structure theorems for subgroups of $PSL_2(\mathbb{C})$. We state these results but first we need some terminology. A subgroup G of $PSL_2(\mathbb{C})$ is *elliptic* if each non-trivial element of G is elliptic. If H is any group then the *rank of H* denoted $r(H)$ is the minimal cardinal number r of a generating system for H. A generating system X of H is a *minimal generating system* if X has cardinal number $r(H)$.

Theorem 1. ([24]) *Let G be a non-elementary and non-elliptic subgroup of $PSL_2(\mathbb{C})$. Then G can be generated by a minimal generating system which contains either only hyperbolic elements or only loxodromic elements.*

Theorem 2. ([8]) *Let G be a non-elementary and non-elliptic subgroup of $PSL_2(\mathbb{C})$. Assume that $trU \in \mathbb{R}$ for all $U \in G$. Then there is a $T \in PSL_2(\mathbb{C})$ with $TGT^{-1} \subset PSL_2(\mathbb{R})$.*

Theorem 3. ([25]) *Suppose $A, B \in PSL_2(\mathbb{C})$ with $G = < A, B >$ and assume that G is non-elementary and non-elliptic. Then there is a generating pair $\{U, V\}$ of G which is Nielsen equivalent to $\{A,B\}$ such that $< U^n, V^n >$ is a discrete free group of rank 2 for some large integer n.*

For the proof of Theorem 2 we also used the following result which is quite useful and the proof of which is computational.

Proposition 5. *Let A,B,C,X be four elements of $GL_2(\mathbb{C})$ with $AB = C$. For A we write $A = \begin{pmatrix} a_1 & a_2 \\ a_3 & a_4 \end{pmatrix}$ and use similar notation for B,C and X. Let $\overline{x} = (x_1, x_2, x_3, x_4)^t \in \mathbb{C}^4$, let $\overline{r} = (trX, trAX, trBX, trCX)^t$ and let*

$$M = \begin{pmatrix} 1 & 0 & 0 & 1 \\ a_1 & a_3 & a_2 & a_4 \\ b_1 & b_3 & b_2 & b_4 \\ c_1 & c_3 & c_2 & c_4 \end{pmatrix} \in \mathbb{C}^{4,4}.$$

Then $M\overline{x} = \overline{r}$ and $detM = (tr[A, B] - 2)\, detA\, detB$.

A straightforward application of Proposition 5 also leads to the following.

Proposition 6. *Let $A_1, \ldots, A_n, B_1, \ldots, B_n \in SL_2(\mathbb{C}), n \geq 3$. Suppose $tr[A_1, A_2] \neq 2$, $trA_i = trB_i$ for $i = 1, \ldots, n$ and $trA_jA_k = trB_jB_k$ for all j,k with $1 \leq j < k \leq n$. Then there exists a $C \in SL_2(\mathbb{C})$ with $CA_iC^{-1} = B_i^{\epsilon}, \epsilon = \pm 1$, for $i = 1, \ldots, n$.*

Finally for our further classification we need the listing and numbering of presentations in Theorem A which we rewrite below.

Listing. Suppose $G = < A, B >$ is two-generator Fuchsian group. Then we say G is of *type*

(1.1) if $G = < A, B; >$ G is a free group on A, B;

(1.2) if $G = < A, B; A^p = 1 >$ for $2 \leq p$;

(1.3) if $G = < A, B; A^p = B^q = 1 >$ for $2 \leq p \leq q$ and $p + q \geq 5$;

(1.4) if $G =< A, B; A^p = B^q = (AB)^r = 1 >$ for $2 \leq p \leq q \leq r$ and $(1/p) + (1/q) + (1/r) < 1$;

(1.5) if $G =< A, B; [A, B]^p = 1 >$ for $2 \leq p$;

(1.6) if $G =< A, B, C; A^2 = B^2 = C^2 = (ABC)^p = 1 >$ for $p = 2k + 1 \geq 3$.

Here, as mentioned in the introduction, a Fuchsian group means a non-elementary Fuchsian group. The elementary Fuchsian groups are either cyclic or isomorphic to the infinite dihedral group $D_\infty =< A, B; A^2 = B^2 = 1 >$ {see [3] or [6] }.

Group theoretically, Fuchsian groups of types (1.1),(1.2) or (1.3) are free products of the cyclic subgroups $< A >$ and $< B >$. A group of type (1.4) is called a (p, q, r) - *triangle group.*

Theorem A was proved independently by Peczynski, Rosenberger and Zieschang [20] and Purzitsky [22]. The proof of Peczynski, Rosenberger and Zieschang is combinatorial and uses Nielsen cancellation methods in free products with amalgamation [36]. Our results in the next section are more direct and will provide another proof along the lines of the Purzitsky proof and depend on the trace properties.

3. The case $tr[A, B] > 2$

Our results depend on the trace of the commutator of the generators A, B. We first consider the case where $tr[A, B] > 2$. Throughout the rest of the paper we use the numbering notation of the listing above.

Lemma 5. *Suppose $U, V \in PSL_2(\mathbb{R})$ with $tr[U, V] > 2$. Suppose $G =< U, V >$ is non-elementary and assume that each elliptic element in G has finite order. Then there is an extended Nielsen transformation from (U, V) to a pair (R,S) which satisfies {after a suitable choice of signs}*

(1) $0 \leq trR \leq trS \leq |trRS|$;

(2) $trR = 2\cos(\pi/p)$ *or* $trR \geq 2$;

(3) $trS = 2\cos(\pi/q)$ *or* $trS \geq 2$ *and*

(4) $trRS = -2\cos(\pi/r)$ *or* $trRS \leq -2$,

where $p, q, r \in \mathbb{N} \setminus \{1\}$.

PROOF. If $R \in G, R \neq \pm I$, is elliptic then $trR = \pm 2\cos(m\pi/n), 1 \leq m \leq n/2$, $\gcd(m, n) = 1$. Let $x = trA$, $y = trB$, $z = trAB$. From Lemma 2 we may assume that $0 \leq x \leq y \leq |z|, z < 0$ {possibly after a suitable choice of signs}. If $x \geq 2$ there is nothing to prove so assume that $0 \leq x < 2$, that is $x = \pm 2\cos(m\pi/n), 1 \leq m \leq n/2$, $\gcd(m, n) = 1$. With respect to the remarks after Lemma 3 we may assume $x = \lambda_n = 2\cos(\pi/n), n \geq 2$ (possibly after an extended Nielsen transformation). Therefore let $x = \lambda_n$.

If $y \geq 2$ there is nothing to prove so, as above, assume $0 \leq y < 2$, that is $y = \pm 2\cos(t\pi/s)$, $\gcd(t,s) = 1$. Then we get an extended Nielsen transformation from $\{A, B\}$ to a pair $\{U, V\}$ with $trU = \lambda_n, trV = \lambda_s = 2\cos(\pi/s), s \geq 2$ and $trUV < 2$, (possibly after a suitable choice of signs), because if $trU = \lambda_n, trV = \lambda_s$ and $trUV \geq 2$ then $trUV^{-1} = \lambda_n\lambda_s - trUV < 2$.

If $trUV \leq -2$ then there is nothing to prove so assume that $|trUV| < 2$, that is $trUV = \pm 2\cos(m\pi/k), 1 \leq m \leq k/2$, $\gcd(k,m) = 1$. If $m = 1$ and $trUV = -2\cos(\pi/k)$ then there is nothing more to prove. If $m = 1$ and $trUV = 2\cos(\pi/k)$ then we may assume that $\lambda_n \leq \lambda_s \leq \lambda_k = 2\cos(\pi/k)$, possibly after a change of generators. From this we get that $trUV^{-1} = \lambda_n\lambda_s - \lambda_k < \lambda_k$, and now we consider the generating pair $\{U, V^{-1}\}$ to continue as above.

Now let $m \geq 2$ Then we regard the generating pair $\{U, W\}$ with $W = UV$. Since $\gcd(m,k) = 1$ there is a $W_1 \in G$ with $W = W_1^m$ and $trW_1 = \lambda_k = 2\cos(\pi/k)$, (possibly after a suitable choice of signs), and we have $G = < U, W > = < U, W_1 >$. Further

$$tr[U,V] - 2 = tr[U,W] - 2 = S_m^2(\lambda_k)(tr[U,W_1] - 2).$$

Here it is $S_m^2(\lambda_k) > 1$ since $m \geq 2$ and hence $2 < tr[U,W_1] < tr[U,W]$. We now regard the generating pair $\{U, W_1\}$ and continue as above.

Now if R is any element of G then $trR \in \mathbb{Q}(\lambda_n, \lambda_s, \lambda_k)$ and there are only finitely many $\lambda_h = 2\cos(\pi/h), h \geq 2$ with $\lambda_h \in \mathbb{Q}(\lambda_n, \lambda_s, \lambda_k)$, and there are only finitely many $j \in \mathbb{N}$ with $1 \leq j \leq h/2$, gcd(j,h) = 1. Therefore we get an extended Nielsen transformation from $\{A, B\}$ to a pair $\{R, S\} \in E_G^\star$ with $2 < tr[R,S] \leq tr[C,D]$ for all $(C,D) \in E_G^\star$. (Recall that $E_G^\star = \{(U,V); U, V \in PSL_2(\mathbb{R})$ and $(U,V) \overset{eN}{\sim} (A,B)\}$.)

Now we may assume that $trR = \lambda_p = 2\cos(\pi/p), p \geq 2, trS = \lambda_q = 2\cos(\pi/q), q \geq 2$, and $|trRS| = \lambda_r = 2\cos(\pi/r), r \geq 2$ with $\lambda_p \leq \lambda_q \leq \lambda_r$ (possibly after a change of the generators and a suitable choice of signs). If $trRS = -\lambda_r$ then there is nothing more to prove. If $trRS = +\lambda_r$ then $trRS^{-1} = \lambda_p\lambda_q - \lambda_r$. Because $tr[R,S] = tr[R,S^{-1}]$ and because of the minimality of $tr[R,S]$ we get that $trRS^{-1} = \lambda_p\lambda_q - \lambda_r = \pm\lambda_h < \lambda_r, \lambda_h = 2\cos(\pi/h), h \geq 2$ for otherwise there would be a pair $(E,F) \in M_G^\star$ with $2 < tr[E,F] < tr[R,S]$ because RS^{-1} has finite order contradicting the minimality of $tr[R,S]$. Therefore we may assume that $trRS = -\lambda_r$ and Lemma 5 is proven. □

We next mention the following theorem {see T. Jorgensen[4] and Rosenberger [28],[30] }. However we provide a different proof based on our trace techniques. Jorgensen's proof used his well-known inequality

$$|(trA)^2 - 4| + |tr[A,B] - 2| \geq 1$$

for discrete non-elementary two-generator subgroups $< A, B > \subset PSL_2(\mathbb{C})$ together with the result of Selberg [31] that finitely generated linear groups contain a torsion-free subgroup of finite index and a result of C.L. Siegel[32] concerning a discreteness condition for torsion-free subgroups of $PSL_2(\mathbb{R})$.

Theorem 4. ([4],[28],[30]) *Let G be a non-elementary subgroup of $PSL_2(\mathbb{R})$. Then G is discrete if and only if each cyclic subgroup of G is discrete .*

PROOF. If G is discrete then certainly each cyclic subgroup is discrete and in particular each cyclic subgroup generated by an elliptic element is finite. Now let G have the property that each cyclic subgroup of G is discrete. Then each elliptic element in G has finite order. Let U, V be two elements of G with $tr[U, V] \neq 2$ (such U and V exist since G is non-elementary). If $tr[U, V] > 2$ then $< U, V >$ is isomorphic to the infinite dihedral group or $< U, V >$ is non-elementary. If $tr[U, V] > 2$ and $< U, V >$ is non-elementary then after applying extended Nielsen transformations we obtain by Lemma 5 that $tr[U, V] \geq (2\cos(\pi/7))^2 - 1$. If $tr[U, V] < 2$ then $tr[U, VU^{-1}V^{-1}] > 2$ and therefore G must be discrete. □

Lemma 6. *Suppose $U, V \in PSL_2(\mathbb{R})$ and suppose $G =< U, V >$ is non-elementary. Suppose further that*

(1) $0 \leq trU \leq trV \leq |trUV|$;

(2) $trU = 2\cos(\pi/p)$ *or* $trU \geq 2$;

(3) $trV = 2\cos(\pi/q)$ *or* $trV \geq 2$ *and*

(4) $trUV = -2\cos(\pi/r)$ *or* $trUV \leq -2$,

where $p, q, r \in \mathbb{N}/\{1\}$. Then G is discrete. Moreover, G is of type (1.1), (1.2), (1.3) or (1.4).

PROOF. We note that $tr[U, V] > 2$. The conditions in Lemma 6 just describe a pair of canonical generators for G from which we may construct a standard picture for a fundamental domain for G. We demonstrate this for the case where $trU = 2\cos(\pi/p), trV = 2\cos(\pi/q)$ and $trUV = -2\cos(\pi/r)$. After a suitable conjugation we may assume that $U = \begin{pmatrix} \cos(\pi/p) & -\sin(\pi/p) \\ \sin(\pi/p) & \cos(\pi/p) \end{pmatrix}$ and $V = \begin{pmatrix} \cos(\pi/q) & b \\ c & \cos(\pi/q) \end{pmatrix}$ with $b < 0, c > 0, c < |b|, bc = -\sin^2(\pi/q)$ and $trUV = -2\cos(\pi/r)$. The fixed points (FP's) in the hyperbolic plane $\mathcal{H}^*$ of U, V, UV and VU and the resulting non-Euclidean triangle are pictured in Figure 1.

This diagram is by the Poincare Theorem [12] on the generation of Fuchsian groups a standard picture for a fundamental domain for the (p, q, r)-triangle group G. Hence G is discrete in the case $trU = 2\cos(\pi/p), trV = 2\cos(\pi/q)$ and $trUV = -2\cos(\pi/r)$ and G is a (p, q, r)-triangle group. The other non-cocompact cases are similar and we omit the details. □

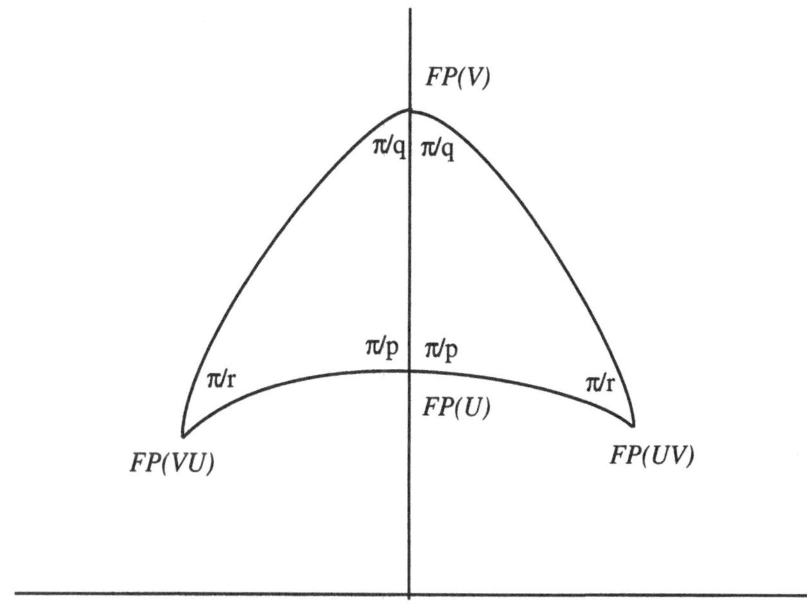

Figure 1. Non-euclidean triangle

Lemmas 5 and 6 combined give us Theorem C.

Theorem C. *Suppose $U, V \in PSL_2(\mathbb{R})$ with $tr[U,V] > 2$ and suppose $G = < U, V >$ is non-elementary. Then G is discrete if and only if there is an extended Nielsen transformation from (U,V) to a pair (R,S) which satisfies {after a suitable choice of signs}*

(1) $0 \leq trR \leq trS \leq |trRS|$;

(2) $trR = 2\cos(\pi/p)$ *or* $trR \geq 2$;

(3) $trS = 2\cos(\pi/q)$ *or* $trS \geq 2$ *and*

(4) $trRS = -2\cos(\pi/r)$ *or* $trRS \leq -2$,

where $p, q, r \in \mathbb{N} \setminus \{1\}$. Moreover, if G is discrete then G is of type (1.1), (1.2), (1.3) or (1.4).

Corollary 1. *Suppose $U, V \in PSL_2(\mathbb{R})$ with $tr[U,V] > 2$ and suppose $G =< U, V >$ is non-elementary. Then G is discrete if and only if $M_G^\star$ is a discrete subset of $\mathbb{R}$. $\{M_G^\star = \{trU; (U,V) \in E_G^\star$ for some $V \in G\}.\}$*

PROOF. If G is discrete then $M_G^\star$ is also discrete from the construction in the proof of Lemma 6. Now suppose $M_G^\star$ is a discrete subset of $\mathbb{R}$. Let $U \in G$ with $trU \in M_G^\star$ and $|trU| < 2$. U must have finite order for otherwise the set $\{trU^nV; n \in \mathbb{Z}\}$ would have an accumulation point for each V with

$(U, V) \in E_G^\star$. Hence Lemma 5 and Lemma 6 together show that G is discrete. □

J.Lehner [11] has shown the following: Suppose $A, B \in PSL_2(\mathbb{R})$ with $trA = 2\cos(\pi/p), trB = 2\cos(\pi/q)$ and $trAB = \pm 2\cos(\pi/r)$ where $p, q, r \in \mathbb{N}$ with $2 \leq p \leq q \leq r$ and $(1/p) + (1/q) + (1/r) < 1$. Then if $p \geq 3$ and $< A, B >$ is non-elementary then $< A, B >$ is a (p, q, r)-triangle group if and only if $trAB = -2\cos(\pi/r)$.

Corollary 2. *Let $U, V \in PSL_2(\mathbb{R})$ with $tr[U, V] > 2$. Let G be a two-generator Fuchsian group of type (1.2),(1.3) or (1.4) with A,B the generators given in Theorem A. Then $(U, V) \overset{eN}{\approx} (A, B)$.*

PROOF. This is clear if G is of type (1.2) or (1.3). Now suppose G is of type (1.4). From the proofs of Lemma 5 (recall that a (p, q, r)-triangle group is not isomorphic to a free product of cyclics) and Lemma 6 we may assume (after an extended Nielsen transformation) that

(1) $trU = trA = 2\cos(\pi/p)$;

(2) $trV = trB = 2\cos(\pi/q)$;

(3) $trUV = trAB = -2\cos(\pi/r)$,

with $2 \leq p \leq q \leq r$ and $(1/p) + (1/q) + (1/r) < 1$. The map $A \to U$, $B \to V$ defines an automorphism ϕ of G. We consider the Teichmuller Space of G, here described by the trace point $(2\cos(\pi/p), 2\cos(\pi/q), 2\cos(\pi/r))$. Up to the inversion $A \to A^{-1}, B \to B^{-1}$ and the inner automorphisms, ϕ can be regarded just as a permutation on the triple $(2\cos(\pi/p), 2\cos(\pi/q), 2\cos(\pi/r))$. Hence $(U, V) \overset{eN}{\approx} (A, B)$. □

4. The case $tr[A, B] < 2$

Note that if $A, B \in PSL_2(\mathbb{C})$ with $tr[A, B] = 2$ then A and B (regarded as linear fractional transformations) have a common fixed point. Now let $A, B \in PSL_2(\mathbb{R})$ with $tr[A, B] < 2$. This means that A and B are hyperbolic with axes intersecting in exactly one point τ. Here we now follow an idea of N. Purzitsky [22]. Let E_1 be the elliptic transformation of order 2 fixing τ. Observe that $E_1AE_1 = A^{-1}$ and $E_1BE_1 = B^{-1}$. Let $E_2 = E_1A$ and $E_3 = BE_1$. Then E_2 and E_3 are of order 2 and $< A, B >$ is of index less than or equal to 2 in $< E_1, E_2, E_3 >$, and we have $[A, B] = (E_1E_2E_3)^2$. Further $< A, B >$ is a Fuchsian group if and only if $< E_1, E_2, E_3 >$ is also a Fuchsian group. Applying the appropriate extension of Poincare's theorem [12] it follows that $< E_1, E_2, E_3 >$ is discrete if $|tr(E_1E_2E_3)| \geq 2$. Hence we have the following.

Lemma 7. *Let $A, B \in PSL_2(\mathbb{R})$ with $tr[A, B] < 2$. Then $< A, B >$ is a discrete free group of rank two if and only if $tr[A, B] \leq -2$.*

Now let $-2 < tr[A,B] < 2$. If $< A, B >$ is discrete then necessarily $[A,B]$ has finite order. Then also $E_1E_2E_3$ has finite order. Hence let $trE_1E_2E_3 = \pm 2\cos(p\pi/q)$ with $p, q \in \mathbb{N}, 2 \leq q, 1 \leq p \leq q/2$ and $\gcd(p,q) = 1$. Observe that $tr[A,B] = \pm 2\cos(2p\pi/q)$.

For the following we assume first that $< E_1, E_2, E_3 >$, and hence $< A, B >$, is discrete. We construct the following non-Euclidean triangle (after possibly a suitable conjugation). Let z_0 be the fixed point(FP) of $E_1E_2E_3$, let $z_1 = E_3(z_0)$ and $z_2 = E_2(z_1)$. We observe that z_1 is on the geodesic joining z_0 to the fixed point of E_3. Similarly z_2 is on the geodesic joining z_1 to the fixed point of E_2 and z_0 is on the geodesic joining z_2 to the fixed point of E_1. Thus after a suitable conjugation we have the triangle T pictured in Figure 2.

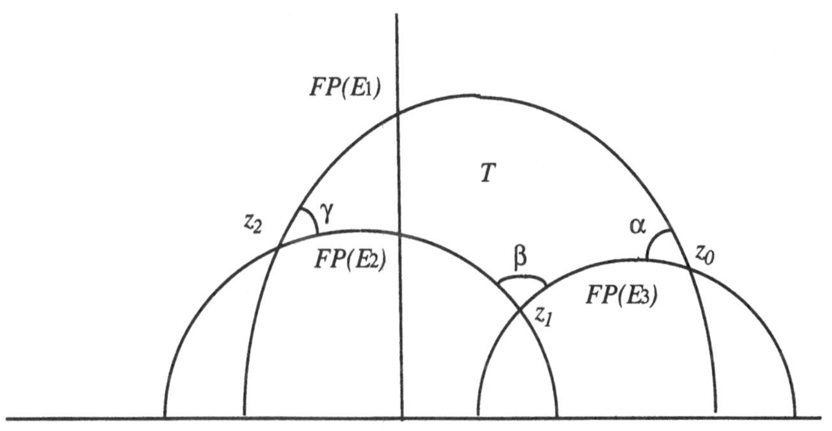

Figure 2. Non-euclidean triangle

Now we use the following theorem of Knapp [10].

Theorem 5. ([10]) *Let P be an open simple, finite sided geodesic polygon in $\mathcal{H}^+$, the hyperbolic plane, whose closure $\overline{P}$ is compact in $\mathcal{H}^+$. Suppose that all the sides of P are matched in disjoint pairs by analytic automorphisms of $\mathcal{H}^+$. Say $L_a(s_a) = s'_a$ with $\{s_a, s'_a\}$ a matched side pair. As an orientation condition on the transformation L_a suppose that for each x within the side s_a, each sufficiently small disc N centered on x satisfies the inclusion condition $N \subset s_a \cup P \cup L_a^{-1}(P)$. Finally suppose that the group H generated by the transformations L_a is discrete. Then*

(1) $\mathcal{H}^+ \setminus H(\partial P)$ is open and every point in it is equivalent to members of P under exactly t elements of H where t is a finite positive integer independent of the point;

(2) moreover if Q is a standard open fundamental polygon for H, then $\mu(P) = t\mu(Q)$ where μ is the hyperbolic area.

We now return to our groups $< A, B >$ and $< E_1, E_2, E_3 >$ and the triangle T. By cutting and pasting of T we have $\alpha + \beta + \gamma = 2p\pi/q$. If $< E_1, E_2, E_3 >$ has signature $(g; \gamma_1, \ldots, \gamma_n; 0), \gamma_i \geq 2$ then by the area formula and Knapp's theorem we obtain with $t \geq p$:

$$1 - \frac{2p}{q} = 2t(2g - 2 + \sum_{i=1}^{n}(1 - \frac{1}{\gamma_i})).$$

Since $< E_1, E_2, E_3 >$ is generated by three elements of order 2 we see that $g = 0$ and $n = 3$ or $n = 4$. If $n = 4$ then it is easy to check that we must have $t = p = 1$ and $< E_1, E_2, E_3 >$ has signature $(0; 2, 2, 2, q; 0)$ with $q \geq 3$. Hence let $n = 3$ and

$$1 - \frac{2p}{q} = 2t(2g - 2 + \sum_{i=1}^{3}(1 - \frac{1}{\gamma_i}))$$

with $t \geq p$. Then by Poincare's Theorem we must have $p \geq 2$ which implies that $q \geq 5$ and $t \geq 2$. Further $t \geq 2$ imples that

$$\frac{p}{2q} \leq \frac{1}{\gamma_1} + \frac{1}{\gamma_2} + \frac{1}{\gamma_3} - \frac{3}{4}.$$

Without loss of generality we may assume that $2 \leq \gamma_1 \leq \gamma_2 \leq \gamma_3 < \infty$. The elliptic element $E_1E_2E_3$ has order q and must be conjugate to some power of one of the three elliptic generators in the standard presentation for a $(\gamma_1, \gamma_2, \gamma_3)-$ triangle group. Hence, $q \leq \gamma_3$ and we get

$$0 \leq \frac{p}{2\gamma_3} - \frac{1}{\gamma_3} \leq \frac{p}{2q} - \frac{1}{\gamma_3} \leq \frac{1}{\gamma_1} + \frac{1}{\gamma_2} - \frac{3}{4}$$

with $p \geq 2$. This gives the possibilites:

(a) $\gamma_1 = 2$ and $\gamma_2 = 4$;

(b) $\gamma_1 = 2$ and $\gamma_2 = 3$.

In (a) we must have $\gamma_3 = q$ and $p = 2$, and therefore $t = 2$. So we consider (b) with $\gamma_1 = 2$ and $\gamma_2 = 3$. None of the fixed points $z_0, z_1, z_2, FP(E_1), FP(E_2)$, $FP(E_3)$ can lie in the orbit of an elliptic element of order 3 in the $(2, 3, \gamma_3)-$ triangle group (recall that $q \geq 5$). This implies that the above fixed points lie in at most two orbits, and therefore we get that $t \geq 3$. Moreover t must be a multiple of 3 (consider the fixed points of order 3 in T).

We have $\gamma_3 = qk$ for some $k \in \mathbb{N}$. From the area formula we now get

$$1 - \frac{2p}{q} = 2t(1 - \frac{1}{2} - \frac{1}{3} - \frac{1}{qk}) = 2t(\frac{1}{6} - \frac{1}{qk}) \geq 1 - \frac{6}{qk},$$

that is $3 \geq pk$ since $t \geq 3$. Therefore $k = 1$ since $p \geq 2$ and then $q = \gamma_3$ and $2 \leq p \leq 3$.

If $p = 2$ then $t = (3q - 12)/(q - 6)$. Recall that $q \geq 7$ since $q = \gamma_3$ and we get the possible solutions $(t, q) = (9, 7)$ and $(t, q) = (6, 8)$ since t is a multiple of 3. However $(t, q) = (6, 8)$ cannot occur because $p = 2$ and $\gcd(p, q) = 1$. Therefore we have $(t, q) = (9, 7)$. If $p = 3$ we get $t = 3$ and $q \geq 7$. Therefore altogether the solutions provided $< E_1, E_2, E_3 >$ is discrete are summarized in the following lemma.

Lemma 8. *Let $< E_1, E_2, E_3 >$ be discrete. Then one of the following holds {using the above notation}.*

(1) $t = p = 1$ *and* $< E_1, E_2, E_3 >$ *has signature* $(0; 2, 2, 2, q; 0), q \geq 3$.

(2) $t = p = 2$ *and* $< E_1, E_2, E_3 >$ *has signature* $(0; 2, 4, q; 0)$ *with* $q \geq 5$ *and* $\gcd(2, q) = 1$.

(3) $t = p = 3$ *and* $< E_1, E_2, E_3 >$ *has signature* $(0; 2, 3, q; 0)$ *with* $q \geq 7$ *and* $\gcd(3, q) = 1$.

(4) $t = 9, p = 2$ *and* $< E_1, E_2, E_3 >$ *has signature* $(0; 2, 3, q; 0)$ *with* $q = 7$.

We remark that case (4) was omitted by N.Purzitzky a fact which was discovered by J.P. Matelski [19]. Unfortunately Matelski's paper contains several errors in its cases (4),(5) and (6). $< A, B >$ can be non-discrete and $A^N B^{-1}$ not elliptic for all $N \geq 1$.

Further notice that $< E_1, E_2, E_3 > = < A, B >$ if and only if q is odd and that $< A, B >$ has index 2 in $< E_1, E_2, E_3 >$ if and only if q is even. If $p \geq 2$ and $q = 2q_1 \geq 8$ is even then $< E_1, E_2, E_3 >$ has signature $(0; 2, 3, q; 0)$ and the subgroup $< A, B >$ has index 2 and the signature $(0; 3, 3, q_1; 0)$. If $p = 1$ and $q = 2q_1 \geq 4$ is even then $< E_1, E_2, E_3 >$ has signature $(0; 2, 2, 2, q; 0)$ and the subgroup $< A, B >$ has index 2 and the signature $(1; q_1; 0)$.

We remark that the results mentioned above, together with Lemma 5, Lemma 6, Lemma 7 and Lemma 8 provide an independent proof of Theorem A which we now restate.

Theorem A. *A two-generator Fuchsian group has one and only one of the following descriptions in terms of generators and relations:*

(1.1) $G = < A, B; > $ *G is a free group on A,B;*

(1.2) $G = < A, B; A^p = 1 >$ *for* $2 \leq p$;

(1.3) $G = < A, B; A^p = B^q = 1 >$ *for* $2 \leq p \leq q$ *and* $p + q \geq 5$;

(1.4) $G = < A, B; A^p = B^q = (AB)^r = 1 >$ *for* $2 \leq p \leq q \leq r$ *and* $(1/p) + (1/q) + (1/r) < 1$;

(1.5) $G = < A, B; [A, B]^p = 1 >$ *for* $2 \leq p$;

(1.6) $G = < A, B, C; A^2 = B^2 = C^2 = (ABC)^p = 1 >$ *for* $p = 2k + 1 \geq 3$.

We now want to determine how to realize the cases and situations in Lemma 8. For that we consider elements E_1, E_2, E_3 in $PSL_2(\mathbb{R})$ with $E_1^2 =$

$E_2^2 = E_3^2 = 1$ and $trE_1E_2E_3 = \pm 2\cos(p\pi/q)$ with $p, q \in \mathbb{N}, 2 \leq q, 1 \leq p \leq (q/2)$ and $\gcd(p,q) = 1$. With $A = E_1E_2$ and B $= E_3E_1$ we have $[A,B] = (E_1E_2E_3)^2$ and $tr[A,B] = \pm 2\cos(p\pi/q)$. Especially A and B are hyperbolic elements, We may construct the triangle T as in the proof of Lemma 8, independently of whether $< E_1, E_2, E_3 >$ is discrete or not. However the above considerations show that if $p \geq 2$ then either $< E_1, E_2, E_3 >$ is a triangle group as described in the cases (2),(3) and (4) or is not discrete. Hence if $p \geq 2$ we just have to realize the cases (2),(3),(4) in the triangle group belonging to them.

If $p = 1$, the triangle T then gives a fundamental domain for $< E_1, E_2, E_3 >$ and then Poincare's Theorem gives us the following result.

Lemma 9. *Let $E_1, E_2, E_3 \in PSL_2(\mathbb{R})$ with $E_1^2 = E_2^2 = E_3^2 = 1$. Then $< E_1, E_2, E_3 >$ is a Fuchsian group with signature $(0; 2, 2, 2, q; 0)$ if and only if $trE_1E_2E_3 = \pm\cos(\pi/q)$.*

Moreover if q is even, $q = 2n \geq 4$ then $< A, B >=< A, B; [A,B]^n = 1 >$ while if q is odd $q = 2k + 1 \geq 3$, then $< A, B >=< E_1, E_2, E_3; E_1^2 = E_2^2 = E_3^2 = (E_1E_2E_3)^q = 1 >$.

Lemma 10. *Let $A, B \in PSL_2(\mathbb{R})$. Let $G =< A, B >=< A, B; [A,B]^n = 1 >, n \geq 3$, be a Fuchsian group. Then $tr[A,B] \neq 2\cos(\pi/n)$.*

PROOF. Let $tr[A,B] = 2\cos(\pi/n) = \lambda, n \geq 3$ and let $trA = \mu$. From Lemma 2 we may assume that $2 < \mu < 3$ (we may choose $trA, trB \geq 0$). Let $C = [A,B]$. Then $trAC = \mu(\lambda - 1) \geq 0$. Since G is discrete and has a presentation $< A, B >=< A, B; [A,B]^n = 1 >$ we see that $\lambda \geq 1 + 2/\mu$. Similarly $tr[A^2, B] = \mu^2(\lambda - 2) + 2 < 2$. Hence $\mu^2(\lambda - 2) \leq -4$. Thus altogether $2 - 4/\mu^2 \geq \lambda \geq 1 + 2/\mu$ which implies $\mu \geq \sqrt{5} + 1 > 3$. Since $\mu < 3$ we have $tr[A,B] \neq 2\cos(\pi/n)$. □

Lemma 11. *Let $A, B \in PSL_2(\mathbb{R})$. Let $G =< A, B >=< E_1, E_2, E_3; E_1^2 = E_2^2 = E_3^2 = (E_1E_2E_3)^q = 1 >$, with $q = 2k+1 \geq 3$ be a Fuchsian group. Then $tr[A,B] \neq 2\cos(2\pi/q)$.*

PROOF. Let $tr[A,B] = 2\cos(2\pi/q) = \lambda, q = 2k + 1 \geq 1$. If $q = 3$ then $tr[A,B] = 2\cos(2\pi/3) = -2\cos(\pi/3)$ and G must have a presentation $< A, B; [A,B]^3 = 1 >$ by Lemma 7 and Lemma 9 which is impossible.

Therefore $q \geq 5$. In the generators A and B we have a presentation

$$< A, B; [A,B]^q = ([A,B]^{(q-1)/2}A)^2 =$$
$$([A,B]^{(q-1)/2}AB)^2 = ([A,B]^{(q-1)/2}ABA)^2 = 1 > .$$

Let $\{U, V\}$ be a pair which is Nielsen equivalent to $\{A, B\}$. Analagously as for A and B we may construct F_1, F_2, F_3 such that $U = F_1F_2$ and V $= F_3F_1$

and $G =< F_1, F_2, F_3; F_1^2 = F_2^2 = F_3^2 = (F_1F_2F_3)^q = 1 >$. Therefore by Lemma 2 we may assume that $2 < trA < 3$ and we may argue analagously as in Lemma 10. □

Therefore from Lemmas 9, 10 and 11 we get.

Lemma 12. *Let $A, B \in PSL_2(\mathbb{R})$. Then:*

(1) $< A, B >$ is a Fuchsian group with signature $(1; n; 0), n \geq 2$ if and only if $tr[A, B] = -2\cos(\pi/n)$;

(2) $< A, B >$ is a Fuchsian group with signature $(0; 2, 2, 2, q; 0), q = 2k + 1 \geq 3$ if and only if $tr[A, B] = -2\cos(2\pi/q)$.

Lemma 13. *Let $A, B \in PSL_2(\mathbb{R})$ and $G =< A, B >$ a cocompact Fuchsian group. Suppose $\{U, V\}$ is a generating pair for G. Then:*

(a) If $G =< A, B; [A, B]^n = 1 >$ then $(U, V) \overset{N}{\sim} (A, B)$.

(b) If $G =< E_1, E_2, E_3; E_1^2 = E_2^2 = E_3^2 = (E_1E_2E_3)^q = 1 >$, with $q = 2k + 1 \geq 3$, then $(U, V) \overset{N}{\sim} (E_1E_2, E_3E_1)$.

(c) If $G =< A, B; A^2 = B^3 = (AB)^q = 1 >, q \geq 8$, $\gcd(q, 2) = 1$ and $tr[U, V] < 2$, then $(U, V) \overset{N}{\sim} (ABAB^2, B^2ABA)$.

(d) If $G =< A, B; A^2 = B^4 = (AB)^q = 1 >, q \geq 5$, $\gcd(q, 2) = 1$ and $tr[U, V] < 2$, then $(U, V) \overset{N}{\sim} (AB^2, B^3AB^3)$.

(e) If $G =< A, B; A^3 = B^3 = (AB)^q = 1 >, q \geq 4$, $\gcd(q, 3) = 1$ and $tr[U, V] < 2$, then $(U, V) \overset{N}{\sim} (AB^2, B^2A)$.

PROOF. (a) Because G is not a free product of cyclics or a cocompact triangle group we must have $tr[U, V] < 2$. By Lemma 12 we have $tr[U, V] = -2\cos(\pi/n)$. Hence $[U, V]$ is conjugate in G to $[A, B]^\epsilon, \epsilon = \pm 1$, independent of n. Therefore if $F =< a, b >$ is free on a and b and $\phi : F \to G$ is the canonical epimorphism, we have elements $u, v \in F$ such that $[u, v]$ is conjugate within F to $[a, b]^\epsilon, \epsilon = \pm 1$ and $U = \phi(u), V = \phi(v)$. By Nielsen's Theorem {see[12]},$(u, v) \overset{N}{\sim} (a, b)$ in F and hence $(U, V) \overset{N}{\sim} (A, B)$ in G.

(b) As in case (a) we must have $tr[U, V] < 2$. By Lemma 12 we have $tr[U, V] = -2\cos(2\pi/q)$. Hence [U,V] is conjugate within G to $(E_1E_2E_3)^{2\epsilon}, \epsilon = \pm 1$ independent of q. Therefore if $F =< e_1, e_2, e_3; e_1^2 = e_2^2 = e_3^2 = 1 >$ is the free product on the three elements e_1, e_2, e_3 of order two and $\phi : F \to G$ is the canonical epimorphism, we have elements $u, v \in F$ such that $[u, v]$ is conjugate within F to $(e_1e_2e_3)^{2\epsilon}, \epsilon = \pm 1$ and $U = \phi(u), V = \phi(v)$. If we regard F as a Fuchsian group with finite area we see that $< u, v >$ is the only two-generator free subgroup of genus 1 and index 2 in F. Therefore $(u, v) \overset{N}{\sim} (e_1e_2, e_3e_1)$ in F and hence $(U, V) \overset{N}{\sim} (E_1E_2, E_3E_1)$ in G.

(c) Again we must have $tr[U, V] < 2$. From Lemma 8 and the remarks following Lemma 8 we have $tr[U, V] = \pm 2\cos(6\pi/q)$. Hence $[U, V]$ is conjugate within G to $(AB)^{6\epsilon}, \epsilon = \pm 1$ independent of q. Therefore if $F =< a, b; a^2 = b^3 = 1 >\cong PSL_2(\mathbb{Z})$ is the free product on a of order 2 and b of order 3 and $\phi : F \to G$ is the canonical epimorphism, we have elements $u, v \in F$ such that $[u, v]$ is conjugate within F to $(ab)^{6\epsilon}, \epsilon = \pm 1$ and $U = \phi(u), V = \phi(v)$. If we regard F as a Fuchsian group with finite area, in fact as the classical modular group, we see that $< u, v >$ is the only two-generator free subgroup of genus 1 and index 6 in F. In fact $< u, v >$ is the commutator subgroup F' of F. Therefore $(u, v) \overset{N}{\sim} (abab^2, b^2aba)$ in F and hence $(U, V) \overset{N}{\sim} (ABAB^2, B^2ABA)$ in G.

The cases (d) and (e) are handled in an analagous manner. □

Corollary 3. *Let $G =< A, B; [A, B]^n = 1 >, n \geq 2$. Then the following statements hold:*

(1) Two elements $U, V \in G$ generate G if and only if $[U, V]$ is conjugate in G to $[A, B]^\epsilon, \epsilon = \pm 1$.

(2) Each automorphism of G is induced by an automorphism of the free group of rank 2.

(3) Each automorphism of the free group of rank 2 induces an automorphism of G.

Corollary 4. *Let $G =< E_1, E_2, E_3; E_1^2 = E_2^2 = E_3^2 = (E_1E_2E_3)^q = 1 >$, $q = 2k + 1 \geq 3$ Then the following statements hold:*

(1) Two elements $U, V \in G$ generate G if and only if $[U, V]$ is conjugate in G to $[E_1E_2, E_3E_1]^\epsilon, \epsilon = \pm 1$.

(2) Each automorphism of G is induced by an automorphism of the free group of rank 2.

(3) Each automorphism of the free group of rank 2 induces an automorphism of G.

The first statement in each of the corollaries follows easily from the proof of Lemma 14 while the other two statements follow by a simple calculation {see the presentation given in the proof of Theorem 12}.

Corollary 5. *Let $A, B \in PSL_2(\mathbb{R})$ and $G =< A, B >$ a cocompact triangle group. Suppose $\{U, V\}$ is a generating pair for G with $tr[U, V] < 2$ and $0 \leq trU, trV$. Then:*

(a) If $G =< A, B; A^2 = B^3 = (AB)^q = 1 >, q \geq 8$, $\gcd(q, 6) = 1$ then $tr[U, V] = -2\cos(6\pi/q)$, and (U, V) is Nielsen equivalent in a trace minimizing manner to a pair (R, S) which satisfies $trR = trS = trRS$.

(b) If $G =< A, B; A^2 = B^4 = (AB)^q = 1 >, q \geq 5$, $\gcd(q, 2) = 1$ then $tr[U, V] = -2\cos(4\pi/q)$, and (U, V) is Nielsen equivalent in a trace minimizing manner to a pair (R, S) which satisfies $trR = trS$ and $trRS = \frac{1}{2}(trR)^2$.

(c) *If* $G =< A, B; A^3 = B^3 = (AB)^q = 1 >, q \geq 4$, $\gcd(q, 3) = 1$ *then* $tr[U, V] = -2\cos(3\pi/q)$, *and* (U, V) *is Nielsen equivalent in a trace minimizing manner to a pair* (R, S) *which satisfies* $trR = trS = trRS$.

PROOF. (a) By Lemma 13 group theoretically we may assume that $U = ABAB^2, V = B^2ABA$. Now we regard A and B as the standard generators for G with $trA = 0, trB = 1$ and $trAB = -2\cos(\pi/q)$. Then $2 < trU = 4\cos^2(\pi/q) - 1 < 3$ and $trU = trV = trUV$ (recall that we chose $0 \leq trU, trV$). From this we obtain $tr[U, V] = 3(4\cos^2(\pi/q) - 1) - (4\cos^2(\pi/q))^3 - 2 = -2\cos(6\pi/7)$ completing part (a).

The other two cases are handled in an analagous manner. □

In [5], [22] and [23] proofs for the various cases in Lemma 13 were given based on the Nielsen cancellation method. The proof is not difficult this way but our direct method avoided this method. We mention however that S.Pride proved the following theorem [21] on the cancellation method in HNN groups.

Theorem 6. ([21]) *Let* $G =< a, b; R^m(a, b) = 1 >, 2 \leq m$, *where R(a,b) is cyclically reduced, not a proper power and not a primitive element in the free group on a,b. If (u,v) is a generating pair for* G *then* $(u, v) \overset{N}{\sim} (a, b)$.

Lemma 12 and Lemma 13 together with Lemma 2 show that the set of conjugacy classes in $PGL_2(\mathbb{R})$ of the groups with signature $(1; n; 0), n \geq 2$, or $(0; 2, 2, 2, q; 0), q = 2k + 1 \geq 2$ are in one to one correspondence with the spaces

$$\{(x, y, z) \in \mathbb{R}^3; x^2 + y^2 + z^2 - xyz = 2 - 2\cos(\pi/n), 2 < x \leq y \leq z \leq (1/2)xy\}$$

and

$$\{(x, y, z) \in \mathbb{R}^3; x^2 + y^2 + z^2 - xyz = 2 - 2\cos(2\pi/q), 2 < x \leq y \leq z \leq (1/2)xy\}$$

respectively. Analagously, with the help of Lemma 7, we have that the set of conjugacy classes in $PGL_2(\mathbb{R})$ of thc fundamental group of a once punctured torus, that is up to isomorphism the group $< A, B, P; [A, B]P = 1 >$ with $A, B, P \in PSL_2(\mathbb{R})$, P parabolic and $tr[A, B] = -2$ is in one to one correspondence with the space

$$\{(x, y, z) \in \mathbb{R}^3; x^2 + y^2 + z^2 = xyz, 2 < x \leq y \leq z \leq (1/2)xy\}.$$

A complete discussion of the possibilities of a commutator to be a proper power in a free product of groups can be found in the work of Comerford, Edmonds and Rosenberger [2].

Some of the questions which arise from the proof of Lemma 13 play a key role in the characterization of the normal subgroups of genus one in free products of finitely many finite cyclic groups {see [7]}. In the special case of the Hecke group $G(\sqrt{2}) \cong < a, b; a^2 = b^4 = 1 >$ we get the following: If $b_1(n)$ is the number of normal subgroups of $G(\sqrt{2})$ of index 4n and genus one, $n \in \mathbb{N}$, then $b_1(n) = \frac{1}{4} card\{(x,y) \in \mathbb{Z}^2; x^2 + y^2 = n\}$.

Lemma 13 and Corollary 5 did not handle the case of a (2,3,7)-triangle group. We do this now. Here however we need a different argument because of the additional last case in Lemma 7. Fortunately for this case we may use some arithmetic arguments. By Takeuchi [31] the (2,3,7)-triangle group is arithmetic. For our purposes we may take the following theorem [14] as a definition of an arithmetic Fuchsian group. This theorem is a reformulation and slight extension of Takeuchi's theorem [34].

Theorem 7. ([14]) *Let Γ be a Fuchsian group of the first kind. Then Γ is an arithmetic Fuchsian group if and only if*

(1) $k = \mathbb{Q}(tr\gamma | \gamma \in \Gamma)$ is a totally real algebraic number field and the set of traces $tr\Gamma \subset O_{k_1}$ the ring of integers in k_1, and

(2) for every $\mathbb{Q}$ - isomorphism $\sigma : k_1 \to \mathbb{R}$ such that $\sigma|_{k_2} \neq Id, k_2 = \mathbb{Q}((tr\gamma)^2)|\gamma \in \Gamma)$, then $|\sigma(tr\gamma)| < 2$ for all $\gamma \neq \pm I$.

Lemma 14. *Let $G =< A, B; A^2 = B^3 = (AB)^7 = 1 >\subset \mathbb{R}$ be a $(2,3,7)$-triangle group. Suppose (U,V) is a generating pair for G with $tr[U,V] < 2$ and $0 \leq trU, trV$. Then one and only one of the following cases holds:*

(1) $(U,V) \overset{N}{\sim} (ABAB^2, B^2ABA), tr[U,V] = -2\cos(6\pi/7)$, and (U,V) is Nielsen equivalent in a trace minimizing manner to a pair (R,S) which satisfies $trR = trS = trRS$;

(2) $(U,V) \overset{N}{\sim} (AB^2ABAB^2AB^2AB, B^2ABAB^2ABABA)$, $tr[U,V] = -2\cos(4\pi/7)$, and (U,V) is Nielsen equivalent in a trace minimizing manner to a pair (R,S) which satisfies $trS = trRS = trR + 1$.

PROOF. G is an arithmetic Fuchsian group. We have $k_1 = \mathbb{Q}(2\cos(\pi/7)) = k_2$ and $O_{k_1} = \mathbb{Z}[2\cos(\pi/7)]$ in the notation of Theorem 7. k_1 is a totally real field of degree 3 over $\mathbb{Q}$ and moreover $\mathbb{Q} \subset k_1$ is Galois. Hence we must consider only three -automorphisms of k_1, given by the identity and $\sigma_1 : k_1 \to k_1, \sigma_1(2\cos(\pi/7)) = -2\cos(2\pi/7)$ and $\sigma_2 : k_1 \to k_1, \sigma_2(2\cos(\pi/7)) = 2\cos(3\pi/7)$. Let $\lambda_1 = 2\cos(\pi/7), \lambda_2 = -2\cos(2\pi/7), \lambda_3 = 2\cos(3\pi/7)$.

Let $\{U,V\}$ be a generating pair for G with $tr[U,V] < 2$ and $0 \leq trU, trV$. As we noted earlier we must have $2 < trU, trV$ and $|tr[U,V]| < 2$. By Lemma 2 we may assume that $2 < trU < 3$ and $trU \leq trV \leq trUV \leq (1/2)trUtrV$. Let $x = trU$, $y = trV$ and $z = trUV$. By Lemma 8 and the remarks after Lemma 8 we have $tr[U,V] = \pm 2\cos(4\pi/7)$ or $tr[U,V] = \pm 2\cos(6\pi/7)$. Suppose first that $tr[U,V] = \pm 2\cos(4\pi/7)$, that is,

$$x^2 + y^2 + z^2 - xyz = 2 - 2\epsilon\cos(4\pi/7), \epsilon = \pm 1. \qquad (*)$$

Let $\gamma_\epsilon = 2 - 2\epsilon\cos(4\pi/7)$. We have $2 < x < 3$ and $|\sigma_i(x)| < 2$ for $i = 1, 2$. Now $|\sigma_i(x)| < 2$ implies $|\sigma_i(y)| < 2$ for $i = 1, 2$.

Since k_1 is totally real, solving the equation (*) for z shows that we must have $(\sigma_i(x)^2-4)(\sigma_i(y)^2-4) \geq 4(4-\sigma_i(\gamma_\epsilon))$, and hence $\sigma_i(x)^2 \leq \sigma_i(\gamma_\epsilon), i = 1, 2$. Analagously $\sigma_i(y)^2 \leq \sigma_i(\gamma_\epsilon)$ and $\sigma_i(z)^2 \leq \sigma_i(\gamma_\epsilon)$ for $i = 1, 2$.

Now if we set $x^2 = a_0 + a_1\lambda_1 + a_2\lambda_1^2$ with $a_j \in \mathbb{Z}, j = 0, 1, 2$, we get the following inequalities for x^2

(1) $4 < a_0 + a_1\lambda_1 + a_2\lambda_1^2 < 9$

(2) $0 < a_0 + a_1\lambda_2 + a_2\lambda_2^2 \leq \sigma_1(\gamma_\epsilon) = 2 + \epsilon\lambda_1$

(3) $0 < a_0 + a_1\lambda_3 + a_2\lambda_3^2 \leq \sigma_2(\gamma_\epsilon) = 2 + \epsilon\lambda_2$

which we must solve for a_0, a_1, a_2.

From the equation (*) and the inequalities between x, y, z we have

$$4\left(\frac{x^2 - \gamma_\epsilon}{x^2 - 4}\right) \leq y^2 \leq \frac{x^2 - \gamma_\epsilon}{x - 2}.$$

Having the solutions a_0, a_1, a_2 for x^2 we may get the possibility $y^2 = b_0 + b_1\lambda_1 + b_2\lambda_1^2$ with $b_j \in \mathbb{Z}, j = 0, 1, 2$, from these inequalities together with $0 < \sigma_i(y^2) \leq \sigma_i(\gamma_\epsilon)$ for $i = 1, 2$. Having the possible solutions a_0, a_1, a_2 for x^2 and the possible solutions b_0, b_1, b_2 for y^2 we can get z from the equation (*) and the inequalities $2 < x \leq y \leq z \leq (1/2)xy$. Moreoever we have to consider the fact that $k_1 = k_2$, which means especially, $x, y \in k_1$ if $x^2, y^2 \in k_1$. We note that this remark here eliminates the (2,4,7)-triangle group.

With the above in mind we can obtain only the following two possibilities for $(x, y, z) \in k_1^3$:

(i) $(x, y, z) = (\lambda_1^2 - 1, \lambda_1^2, \lambda_1^2)$ if $\epsilon = 1$ and

(ii) $(x, y, z) = (\lambda_1 + 1, \lambda_1 + 1, \lambda_1 + 1)$ if $\epsilon = -1$.

Suppose first that $(x, y, z) = (\lambda_1^2 - 1, \lambda_1^2, \lambda_1^2)$ and $\epsilon = 1$. Then $\gamma_1 = 2 - 2\cos(4\pi/7)$. Let $R = AB^2ABAB^2AB^2AB$ and $S = B^2ABAB^2ABABA$. We have $trRS = \lambda_1^2 - 1, trR^{-1} = trS = \lambda_1^2$. Therefore we have realized the triple $(\lambda_1^2 - 1, \lambda_1^2, \lambda_1^2)$ as a trace triple for a generating pair for G. {Recall that $tr[R, S] = 2\lambda_1^4 + (\lambda_1^2 - 1)^2 - \lambda_1^4(\lambda_1^2 - 1) - 2 = -2\cos(4\pi/7)$.}

Since there is exactly one possible triple for $\epsilon = 1$ since two (2,3,7)-triangle groups are conjugate within $PSL_2(\mathbb{R})$ and since a (2,3,7)-triangle group is not a proper subgroup of any other Fuchsian group [12], $\{U, V\}$ with $tr[U, V] = -2\cos(4\pi/7)$ is Nielsen equivalent to $\{R, S\}$.

Now suppose $(x, y, z) = (\lambda_1 + 1, \lambda_1 + 1, \lambda_1 + 1)$ and $\epsilon = -1$. Then $\gamma_{-1} = 2 + 2\cos(4\pi/7) = 2 - 2\cos(3\pi/7)$. This triple belongs to a (3,3,7)-triangle group $H = < C, D; C^3 = D^3 = (CD)^7 = 1 >$ because $|trCD^2| = |trD^2C| = |trCDC| = \lambda_1 + 1$. Recall that we also have $\mathbb{Q}(tr\gamma; \gamma \in H) = \mathbb{Q}((tr\gamma)^2; \gamma \in H) = k_1 = \mathbb{Q}(2\cos(\pi/7) = k_2$ and hence the above triple does not occur in connection with the (2,3,7)-triangle group G.

The case $tr[U,V] = \pm 2\cos(6\pi/7)$ may be considered in an analagous manner. In fact this case corresponds to case (c) in Lemma 13 and case (a) in Corollary 5. This completes the proof of Lemma 14. □

We remark that the two-generator arithmetic Fuchsian groups were completely determined by Takeuchi [33] and [35] and C. Maclachlan and G. Rosenberger [14]. As a consequence Maclachlan and Rosenberger obtained a description of the commensurability classes of two-generator Fuchsian groups [15].

With the proof of Lemma 14 together with the other results in this section and the last section we obtain Theorem B which we now restate.

Theorem B. *Let $G = \langle U, V \rangle$ be a two-generator Fuchsian group.*

(2.1) If G is of type (1.1) or (1.5) then $(U,V) \overset{N}{\sim} (A,B)$.

(2.2) If G is of type (1.6) then $(U,V) \overset{N}{\sim} (AC,CA)$.

(2.3) If G is of type (1.2) or (1.3) then $(U,V) \overset{eN}{\sim} (A,B)$.

(2.4) If G is of type (1.4) then one and only one of the following cases holds

- *(a) $(U,V) \overset{eN}{\sim} (A,B)$;*
- *(b) G is a $(2,3,r)$ triangle group with $\gcd(r,6) = 1$ and $(U,V) \overset{N}{\sim} (ABAB^2, B^2ABA)$;*
- *(c) G is a $(2,4,r)$ triangle group with $\gcd(r,2) = 1$ and $(U,V) \overset{N}{\sim} (AB^2, B^3AB^3)$;*
- *(d) G is a $(3,3,r)$ triangle group with $\gcd(r,3) = 1$ and $(U,V) \overset{N}{\sim} (AB^2, B^2A)$;*
- *(e) G is a $(2,3,7)$ triangle group with $\gcd(r,6) = 1$ and $(U,V) \overset{N}{\sim} (AB^2ABAB^2AB^2AB, B^2ABAB^2ABABA)$.*

Statements (2.1),(2.2) and (2.3) follow from Lemma 13 and general properties of free products of cyclics and (2.4) follows from Lemmas 13 and 14 and the corollaries.

All Nielsen equivalence classes of generating pairs - up to the obvious ones - in (2.3) and (2.4) can be obtained from the following possibilites for generating pairs of (p,q,r) - triangle groups consisting of two elements of finite order whose product is also of finite order. That this gives a complete list can be derived from the work of Knapp [10].

(1) The $(2,3,r)$-triangle group with $\gcd(r,3) = 1$ can be generated by $U = ABABA$ and $V = B$ with $UV = (AB)^3$.

(2) The $(2,3,r)$-triangle group with $\gcd(r,2) = 1$ can be generated by $U = A$ and $V = BAB$ with $UV = (AB)^2$.

(3) The $(2,3,r)$-triangle group with $\gcd(r,2) = 1$ can be generated by $U = (AB)^{-1}$ and $V = BABABAB^2$ with $UV = B^2(AB)^4B$.

(4) The $(2,3,7)$-triangle group can be generated by $U = ABAB^2ABA$ and $V = B$ with $UV = ABAB^2ABAB = ABAB^2ABAB(AB)^{-7} = ABABAB^2(AB)BAB^2AB^2A$.

(5) The $(2,p,q)$-triangle group with $\gcd(2,q) = 1$ can be generated by $U = ABA$ and $V = B$ with $UV = (AB)^2$.

(6) The $(2,p,q)$-triangle group with $\gcd(2,p) = 1$ can be generated by $U = BA$ and $V = AB$ with $UV = B^2$.

The Nielsen inequivalence of two generating pairs $(U_1, V_1), (U_2, V_2)$ of a (p,q,r)-triangle group with U_i, V_i and U_iV_i of finite order can be obtained form the following proposition which is derived directly from work of Kern-Isbrenner and Rosenberger[9].

Proposition 7. *Let $r, r_1, r_2, s, s_1, s_2 \in \mathbb{N}$ such that $r, s \geq 2$ and $1 \leq r_i \leq r/2, gcd(r, r_i) = 1, 1 \leq s_i \leq s/2, gcd(s, s_i) = 1$, for $i = 1,2$. Let $\sin(r_1\pi/r)\sin(s_1\pi/s) = \sin(r_2\pi/r)\sin(s_2\pi/s)$. If $r \neq s$ then $r_1 = r_2$ and $s_1 = s_2$. If $r = s$ then $r_1 = r_2$ and $s_1 = s_2$ or $r_1 = s_2$ and $r_2 = s_1$.*

Now we turn our attention to Theorem D. Most has been done already. To complete the result we need the following.

Lemma 15. *Suppose $U, V \in PSL_2(\mathbb{R})$ with $0 \leq trU, trV$. Assume that one of the following cases holds:*

(i) $tr[U,V] = -2\cos(6\pi/q), q \geq 7, \gcd(q,6) = 1$ *and* $trU = trV = trUV$.

(ii) $tr[U,V] = -2\cos(4\pi/q), q \geq 5, \gcd(q,2) = 1$ *and* $trU = trV, trUV = (1/2)(trU)^2$.

(iii) $tr[U,V] = -2\cos(3\pi/q), q \geq 4, \gcd(q,3) = 1$ *and* $trU = trV = trUV$.

(iv) $tr[U,V] = -2\cos(4\pi/q), q = 7,$ *and* $trV = trUV = trU + 1$.

Then $< U, V >$ is discrete.

PROOF. (i) Let $tr[U,V] = -2\cos(6\pi/q), q \geq 7, \gcd(q,6) = 1$, and $trU = trV = trUV$. Let $x = trU$. Then $0 = tr[U,V] + 2\cos(6\pi/q) = -x^3 + 3x^2 - 2 + 2\cos(6\pi/q)$. The polynomial $f(x) = -x^3 + 3x^2 - 2 + 2\cos(6\pi/q)$ has three real zeros and only one of these zeros has an absolute value greater than 2. This is the zero $\lambda^2 - 1$ where $\lambda = 2\cos(\pi/q)$. Therefore $trU = trV = trUV = \lambda^2 - 1$. On the other hand in the $(2,3,q)$-triangle group, $q \geq 7$, $\gcd(q,6) = 1$, $G =< A, B; A^2 = B^3 = (AB)^q = 1 >$ we have the generating pair $\{R, S\}$ with $R = ABAB^2, S = B^2ABA$ and $trR = trS = trRS = \lambda^2 - 1$. We know that $\lambda^2 - 1 > 2$. Therefore $< U, V >$ is conjugate within $PGL_2(\mathbb{R})$ to $G =< R, S >$ and hence $< U, V >$ is discrete because G is discrete.

In case (ii) we must consider the polynomial $f(x) = -(1/4)x^3 + 2x^2 - 2 + 2\cos(4\pi/q)$ with $x = trU$. In case (iii) we must consider the polynomial $f(x) = -x^3 + 3x^2 - 2 + 2\cos(3\pi/q)$ again with $x = trU$. Finally in case (iv) we must consider the polynomial $f(x) = -x^3 + 4x^2 - 2x - 1 + 2\cos(4\pi/7)$ with $x = trV$. In each case $f(x)$ has three real zeros only one of which has an absolute value greater than 2 namely the zero $2\sqrt{2}\cos(\pi/q), 2\cos(\pi/q)$ and $(2\cos(\pi/7))^2$ respectively. Once we have this zero we may argue analagously as in case (i). □

This then completes all the material for Theorem D and the corollary which we now restate.

Theorem D. *Suppose* $U, V \in PSL_2(\mathbb{R})$ *with* $0 \leq trU, 0 \leq trV$ *and* $tr[U,V] < 2$ *and suppose* $G =< U, V >$. *Then* G *is discrete if and only if one of the following cases holds:*

(3.1) $tr[U,V] \leq -2$;

(3.2) $tr[U,V] = -2\cos(\pi/p), p \in \mathbb{N} \setminus \{1\}$;

(3.3) $tr[U,V] = -2\cos(2\pi/p), p \in \mathbb{N} \setminus \{1\}$ *and* $\gcd(p, 2) = 1$;

(3.4) $tr[U,V] = -2\cos(6\pi/r), r \in \mathbb{N}, r \geq 5$ *with* $\gcd(r, 2) = 1$ *and* (U, V) *is Nielsen equivalent in a trace minimizing manner to a pair* (R, S) *which satisfies* $trR = trS = trRS$;

(3.5) $tr[U,V] = -2\cos(4\pi/r), r \in \mathbb{N}, r \geq 7$ *with* $\gcd(r, 6) = 1$ *and* (U, V) *is Nielsen equivalent in a trace minimizing manner to a pair* (R, S) *which satisfies* $trR = trS$ *and* $trRS = \frac{1}{2}(trR)^2$;

(3.6) $tr[U,V] = -2\cos(3\pi/r), r \in \mathbb{N}, r \geq 4$ *with* $\gcd(r, 3) = 1$ *and* (U, V) *is Nielsen equivalent in a trace minimizing manner to a pair* (R, S) *which satisfies* $trR = trS = trRS$;

(3.7) $tr[U,V] = -2\cos(4\pi/7)$ *and* (U, V) *is Nielsen equivalent in a trace minimizing manner to a pair* (R, S) *which satisfies* $trS = trRS = trR + 1$.

Moreover, if G *is discrete then* G *is of type (1.1) in case (3.1), of type(1.5) in case (3.2), of type (1.6) in case (3.3), a* $(2,3,r)$ *triangle group in case (3.4), a* $(2,4,r)$ *triangle group in case (3.5), a* $(3,3,r)$ *triangle group in case (3.6), and a* $(2,3,7)$ *triangle group in case (3.7).*

Corollary. *Suppose* $G =< U, V > \subset PSL_2(\mathbb{R})$ *is a Fuchsian group. Then* $|tr[U,V] - 2| \geq 2 - 2\cos(\pi/7)$.

References

[1] A.F. Beardon, *The Geometry of Discrete Groups* (Springer-Verlag, 1983).

[2] L. Comerford, C.Edmonds and G.Rosenberger, Commutators as powers in free products of groups, *Proc. Amer. Math. Soc.*, to appear.

[3] L. Ford, *Automorphic Functions* (Chelsea, New York, 1929).

[4] T.Jorgensen, A note on subgroups of $SL_2(\mathbb{C})$, *Quart. J. Math.* Ser. II **28**(1977), 209–212.

[5] R.N. Kalia and G.Rosenberger, Automorphism of the Fuchsian Groups of type (0;2,2,2,q;0), *Comm. Algebra* **6**(1978), 1115–1129.

[6] S. Katok, *Fuchsian Groups* (University of Chicago Press, Chicago, 1993).

[7] G. Kern-Isberner and G.Rosenberger, Normalteiler von Geschlecht eins in freien Produkten endlicher zyklischer Gruppen, *Resultate Math.* **11**(1987), 272–278.

[8] G. Kern-Isberner and G.Rosenberger, Einige Bemerkungen uber Untergruppen der $PSL_2(\mathbb{C})$, *Resultate Math.* **6**(1983), 40–47.

[9] G. Kern-Isberner and G.Rosenberger, Uber Diskretheitsbedingungen und die diophantische Gelichung $ax^2 + by^2 + cz^2 = dxyz$, *Arch. Math.* **34**(1980), 481–493.

[10] A.W. Knapp, Doubly generated Fuchsian groups, *Michigan Math. J.* **15**(1968), 289–304.

[11] J.Lehner, On (p,q,r)- groups, *Bull. Inst. Math. Soc. Academica Sinica* **6**(1978), 419–422.

[12] J.Lehner, *Discontinuous Groups and Automorphic Functions* (American Mathematical Society, Math Surveys No. VIII, 1964).

[13] R.C. Lyndon and P.E. Schupp, *Combinatorial Group Theory* (Springer-Verlag, 1977).

[14] C.Maclachlan and G.Rosenberger Two-Generator Arithmetic Fuchsian Groups I. *Math. Proc. Cambridge Philos. Soc.* **93**(1983), 383–391; II., *Math. Proc. Cambridge Philos. Soc.*.

[15] C.Maclachlan and G.Rosenberger, Commensurability Classes of Two-Generator Arithmetic Fuchsian Groups, in *Discrete Groups and Geometry* (London Math. Soc., 1992), 171–190.

[16] W. Magnus, The uses of 2×2 matrices in combinatorial group theory, *Resultate Math.* **4**(1981), 171–192.

[17] W. Magnus, Two generator subgroups of $PSL_2(\mathbb{C})$, *Nachr. Akad. Wiss. Gottingen II. Math. Phys. Kl.* **7**(1975), 81- 94.

[18] W. Magnus, A. Karrass and D. Solitar, *Combinatorial Group Theory* (Wiley 1966, Second Edition, Dover Publications, New York, 1976).

[19] J.P. Matelski, The classification of discrete 2-generator subgroups of $PSL_2(\mathbb{R})$, *Israel J. Math.* **42**(1982), 309–317.

[20] N. Peczynski, G. Rosenberger and H. Zieschang, Uber Erzeugende ebener diskontinuierlicher Gruppen, *Invent. Math.* **29**(1975), 161–180.

[21] S.J. Pride, The isomorphism problem for two generator one-relator groups with torison is solvable, *Trans. Amer. Math. Soc.* **227**(1977), 109–139.

[22] N. Purzitsky, All two-generator Fuchsian groups, *Math. Z.* **147**(1976), 87 - 92.

[23] N. Purzitsky and G. Rosenberger, Two-generator Fuchsian groups of genus one, *Math. Z.* **128**(1972), 245–251; Correction: *Math Z.* **132**(1973), 261–262.

[24] G.Rosenberger, Minimal generating systems of a subgroup of $SL_2(\mathbb{C})$, *Proc. Edinburgh Math. Soc.* **31**(1988), 261–265.

[25] G.Rosenberger, Some remarks on a paper of A. Majeed on the freeness of the groups $< a^n, b^n >$ for some integer n; $a, b \in SL_2(\mathbb{C})$, *Results in Math.* **11**(1987), 267–271.

[26] G.Rosenberger, All generating pairs of all two-generator Fuchsian groups, *Arch. Math.* **46**(1986), 198–204.

[27] G.Rosenberger, Some remarks on a paper of A.F. Beardon and P.L. Waterman about strongly discrete subgroups of $SL_2(\mathbb{C})$, *J. London Math. Soc.* **27**(1983), 39–42.

[28] G.Rosenberger, *Zum Rang und Isomorphieproblem fur freie Produkte mit Amalgam* (Habilitationsschrift, Hamburg 1974).

[29] G.Rosenberger, Applications of Nielsen's Reduction Method in the Solution of Combinatorial Problems in Group Theory (in London Math. Soc. Lecture Notes **36**, 1979), 339–358.

[30] G.Rosenberger, Eine Bemerkung zu einer Arbeit von T. Jorgensen, *Math Z.* **165**(1979), 261–265.

[31] A. Selberg, On discontinuous groups in higher dimensional symmetric spaces, *Contributions to Function Theory Tata Institute Bombay* (1960), 147-164.

[32] C.L. Siegel, Bemerkung zu einem Satze von Jakob Nielsen, *Mat. Tidsskr. B* (1950), 66–70.

[33] K.Takeuchi, Arithmetic Triangle Groups, *J. Math. Soc. Japan* **29**(1977), 91–106.

[34] K.Takeuchi, A Characterization of Arithmetic Fuchsian Groups, *J. Math. Soc. Japan* **27**(1975), 600–612.

[35] K.Takeuchi, Arithmetic Fuchsian Groups with signature $(1; e)$, *J. Math. Soc. Japan* **35**(1983), 381–407.

[36] H. Zieschang, Uber die Nielsensche Kurzungsmethode in freien Produkten mit Amalgam, *Invent. Math.* **10**(1970), 4–37.

PARAMETRIC WORDS AND MODELS OF THE ELEMENTARY THEORY OF NON-ABELIAN FREE GROUPS

ANTHONY M. GAGLIONE*[1] and DENNIS SPELLMAN†

*U.S. Naval Academy, Annapolis, MD 21402, U.S.A.
†Philadelphia, PA 19124, U.S.A.

Abstract

A group is n-free if every subgroup generated by n or fewer distinct elements is free. In [8], the authors observed that if G is a finitely generated model of the elementary theory of the non-Abelian free groups, then G is 2-free. The main result here is that such a group is 3-free. The principal tool used is a characterization, due to Hmelevskii [10], of the set of nontrivial solutions in a free group of a three variable word equation without coefficients.

1. Introduction and preliminaries

We start by giving a convention and definition which are used throughout this paper.

Convention. The trivial group $\{1\}$ is free of rank zero.

Definition 1. Let n be a positive integer. The group G is *n-free* provided every subgroup of G generated by n or fewer distinct elements is free.

Clearly, every n-free group is m-free for all integers m with $1 \leq m \leq n$. The 1-free groups are precisely the torsion free groups.

Lemma 1. (Harrison [9]) *Let G be a group. Then the following three properties are pairwise equivalent.*

(i) The relation of commutativity is transitive on the non-identity elements of G.

(ii) The centralizer in $G, Z_G(g)$, of every non-identity element $g \neq 1$ in G is Abelian.

(iii) Every pair of distinct maximal Abelian subgroups $M_1 \neq M_2$ in G has a trivial intersection; $M_1 \cap M_2 = \{1\}$.

---

[1]The research of this author was partially supported by the Naval Research Laboratory, Radar Division, Identification Systems Branch.

Definition 2. If G satisfies any one of the three conditions (and therefore all three) of Lemma 1, then G is *commutative transitive.*

As observed in [5], if G is 2-free, then for all $g \in G - \{1\}$ one has $Z_G(g)$ is a locally cyclic group (necessarily torsion free). In particular, G is commutative transitive. Thus, this conclusion also applies to n-free groups for all $n \geq 2$.

Definition 3. The group G is $\mathbb{Z}$-*commutative transitive* provided for every $g \in G - \{1\}$, we have $Z_G(g)$ is infinite cyclic.

Clearly, $\mathbb{Z}$-commutative transitive groups are commutative transitive by condition (ii) of Lemma 1. Observe that $\mathbb{Z}$-commutative transitive groups are torsion free.

Definition 4. (cf. p. 242, [11]) Let G be a torsion free group. Then G is an *R-group* provided that for every positive integer n and every element $g \in G$ the equation $x^n = g$ has at most one solution in G (i.e., nth roots, when they exist in G, are unique).

It is an easy exercise to verify that every torsion free, commutative transitive group is an R-group. In particular, every $\mathbb{Z}$-commutative transitive group is an R-group.

Definition 5. (cf. p. 243, [11]) Let G be a $\mathbb{Z}$-commutative transitive group and let P be a subgroup of G. Then P is *isolated* in G provided that for every positive integer n and every element $g \in G$ it is the case that $g \in P$ whenever $g^n \in P$.

Since the intersection of any family of isolated subgroups of G is again isolated in G, we must have given any subgroup H in G a least isolated subgroup $I(H)$ in G such that $I(H)$ contains H. $I(H)$ is the *isolator* of H in G.

Definition 6. Let G be a $\mathbb{Z}$-commutative transitive group and let C be a subgroup of G. Then C is *centralizer closed* in G provided that for any nontrivial element g of C, we must have $Z_G(g) \subseteq C$.

Let ω be the first limit ordinal. Let H be a subgroup of the $\mathbb{Z}$-commutative transitive group G. We construct an ω-chain

$$I_0(H) \subseteq I_1(H) \subseteq \ldots \subseteq I_n(H) \subseteq \ldots$$

of subgroups of G as follows:

$$\begin{cases} I_0(H) = H \\ I_{n+1}(H) \text{is the subgroup of } G \text{ generated} \\ \text{by } Z_G(g) \text{ as } g \text{ varies over } I_n(H) - \{1\}. \end{cases}$$

Since $1 \in I_{n+1}(H)$ and $g \in Z_G(g) \subseteq I_{n+1}(H)$ whenever $g \in I_n(H) - \{1\}$ we see that the groups $I_n(H)$ are indeed nested. We then claim that the isolator, $I(H)$, is the direct limit

$$I(H) = \bigcup_{n<\omega} I_n(H).$$

Indeed, we have

Lemma 2. *Let G be a $\mathbb{Z}$-commutative transitive group and let H be a subgroup of G. Then H is centralizer closed in G if and only if H is isolated in G. In particular, $\bigcup_{n<\omega} I_n(H)$ is the least centralizer closed subgroup of G containing H (and so it is $I(H)$).*

PROOF. Suppose that H is centralizer closed in G. Let $g^n \in H$ where $g \in G$ and n is a positive integer. Suppose first that $g^n = 1$. Since G is torsion free, this implies $g = 1 \in H$. Now suppose $g^n = h \neq 1$. Then $g \in Z_G(h) \subseteq H$. Thus every centralizer closed subgroup is isolated. Now suppose that H is isolated in G. Let $h \in H - \{1\}$. Since G is $\mathbb{Z}$-commutative transitive, $Z_G(h) =< h_0 >$ is infinite cyclic. Furthermore, we may choose h_0 such that $h = h_0^n$ where $n > 0$. The $h_0 \in H$ so $Z_G(h) \subseteq H$. Thus every isolated subgroup is centralizer closed.

To show that the isolator $I(H)$ is $\bigcup_{n<\omega} I_n(H)$ it suffices to show that this direct limit is the least centralizer closed subgroup of G containing H according to the equivalence just established. Let $1 \neq g \in \bigcup_{n<\omega} I_n(H)$. Then $g \in I_m(H)$ for some $m < \omega$. Therefore, $Z_G(g) \subseteq I_{m+1}(H) \subseteq \bigcup_{n<\omega} I_n(H)$. Thus, $\bigcup_{n<\omega} I_n(H)$ is centralizer closed in G. Suppose C is any centralizer closed subgroup of G containing $H = I_0(H)$. Assume C contains $I_n(H)$. Let $g \in I_n(H) - \{1\}$. Then $Z_G(g) \subseteq C$. Therefore, C contains $I_{n+1}(H)$. It follows that C contains $\bigcup_{n<\omega} I_n(H)$. Thus, $\bigcup_{n<\omega} I_n(H)$ is the least centralizer closed subgroup of G containing H. That is, $I(H) = \bigcup_{n<\omega} I_n(H)$. □

Definition 7. Let n be a positive integer and G be a group. G is *strongly n-free* provided the following two properties are satisfied.

(i) G is $\mathbb{Z}$-commutative transitive, and

(ii) if H is any subgroup of G generated by n or fewer distinct elements, then $I(H)$ is free.

Clearly necessary conditions for G to be strongly n-free are that G be n-free and that the centralizer of every non-trivial element in G be infinite cyclic.

Following Hmelevskii [10] we make the following definition (with some modifications).

Definition 8. Given a set $U = \{u_1, \dots, u_k\}$ and a disjoint infinite sequence $N = \{v_m | m < \omega\}$ of distinct elements $v_p \neq v_q$ if $p \neq q$, we define the set of *parametric words* on the *group variables* U and the *integral parameters* N to be the smallest set of strings containing U as a subset and closed under the following rules of formation.

(1) If S is a parametric word, then $(S)^{v_m}$ is a parametric word whenever $v_m \in N$.

(2) If S is a parametric word, then $(S)^{-1}$ is a parametric word.

(3) If S and T are parametric words, then ST is a parametric word.

We define the *length* $L(W)$ of the parametric word W on the group variables U and integral parameters N as follows:

$$\begin{array}{ll} L(W) = 1 & \text{if } W = u_i \in U. \\ L(W) = L(S) + 1 & \text{if } W = (S)^{v_m}. \\ L(W) = L(S) + 1 & \text{if } W = (S)^{-1}. \\ L(W) = L(S) + L(T) & \text{if } W = ST. \end{array}$$

Convention. We allow exactly one *empty parametric word* 1. We assign the length $L(1) = 0$ to 1.

If W is a parametric word on the group variables $u_1, \ldots, u_k$ and the integral parameters $v_0, \ldots, v_m, \ldots$, (we call this sequence v) and G is a $\mathbb{Z}$-commutative transitive group, $(U_1, \ldots, U_k) \in G^k$ and $n \in \mathbb{Z}^\omega, m \mapsto n_m$, is an infinite sequence of integers, then we may define the value $\mathrm{Subst}_W \begin{pmatrix} u_1, \ldots, u_k, v \\ U_1, \ldots, U_k, n \end{pmatrix}$ in G of the substitution into W, $u_1 \mapsto U_1, \ldots, u_k \mapsto U_k; v_0 \mapsto n_0, \ldots, v_m \mapsto n_m, \ldots$ recursively (cf. Hmelevskii [10]) as follows. If $W = u_i \in U$, then

$$\mathrm{Subst}_W \begin{pmatrix} u_1, \ldots, u_k, v \\ U_1, \ldots, U_k, n \end{pmatrix} = U_i.$$

If $W = (S)^{v_m}$, then

$$\mathrm{Subst}_W \begin{pmatrix} u_1, \ldots, u_k, v \\ U_1, \ldots, U_k, n \end{pmatrix} = \mathrm{Subst}_S \begin{pmatrix} u_1, \ldots, u_k, v \\ U_1, \ldots, U_k, n \end{pmatrix}^{n_m}.$$

If $W = (S)^{-1}$, then

$$\mathrm{Subst}_W \begin{pmatrix} u_1, \ldots, u_k, v \\ U_1, \ldots, U_k, n \end{pmatrix} = \mathrm{Subst}_S \begin{pmatrix} u_1, \ldots, u_k, v \\ U_1, \ldots, U_k, n \end{pmatrix}^{-1}.$$

If $W = ST$, then

$$\mathrm{Subst}_W \begin{pmatrix} u_1, \ldots, u_k, v \\ U_1, \ldots, U_k, n \end{pmatrix}$$

$$= \mathrm{Subst}_S \begin{pmatrix} u_1, \ldots, u_k, v \\ U_1, \ldots, U_k, n \end{pmatrix} \mathrm{Subst}_T \begin{pmatrix} u_1, \ldots, u_k, v \\ U_1, \ldots, U_k, n \end{pmatrix}.$$

Suppose $X = \{x_1, x_2, x_3\}$ is a three element set. Let $R = R(x_1, x_2, x_3)$ be a nonempty freely reduced word on $X \cup X^{-1}$. Then the formal expression "$R(x_1, x_2, x_3) = 1$" is a three-variable *word equation without coefficients.* If $F(X)$ is the free group on X and G is a $\mathbb{Z}$-commutative transitive group, then $(g_1, g_2, g_3) \in G^3$ is a *solution* in G to $R(x_1, x_2, x_3) = 1$ provided R (viewed as an element of $F(X)$) lies in the kernel of the homomorphism $h : F(X) \to G$ determined by $h(x_j) = g_j$, $j = 1, 2, 3$. Following Hmelevskii the solution (g_1, g_2, g_3) is *trivial* if the components g_j, $j = 1, 2, 3$, all lie in a common cyclic subgroup (possibly $g_1 = g_2 = g_3 = 1$). Hmelevskii proves that if G is a free group, then the set of nontrivial solutions in G to the three variable word equation without coefficients $R(x_1, x_2, x_3) = 1$ is exhausted by the values in G of a finite set $(W_{i,1}(u_1, u_2), W_{i,2}(u_1, u_2), W_{i,3}(u_1, u_2))$, $i = 1, \ldots, M$, of triples of parametric words on two groups variables u_1 and u_2 and an infinite sequence $v_0, \ldots, v_m, \ldots$ of integral parameters.

Not every substitution will yield a solution so that the above criterion is a necessary but, in general, insufficient condition to be a nontrivial solution. However, that state of affairs is good enough for our purposes. Following Lyndon and Schupp (see p. 64 of [12]) we may view parametric words on the group variables u_1 and u_2 and the integral parameters $N = \{v_m | m < \omega\}$ as representatives of elements of the free $\mathbb{Z}[N]$-group freely generated (as a $\mathbb{Z}[N]$-group) by u_1 and u_2. Here $\mathbb{Z}[N]$ is the polynomial ring over $\mathbb{Z}$ in the algebraically independent indeterminates $N = \{v_m | m < \omega\}$.

This viewpoint allows us to focus on particularly simple parametric words which will be useful for our purposes. Call a parametric word W on u_1 and u_2 and $N = \{v_m | m < \omega\}$ *special* if the only exponents occurring in W other than constants are individual integral parameters and if additionally no integral parameter occurs more than once in W. An example will illustrate why we may restrict ourselves to special parametric words and still not miss any nontrivial solutions.

Example. If x is of the form $((u_1u_2^{v_1}u_1^{-7})^{v_1^3+2v_1v_2-v_2^2}(u_1^5u_2^{-1})^{v_1})^{v_2}$, then x is certainly also of the form $((u_1u_2^{v_1}u_1^{-7})^{v_2}(u_1^5u_2^{-1})^{v_3})^{v_4}$.

Suppose $(W_1(u_1, u_2), W_2(u_1, u_2), W_3(u_1, u_2))$ is a triple of parametric words on the group variables u_1 and u_2 and the integral parameters $N = \{v_m | m < \omega\}$. Let N_j be the set of integral parameters occurring in $W_j(u_1, u_2)$, $j = 1, 2, 3$. The triple is *special* provided each of the components $W_j(u_1, u_2)$ is special $j = 1, 2, 3$ and also N_1, N_2 and N_3 are pairwise disjoint. Clearly the set of values of finitely many special triples exhausts the nontrivial solutions in a free group to a three variable word equation without coefficients.

Let L_0 be the first-order language containing a binary operation symbol $\cdot$, a unitary operation symbol $^{-1}$, and a constant symbol 1 as its only function and constant symbols and $=$ as its only relation symbol - always to be interpreted as the identity relation. (We allow vacuous quantification in formulas of L_0.)

Definition 9. Let $W(u_1, u_2)$ be a special parametric word on the group variables u_1 and u_2 and the integral parameters $N = \{v_m | m < \omega\}$. We define the *depth* of W to be the integer $d(W) \geq 0$ given by

(1) $d(W) = 0$ if W does not involve any integral parameters.
(2) $d(W) = d(S) + 1$ if $W = (S)^{v_m}$.
(3) If $W = (S)^{-1}$ and $d(S) > 0$, then $d(W) = d(S) + 1$.
(4) If $W = ST$ and $\max\{d(S), d(T)\} > 0$, then $d(W) = d(S) + d(T) + 2$.

Note that if $d(S) = d(T) = 0$, then $d((S)^{-1}) = d(ST) = 0$ but $d((S)^{v_m}) = 1$.

We define a formula $\phi_W(y_1, \ldots, y_{d(W)}, z, u_1, u_2)$ of L_0 as follows:

(1) If $W(u_1, u_2)$ does not involve any integral parameters, then $\phi_W(z, u_1, u_2)$ is $z = W(u_1, u_2)$.
(2) If $W = (S)^{v_m}$ where $d(S) = p$, then $\phi_W(y_1, \ldots, y_{p+1}, z, u_1, u_2)$ is
$$(z = 1) \vee ((y_{p+1} \neq 1) \wedge (zy_{p+1} = y_{p+1}z) \wedge \phi_S(y_1, \ldots, y_{p+1}, u_1, u_2)).$$
(3) If $W = (S)^{-1}$ where $d(S) = p > 0$, then $\phi_W(y_1, \ldots, y_{p+1}, z, u_1, u_2)$ is $(z = y_{p+1}^{-1}) \wedge \phi_S(y_1, \ldots, y_{p+1}, u_1, u_2)$.
(4) If $W = ST$ where $d(S) = p$, $d(T) = q$ and $\max\{p, q\} > 0$, then $\phi_W(y_1, \ldots, y_{p+q+2}, z, u_1, u_2)$ is
$$(z = y_{p+1}y_{p+q+2}) \wedge \phi_S(y_1, \ldots, y_{p+1}, u_1, u_2) \wedge \phi_T(y_{p+2}, \ldots, y_{p+q+2}, u_1u_2).$$

Lemma 3. *Let G be a $\mathbb{Z}$-commutative transitive group and $(U_1, U_2) \in G^2$ be a pair of non-commuting elements of G. Let $W(u_1, u_2)$ be a parametric word on the group variables u_1 and u_2 which does not involve any integral parameters. Then*
$$\mathrm{Subst}_W \begin{pmatrix} u_1, u_2, v \\ U_1, U_2, n \end{pmatrix} = W(U_1, U_2)$$
for all $n \in \mathbb{Z}^\omega$. Furthermore, if H is the subgroup of G generated by U_1 and U_2, then $W(U_1, U_2) \in H$.

The proof of Lemma 3 is an easy induction on $L(W)$. The next two lemmas follow by easy inductions on $d(W)$.

Lemma 4. *Let G be a $\mathbb{Z}$-commutative transitive group and $(U_1, U_2) \in G^2$ be a pair of non-commuting elements of G. Let $W(u_1, u_2)$ be a special parametric word on the group variables u_1 and u_2 and the integral parameters $N = \{v_m | m < \omega\}$. Suppose $d(W) = d$. Then for all $g \in G, \exists y_1 \ldots \exists y_d \phi_W(y_1, \ldots, y_d, g, U_1, U_2)$ is true in G whenever*
$$g \in \left\{ \mathrm{Subst}_W \begin{pmatrix} u_1, u_2, v \\ U_1, U_2, n \end{pmatrix} | n \in \mathbb{Z}^\omega \right\}.$$

Lemma 5. *Let G be a $\mathbb{Z}$-commutative transitive group and $(U_1, U_2) \in G^2$ be a pair of non-commuting elements of G. Let H be the subgroup of G generated by U_1 and U_2. Let $W(u_1, u_2)$ be a special parametric word on the group variables u_1 and u_2 and the integral parameters $N = \{v_m | m < \omega\}$. Suppose $d(W) = d$. Then for all $g \in G$, if $\exists y_1 \ldots \exists y_d \phi_W(y_1, \ldots, y_d, g, U_1, U_2)$ is true in G, then $g \in I(H)$.*

2. Some results which give 3-free groups

Let $R(x_1, x_2, x_3) = 1$ be a word equation without coefficients in the three variables x_1, x_2 and x_3. Let $F_2 =< a_1, a_2; >$ be free of rank 2 and suppose the solutions to $R(x_1, x_2, x_3) = 1$ in F_2 are exhausted by the trivial solutions and the set of values in F_2 of the special triples $(W_{i,1}(u_1, u_2), W_{i,2}(u_1, u_2), W_{i,3}(u_1, u_2))$, $i = 1, \ldots, M$, in the two group variables u_1 and u_2 and the infinite sequence of integral parameters $N = \{v_m | m < \omega\}$. Then letting $\overline{x} = (x_1, x_2, x_3)$ and $\overline{u} = (u_1, u_2)$ by Hmelevskii [10] and Lemma 4, there is a nonnegative integer k and a tuple $\overline{y} = (y_1, \ldots, y_k)$ of variables such that the following universal-existential sentence of L_0 is true in F_2.

$$\begin{array}{l} \forall \overline{x} \exists w \exists \overline{y} \exists \overline{u}((R(\overline{x}) = 1) \rightarrow (((w \neq 1) \wedge \bigwedge_{j=1}^{3}(x_j w = w x_j)) \\ \vee((u_1 u_2 \neq u_2 u_1) \wedge \bigvee_{i=1}^{M} \bigwedge_{j=1}^{3} \phi_{W_{ij}}(y_1, \ldots, y_k, x_j, u_1, u_2)))). \end{array} \tag{$*$}$$

Here if $Y_{ij}, 1 \leq i \leq M, 1 \leq j \leq 3$, is the subset of $\{y_1, \ldots, y_k\}$ involved in $\phi_{W_{ij}}(y_1, \ldots, y_k, x_j, u_1, u_2)$ then the $Y_{ij}, 1 \leq i \leq M, 1 \leq j \leq 3$, are pairwise disjoint. (A *universal-existential* sentence of L_0 is one of the form $\forall \overline{s} \exists \overline{t} \phi(\overline{s}, \overline{t})$ where $\overline{s}$ and $\overline{t}$ are tuples of variables and $\phi(\overline{s}, \overline{t})$ is a formula of L_0 containing no quantifiers and containing at most the variables in $\overline{s}$ and $\overline{t}$.) Sacerdote [14] showed that precisely the same universal-existential sentences of L_0 hold in every non-Abelian free group. Let $\mathrm{Th}(F_2) \cap \forall\exists$ be the set of all those universal-existential sentences of L_0 true in every non-Abelian free group. Since we allow vacuous quantifications every universal sentence of L_0 and every existential sentence of L_0 is a special case of a universal-existential sentence of L_0. Thus, if $\Phi = \mathrm{Th}(F_2) \cap (\forall \cup \exists)$ is the set of all those universal sentences of L_0 and all those existential sentences of L_0 true in every non-Abelian free group and Σ is the set of all the sentences of L_0 true in every non-Abelian free group, then $\Phi \subseteq \mathrm{Th}(F_2) \cap \forall\exists \subseteq \Sigma$. Writing $\mathbb{M}(S)$ for the class of all models of a set S of sentences of L_0, we conclude $\mathbb{M}(\Sigma) \subseteq \mathbb{M}(\mathrm{Th}(F_2) \cap \forall\exists) \subseteq \mathbb{M}(\Phi)$.

Theorem 1. *Every strongly 2-free model of* $\mathrm{Th}(F_2) \cap \forall\exists$ *is 3-free.*

PROOF. Let G be a strongly 2-free model of $\mathrm{Th}(F_2) \cap \forall\exists$. Let K be the subgroup of G generated by x_1, x_2 and x_3. If x_1, x_2 and x_3 satisfy no relation

$R(x_1, x_2, x_3) = 1$ (R not freely equal to the empty word), then K is free. Suppose x_1, x_2 and x_3 satisfy the nontrivial relation $R(x_1, x_2, x_3) = 1$. Then (*) is true in G. We then have either of the following two cases. *Case* (1): x_1, x_2 and x_3 all lie in an Abelian subgroup $< w* > \supseteq < w >$, in which case K is cyclic and torsion free (possibly $K = \{1\}$)-hence free of rank ≤ 1. *Case* (2): There are non-commuting elements u_1 and u_2 of G such that by Lemma 5 x_1, x_2 and x_3 all lie in $I(H)$ where H is the subgroup of G generated by u_1 and u_2. Thus, K is free in that case also since $I(H)$ is free according to Definition 7. □

Theorem 2. *Let G be a $\mathbb{Z}$-commutative transitive model of Φ satisfying the maximal condition for two-generator subgroups (i.e., any ascending chain of two-generator subgroups stabilizes). Then G is strongly 2-free.*

PROOF. Let $X = \{x_1, x_2\}$ and let $w(x_1, x_2)$ be a nonempty, freely reduced word on $X \cup X^{-1}$.

For every such word the universal sentence $\forall x \forall y((xy \neq yx) \to (w(x, y) \neq 1)$ of L_0 is true in every non-Abelian free group. It follows that every non-commuting pair of elements of G freely generates a subgroup; moreover, since G is $\mathbb{Z}$-commutative transitive, G is a 2-free group in which the centralizer of every non-trivial element is infinite cyclic. Let $(u_1, u_2) \in G^2$ and let H be the subgroup of G generated by u_1 and u_2. If $u_1 = u_2 = 1$, then $I(H) = \{1\}$ is free of rank zero. If $H \neq \{1\}$ but u_1 and u_2 commute, then $H =< w >$ for some $w \neq 1$ and if $Z_G(w) =< w* >$, then $I(H) =< w* >$ is free of rank one. If u_1 and u_2 do not commute, then H is contained in a maximal two-generator subgroup H^* generated by non-commuting elements u_1^* and u_2^*. We claim that H^* is an isolated subgroup of G. Suppose not. Then there is an integer $n > 1$ and an element $g \in G$ such that $g^n \in H^*$ but $g \notin H^*$. We may assume without loss of generality that g generates its own centralizer in G and that n is the least positive integer such that $g^n \in H^*$. Let K be the subgroup of G generated by u_1^*, u_2^* and g. Then $H^* \subseteq K \subseteq G$ is sandwiched between two models of Φ since $H^* =< u_1^*, u_2^* >$ is free of rank 2. One easily argues as in [6] that K must itself be a model of Φ. Remeslenuikov [13] showed that every finitely generated model of Φ is residually free. Thus, K is a residually free model of Φ. Let $F(X)$ be the free group with basis $X = \{x_1, x_2\}$. Then $w_1(u_1^*, u_2^*) = g^n$ for some nonempty (recall that G is torsion free) freely reduced word $w_1(x_1, x_2)$ on $X \cup X^{-1}$. We claim that $w_1(x_1, x_2)$ is not primitive in $F(X)$. Suppose not. Then there is $w_2(x_1, x_2)$ such that $\{w_1(x_1, x_2), w_2(x_1, x_2)\}$ is also a basis for $F(X)$. But then H^* would be properly contained in the subgroup of G generated by g and $w_2(u_1^*, u_2^*)$. This contradicts the maximality of H^*. Therefore, as claimed, $w_1(x_1, x_2)$ is not primitive in $F(X)$. We further claim that $w_1(x_1, x_2)$ is not a proper power in $F(X)$. Suppose not. Then $w_1(x_1, x_2) = w_0(x_1, x_2)^m$ for some integer

$m \geq 2$. Thus, $g^n = w_0(u_1^*, u_2^*)^m$ in G. By commutative transitivity g and $w_0(u_1^*, u_2^*)$ commute. Thus, $w_0(u_1^*, u_2^*) \in Z_G(g) =< g >$ so $w_0(u_1^*, u_2^*) = g^d$ for some nonzero integer d. Hence, $g^n = g^{dm}$. Since G is torsion free and $g \neq 1$ we conclude $n = dm$. From $n, m > 1$ we conclude that $0 < d < n$; moreover, from $g^{dm} = w_0(u_1^*, u_2^*)^m$ we conclude that $g^d = w_0(u_1^*, u_2^*) \in H^*$ since G is an R-group. (Recall that m-th roots in a torsion free, commutative transitive group are unique.) This contradicts the minimality of n. Therefore, $w_1(x_1, x_2)$ is, as claimed, not a proper power in $F(X)$. But a Theorem of G. Baumslag and A. Steinberg (Proposition 4.12, p. 57 of [12]) asserts that if n and k are integers (not necessarily distinct) each at least two and $F(Y_k)$ is a free group with basis $Y_k = \{y_1, \ldots, y_k\}$ and $w(y_1, \ldots, y_{k-1})$ is neither a primitive element nor a proper power in the free group $F(Y_{k-1})$ on $Y_{k-1} = \{y_1, \ldots, y_{k-1}\}$, then the one-relator quotient $< y_1, \ldots, y_k; y_k^n = w(y_1, \ldots, y_{k-1}) >$ has no free homomorphic images of rank $k-1$. Thus, the group $M =< x_1, x_2, x_3; x_3^n = w_1(x_1, x_2) >$ has no free homomorphic images of rank 2. But every residually free model of Φ is residually a free group of rank 2. (See B. Baumslag [1] and Gaglione and Spellman [7]). Thus there are epimorphisms $M \to K \to F_2 =< a_1, a_2; >$ contrary to the theorem of Baumslag and Steinberg. This contradiction shows that H^* is isolated in G. Therefore, $I(H) \subseteq H^*$ since $I(H)$ is the least isolated subgroup of G containing H. But G is 2-free and H^* is two-generator. Therefore, H^* is free; hence, so is the subgroup $I(H) \subseteq H^*$. Consequently, G is strongly 2-free. □

Theorem 3. *Let G be a $\mathbb{Z}$-commutative transitive, residually free model of Φ. Then G satisfies the maximal condition for two-generator subgroups.*

PROOF. Let $H_0 \subseteq H_1 \subseteq \ldots \subseteq H_n \subseteq \ldots$ be an ω-chain of two-generator subgroups of G. Let $H = \bigcup_{n<\omega} H_n$. Since G is $\mathbb{Z}$-commutative transitive, if H is Abelian it is cyclic and $H_N = H$ for all sufficiently large N. Thus, it suffices to assume that H is non- Abelian. Let $a, b \in H$ with $ab \neq ba$. Put $c = [a, b] = a^{-1}b^{-1}ab$. Then $c \neq 1$. Since G is residually free, there is a free group F and an epimorphism $p : G \to F$ such that $p(c) \neq 1$. Let p_1 be the restriction of p to H. Then $[p_1(a), p_1(b)] = p_1(c) \neq 1$ so that $p_1(H)$ is a non-Abelian free group. Thus, there is a group $F_0 =< f_1, f_2; >$ free of rank 2 and an epimorphism $p_2 : p_1(H) \to F_0$. Let $p_3 = p_2 p_1$. Pick $g_i \in H$ such that $p_3(g_i) = f_i$ for $i = 1, 2$. Let $G =< g_1, g_2 >\subseteq H$. Observe that $G_0 \subseteq H_N$ for all sufficiently large N. Furthermore, since F_0 is free it is projective so that if $K = \ker(p_3)$, then H is the semidirect product $KG_0, K \cap G_0 = 1$ and $G_0 \cong F_0$ is free of rank 2. If $G_0 = H$, then $G_0 = H_N$ for all sufficiently large N. Assume, on the contrary, that there is n sufficiently large so that $G_0 \subsetneqq H_n$. From $G_0 \subseteq H_n \subseteq H$ we get $F_0 = p_3(G_0) \subseteq p_3(H_n) \subseteq p_3(H) = F_0$. Thus, p_3 maps H_n onto F_0. Now H_n is non-Abelian since, for example, it contains G_0. By the argument used in the beginning of the proof of Theorem

2, H_n is free of rank 2 since it is a non-Abelian, two-generator subgroup of a model of Φ. Let p_0 be the restriction of p_3 to H_n. Then p_0 maps H_n onto F_0. Since H_n is free of rank 2 it is Hopfian so that kernel $K \cap H_n$ of the epimorphism $p_0 : H_n \to F_0$ must be trivial. But then p_0 is an isomorphism. It follows that the preimages $p_0^{-1}(f_i) = g_i \in G_0 \subseteq H_n$ freely generate H_n. This contradicts $< g_1, g_2 > \subsetneqq H_n$. Therefore, $H_N = G_0$ for all sufficiently large N. □

Corollary 1. *Every residually free model of Σ is 3-free.*

PROOF. In [6] Gaglione and Spellman proved that every residually free model of Σ is a 2-free group in which the centralizer of every non-trivial element is infinite cyclic. □

Corollary 2. *Every finitely generated model of Σ is 3-free.*

PROOF. We reiterate the result of Remeslennikov [13] asserting that every finitely generated model of Φ is residually free. □

Remark. The proof of Theorem 3 can be modified to show that every finitely generated, fully residually free group satisfies the maximal condition for two-generator subgroups.

If G is Abelian and finitely generated, then it is Noetherian. (Any ascending chain of subgroups stabilizes.) Suppose G is a non-Abelian, finitely generated, fully residually free group. Let $H_0 \subseteq H_1 \subseteq \ldots \subseteq H_n \subseteq \ldots$ be an ω-chain of two-generator subgroups of G. If H is non-Abelian the proof goes through verbatim. Suppose H is Abelian. Remeslennikov [13] shows that if $k \geq 2$ is an integer and G is a non-Abelian, k-generator fully residually free group, then G is a Λ-free group where $\Lambda = (A, \leq)$ is an ordered Abelian group and A is free Abelian of rank not exceeding $3k$. It follows that every Abelian subgroup of G can be embedded in A. In particular, H must be finitely generated so $H_N = H$ for all sufficiently large N.

References

[1] B. Baumslag, Residually free groups, *Proc. London Math. Soc.* **17**(1967), 402–418.

[2] G. Baumslag, Residual nilpotence and relations in free groups, *J. Algebra* **2**(1965), 271–282.

[3] G. Baumslag and A. Steinberg, Residual nilpotence and relations in free groups, *Bull. Amer. Math. Soc.* **70**(1964), 283–284.

[4] C.C. Chang and H.J. Keisler, *Model Theory*, third edition (North-Holland, The Netherlands, 1992).

[5] B. Fine and G. Rosenberger, On restricted Gromov groups, *Comm. Algebra* **20(18)**(1992), 2171–2181.

[6] A.M. Gaglione and D. Spellman, More model theory of free groups, *Houston J. Math.*, to appear.

[7] A.M. Gaglione and D. Spellman, Even more model theory of free groups, in *Proceedings of Infinite Groups and Group Rings*, (Tuscaloosa, AL (1992), World Scientific, 1993, New Jersey), 37–40.

[8] A.M. Gaglione and D. Spellman, Generalizations of free groups: Some questions, *Comm. Algebra*, to appear.

[9] N. Harrison, Real length functions in groups, *Trans. Amer. Math. Soc.* **174**(1972), 77–106.

[10] Ju.I. Hmelevskii, Word equations without coefficients, *Soviet Math. Dokl.* **7**(1966), 1611–1613.

[11] A.G. Kurosh, *The Theory of Groups*, Vol. II, second English edition (Chelsea, New York, 1960).

[12] R.C. Lyndon and P.E. Schupp, *Combinatorial Group Theory* (Springer-Verlag, Germany, 1977).

[13] V.N. Remeslennikov, ∃-Free groups and groups with a length function, preprint.

[14] G.S. Sacerdote, Elementary properties of free groups, *Trans. Amer. Math. Soc.* **178**(1972), 127–138.

[15] A. Steinberg, On free nilpotent quotient groups, *Math. Z.* **85**(1964), 185–196.

THE GROUPS $G(n,l)$ AS FUNDAMENTAL GROUPS OF SEIFERT FIBERED HOMOLOGY SPHERES

LUIGI GRASSELLI[1]

Dipartimento di Matematica, Politecnico di Milano, Piazza Leonardo da Vinci 32, I-20133, Milano, Italy

Abstract

If n and l are odd and coprime, let $G(n,l)$ be the group admitting the following presentation:

$$< a_0, a_1, \ldots, a_{n-1} | a_0 a_1 \ldots a_{n-1} = 1,$$
$$a_i^{-1} a_{i-2}^{-1} a_{i-4}^{-1} \ldots a_{i-(l-3)}^{-1} a_{i-(l-2)} a_{i-(l-4)} \ldots a_{i-1} a_{i+1} = 1,$$
$$i \in \{0, 1, \ldots, n-1\} > .$$

By means of geometric techniques, we prove that the groups $G(n,l)$ are pairwise non-isomorphic, extending results obtained in [C1] and in [JT].

It is well known that PL-manifolds can be represented by edge-coloured graphs (see, for examples, the survey papers [FGG] and [V]).

In [LM] a class of (closed orientable) 3-manifolds $S(n,l,t,c)$, represented by edge-coloured graphs, is introduced and investigated. By direct computation of their homology groups, A. Cavicchioli proves in [C2] that each 3-manifold $M(n,l) = S(n,l,l-1,1)$ of the class $\mathcal{L} = \{M(n,l); n, l \text{ odd and coprime}\}$ is a homology sphere, as conjectured in [LM]. Actually, by means of a technique described in [D1] and in [G], the following presentation $P(n,l)$ for the fundamental group $G(n,l)$ of $M(n,l) \in \mathcal{L}$ is given in [C2]:

$$P(n,l) =< a_0, a_1, \ldots, a_{n-1} | a_0 a_1 \ldots a_{n-1} = 1,$$
$$a_i^{-1} a_{i-2}^{-1} a_{i-4}^{-1} \ldots a_{i-(l-3)}^{-1} a_{i-(l-2)} a_{i-(l-4)} \ldots a_{i-1} a_{i+1} = 1,$$
$$i \in \{0, 1, \ldots, n-1\} >,$$

where the indices are reduced mod n.

The following result is proved in [C1].

Proposition 1. *The class $\mathcal{M} = \{M(3, l = 6q \pm 1), q \in \mathbf{N}\}$ is a countable set of prime non-homeomorphic homology spheres with Heegaard genus two.*

[1]Work performed under the auspices of the G.N.S.A.G.A. of the C.N.R (National Research Council of Italy) and financially supported by M.U.R.S.T. of Italy (project "Geometria Reale e Complessa").

Note that $\mathcal{M} \subset \mathcal{L}$. The main part of the proof is to show that:

Proposition 2. *The fundamental groups $G(3, l = 6q \pm 1)$ of $M(3,l) \in \mathcal{M}$ are pairwise non-isomorphic.*

This is performed by proving the following result.

Proposition 3. *If $Z(3,l)$ denotes the centre of $G(3, l = 6q \pm 1)$, then the factor group $G(3,l)/Z(3,l)$ is the triangle group $D(2,3,l) =< x,y|x^2 = y^3 = (xy)^l = 1 >$.*

Another proof of Proposition 3 (and hence of Proposition 2) is given in [JT]; both proofs are performed by using Tietze transformations on the original presentation $P(3, l = 6q \pm 1)$ in order to obtain a simpler presentation of $G(3,l)$ with only two generators. The following extension of the above result is obtained in [C3], by making use of the extension theory of groups.

Proposition 3′. *If n,l are odd and coprime and $Z(n,l)$ denotes the centre of $G(n,l)$, then the factor group $G(n,l)/Z(n,l)$ is the triangle group $D(2,n,l) = < x,y|x^2 = y^n = (xy)^l = 1 >$.*

This proves that:

Proposition 4. *If $G(n,l), G(n',l')$ are the fundamental groups of the 3-manifolds $M(n,l), M(n',l') \in \mathcal{L}$, then $G(n,l)$ is isomorphic with $G(n',l')$ (and $M(n,l)$ is homeomorphic with $M(n',l')$) if and only if $\{n,l\} = \{n',l'\}$.*

In [D2] it is shown that, for all n and l, the 3-manifold $M(n,l)$ represents the 2-fold covering of the 3-sphere S^3 branched over the torus link $\{n,l\}$, i.e. the Brieskorn manifold $M(n,l,2)$(see [M]). Hence $M(n,l) = M(l,n)$, so that, from now on, we may suppose $n < l$. As a consequence of this fact and of [Mo, Th3.8], A. Donati obtains the following result, which gives a direct proof that each $M(n,l) \in \mathcal{L}$ is a homology sphere.

Proposition 5. *Each $M(n,l) \in \mathcal{L}$ is the Seifert fibered homology sphere*

$$(OoO| - 1; (2,1), (n,p), (l,q)),$$

where p,q are uniquely determined by $2(pl + qn) = nl \pm 1$.

In this setting, the contribution of the present paper is to point out that Proposition 5 allows a deep insight both in the geometrical properties of the homology spheres $M(n,l) \in \mathcal{L}$ and in the algebraic structure of their fundamental groups $G(n,l)$. In fact Seifert fibered homology spheres (and their fundamental groups) are deeply investigated (see, for example, [S], [O], [Mo], [J], [BZ], [K], [BCZ]) and the following results are well known.

Proposition 6.

(a) [S, Th. 12] *A Seifert fibered space M is a homology sphere ($\neq S^3$) if and only if $M = \mathbf{S}(\alpha_1, \alpha_2, \ldots, \alpha_r) = (OoO|b; (\alpha_1, \beta_1), (\alpha_2, \beta_2), \ldots, (\alpha_r, \beta_r))$, with $r \geq 3, \alpha_1 < \alpha_2 < \ldots < \alpha_r$ pairwise coprime and $b, \beta_1, \beta_2, \ldots, \beta_r$ uniquely determined as the solution of the equation:*

$$b\alpha_1 \ldots \alpha_r + \beta_1\alpha_2 \ldots \alpha_r + \alpha_1\beta_2\alpha_3 \ldots \alpha_r + \ldots + \alpha_1\alpha_2 \ldots \alpha_{r-1}\beta_r = 1.$$

(b) [O, §5.2, Th6] *With the exception of the dodecahedral space $\mathbf{S}(2,3,5)$, each $\mathbf{S}(\alpha_1, \alpha_2, \ldots, \alpha_r)$ is "large"; hence, the Seifert fibered homology spheres $\mathbf{S}(\alpha_1, \alpha_2, \ldots, \alpha_r)$ are pairwise non-homeomorphic and have different fundamental groups.*

(c) [M, Th3.7], [BZ, Th12.30] *The Seifert fibered space $(OoO|b; (\alpha_1, \beta_1), \ldots, (\alpha_r, \beta_r))$ is the 2-fold covering of S^3, branched over the Montesinos link $\mathbf{m}(b; \alpha_1/\beta_1, \ldots, \alpha_r/\beta_r)$.*

(d) [K, §7] *Each Seifert fibered space $(OoO|b; (\alpha_1, \beta\alpha_1), (\alpha_2, \beta_2), (\alpha_3, \beta_3))$ has Heegaard genus two.*

(e) [S, Th17] *Each Seifert fibered homology sphere $\mathbf{S}(\alpha_1, \alpha_2, \alpha_3)$ is the $\alpha_{p(1)}$-fold cyclic covering of S^3, branched over the torus knot $\{\alpha_{p(2)}, \alpha_{p(3)}\}$, p being an arbitrary permutation on the set $\{1,2,3\}$.*

(f) [S, Th13] *A homology sphere M can be obtained by Dehn's q-surgery from a torus knot $\{\alpha_1, \alpha_2\}$ ([De]) if and only if $M = \mathbf{S}(\alpha_1, \alpha_2, \alpha_3)$, with $\alpha_3 = |q\alpha_1\alpha_2 - 1|$.*

Since Proposition 5 states that each $M(n,l) \in \mathcal{L}$ is the Seifert fibered homology sphere $\mathbf{S}(2,n,l)$, we obtain, as a corollary of the above results, the following list of properties for the class $\mathcal{L}$.

Proposition 7.

(a) *$\mathcal{L}$ is a countable class of prime non-homeomorphic homology spheres with Heegaard genus two and with different fundamental groups.*

(b) *Each $M(n,l) \in \mathcal{L}$ is the $p(2)$-fold covering of S^3 branched over the torus knot $\{p(n), p(l)\}$, p being an arbitrary permutation on the set $\{2,n,l\}$.*

(c) *$M(n,l) \in \mathcal{L}$ is obtained by Dehn's q-surgery from the torus knot $\{2,n\}$ if and only if $l = |2nq - 1|$.*

Note that Propositions 1,2 and 4 are contained in Proposition 7a, since $\mathcal{M} \subset \mathcal{L}$; moreover, Proposition 7b (resp. 7c) proves that $\mathcal{M}$ is the class of the homology spheres obtained as $(6q \pm 1)$-fold covering of S^3 branched over (resp. obtained by Dehn's q-surgery from) the trefoil knot, with $q \in \mathbf{N}$.

As well as the 3-manifolds $M(n,l)$ of the class $\mathcal{L}$, the algebraic structure of their fundamental groups $G(n,l)$ has been intensively investigated, too. As direct corollaries of [S,§10] or [J,§6], we obtain the following results which contain, in particular, Proposition 3′.

Proposition 8.

(a) A presentation ot the fundamental group $G(n,l)$ of $M(n,l) \in \mathcal{L}$ is given by

$$< x, y, h | x^2 = h, y^n = h^p, (xy)^l = h^q, yhy^{-1} = h >,$$

where p and q are uniquely determined by $2(pl+qn) = nl \pm 1$.

(b) The centre $Z(n,l)$ of $G(n,l)$ is the cyclic normal subgroup $< h >$ generated by h.

(c) The factor group $G(n,l)/Z(n,l)$ is the triangle group $D(2,n,l) =< x, y | x^2 = y^n = (xy)^l = 1 >$.

(d) $G(n,l)$ is infinite if and only if $(n,l) \neq (3,5)$.

Finally, note that, for all n and l, the direct computation of the commutator factor group $H_1(M(n,l))$ of $G(n,l)$ performed in [C3, Prop.5] actually gives the first homology group of the Brieskorn manifold $M(n,l,2)$.

References

[BCZ] M. Boileau, D.J. Collins and H. Zieschang, Scindements de Heegaard des petites varietes de Seifert, *C. R. Acad. Sci. Paris* **305**(1987), 557–560.

[BZ] G. Burde and H. Zieschang, *Knots* (Walter de Gruyter, Berlin–New York, 1985).

[C1] A. Cavicchioli, A countable class of non-homeomorphic homology spheres with Heegaard genus two, *Geom.Dedicata* **20**(1986), 345–348.

[C2] A. Cavicchioli, Lins-Mandel 3-manifolds and their groups: a simple proof of the homology sphere conjecture, in *Atti III Conv. Naz. Topologia, Trieste 1986* (Suppl. Rend.Circ.Mat.Palermo **18**, 1988), 229–237.

[C3] A. Cavicchioli, On some properties of the groups $G(n,l)$, *Ann. Mat. Pura Appl.* **151**(1988), 303–316.

[De] M. Dehn, Uber die Topologie des dreidimensionalen Raumes, *Math. Ann.* **69**(1910), 137–168.

[D1] A. Donati, A calculation method of 3-manifold fundamental groups, *Combin. Inform. System. Sci.* **8,2**(1983), 97–100.

[D2] A. Donati, Lins-Mandel manifolds as branched coverings of S^3, *Discrete Math.* **62**(1986), 21–27.

[FGG] M. Ferri, C. Gagliardi and L. Grasselli, A graph-theoretical representation of PL-manifolds– A survey on crystallizations, *Aeq. Math.* **31**(1986), 121–141.

[G] L. Grasselli, Edge-coloured graphs and associated groups, in *Atti II Conv. Naz. Topologia, Taormina 1984* (Rend.Circ.Mat.Palermo **12**, 1986), 263–269.

[J] W. Jaco, *Lectures on three-manifold topology* (Conf. Board Math. Sci. Am. Math. Soc., Providence, Rhode Island, 1980).

[JT] D.L. Johnson and R.M. Thomas, The Cavicchioli groups are pairwise non-isomorphic, in *Proc. of Groups, St. Andrews 1985* (London Math.Soc. Lecture Note Series **121**, Cambridge Univ.Press, 1986), 220–222.

[K] T. Kobayashi, Structures of the Haken manifolds with Heegaard splittings of genus two, *Osaka J.Math.* **21**(1984), 437–455.

[LM] S. Lins and A. Mandel, Graph-encoded 3-manifolds, *Discrete Math.* **57** (1985), 261–284.

[M] J. Milnor, On the 3-dimensional Brieskorn manifolds $M(p,q,r)$, *Ann. of Math. St.* **84**, Princeton Univ. Press, 1975.

[Mo] J.M. Montesinos, Sobre la representacion de variedades tridimensionales, mimeographed notes, 1977.

[O] P. Orlik, *Seifert manifolds* (Lect. Notes in Math. **291**, Springer Verlag, 1972).

[S] H. Seifert, Topologie dreidimensionaler gefaserter Raume, *Acta Math.* **60** (1933), 147–288; English reprint: Academic Press, London, 1980.

[V] A. Vince, n-Graphs, *Discrete Math.* **72**(1988), 367–380.

LIFTING AUTOMORPHISMS: A SURVEY

C.K. GUPTA* and V. SHPILRAIN†

*University of Manitoba, Winnipeg R3T 2N2, Canada
†Ruhr Universität, 463 Bochum, Germany

1. Introduction

Let $F = F_n$ be the free group of a finite rank $n \geq 2$ with a fixed set $\{x_i; 1 \leq i \leq n\}$ of free generators. If R is a characteristic subgroup of the group F then the natural homomorphism $\epsilon_R : F \rightarrow F/R$ induces the mapping $\tau_R : \mathrm{Aut}F \rightarrow \mathrm{Aut}(F/R)$ of the corresponding automorphism groups. Those automorphisms of the group F/R that belong to the image of τ_R are usually called *tame*. In this survey, we will be concerned with the following general question: *How to determine whether or not a given automorphism of a group F/R is tame?* In a more general situation, when R is an arbitrary normal subgroup of F, one can ask if a given generating system of the group F/R can be *lifted* to a generating system of F (in this case the system will be also called tame). This question has important applications to low-dimensional topology (see, for instance, [LuMo1]).

The questions of lifting automorphisms and generating systems naturally give rise to the following two problems of independent interest:

(1) Finding appropriate necessary and/or sufficient condition(s) for an endomorphism of the group F_n to be an automorphism;

(2) Describing (in one or another way) the group $\mathrm{Aut}(F/R)$ or generating systems of a group F/R.

The starting point for any research in the automorphisms of a free group is the classical result of Nielsen [N2] (see [LS]) who has described $\mathrm{Aut}F_n$ in terms of generators and relations and also has given an algorithm for deciding if n given elements of F_n form a basis of F_n. Nielsen's procedure, however, is not suitable for generating systems of F/R since we face the problem of deciding if, for n given elements $y_1, \dots, y_n$ of F_n, there exist n elements $r_1, \dots, r_n$ of R such that the elements $y_1r_1, \dots, y_nr_n$ generate F_n.

In Section 1 of this survey we will describe three powerful known necessary conditions for an endomorphism of F_n to be an automorphism. It is interesting to note that each of these is based on examination of the special "Jacobian matrix" (to be specified later) of a given endomorphism of F_n. The first approach comes from Bryant *et al.* [BGLM]; it gives a necessary condition for a certain $n \times n$ matrix (over the integral group ring ZF) to be invertible. This condition does not actually use the fact that a matrix under consideration is the Jacobian matrix of some automorphism. In contrast, the second necessary condition which is due to Lustig and Moriah [LuMo2] exploits just this fact: they consider the image of the Jacobian matrix in an appropriate $n \times n$ matrix ring $M_n(K)$ with K a commutative ring with 1, evaluate the determinant of this image and then observe that this determinant should be equal to a trivial unit of the ring K, i.e. to an element of the (multiplicative) group generated by the images of $\pm x_1, \ldots, \pm x_n$. The third approach is due to the second author [Sh2], and it is based on the evaluation of a non-commutative determinant of the Jacobian matrix; this is a generalization of the determinant in the sense of Dieudonné.

These necessary conditions enable one to prove non-tameness of specific automorphisms and generating systems of various groups. We describe the corresponding results in Sections 3 and 4 where we also discuss the question of lifting "partial" generating systems (the so-called primitive systems) and, in particular, of lifting primitive elements. Having obtained a number of results on non-tameness, it becomes more important to fish out some "nice" groups that would have only tame generating systems, or at least those groups whose non-tame generating systems can be reasonably described. The most familiar series of groups of the first kind has been discovered by Bachmuth and Mochizuki [BM2] (another one is given in [BFM]). A good example of a group of the second kind has been given by E. Stöhr [St]; she proved that the automorphism group of the free centre-by-metabelian group $F/[F'', F]$ of rank $n \geq 4$ is generated by tame automorphisms together with a single automorphism which was later proved to be non-tame [BGLM]. Another example with identical feature is due to Bryant and the first author [BG1] where, for $n \geq c-1 \geq 2$, the free nilpotent group $F_{n,c}$ of rank n and class c is shown to be generated by tame automorphisms together with a single non-tame automorphism. A remarkable positive result on lifting primitive systems of the free nilpotent groups has been obtained recently by the first author and N. Gupta [GG]. We give a survey of these and some more positive results in Sections 3 and 4. In Section 5, we mention a few results on lifting automorphisms of certain relatively free groups to free groups of countable infinite rank.

Although after the results of [BM1], [BM2], it has become clear that the rank of a group can play an important role in the questions of tameness, we would like to emphasize a very special case of 2-generator groups. In this case, we have a convenient criterion for an endomorphism of the group F_2 to be an automorphism - the so called Nielsen's commutator test [N1]. Nielsen [N1] has, in addition, proved that every IA-automorphism (i.e. automorphism that is identical modulo the commutator subgroup) of the group F_2 is inner. The applications of these results are also discussed in Section 3. We conclude the survey with a problem section.

For topics other than those covered in our survey, the reader is referred to the survey articles by Rosenberger [Ro] and by Roman'kov [R2].

2. Necessary conditions for tameness

Let ZF be the integral group ring of the free group F of rank n and Δ its augmentation ideal, that is, the kernel of the natural homomorphism $\sigma : ZF \to Z$. When $R \neq F$ is a normal subgroup of F, we denote by Δ_R the ideal of ZF generated by all elements of the form $(r-1), r \in R$. Fox [F] gave a detailed account of the differential calculus in the free group ring. We refer to the book [Gu] for more information, giving only a brief summary here. Fox derivations are the mappings $\partial_i : ZF \to ZF, 1 \leq i \leq n$, having the following basic property: *the augmentation ideal is a free left ZF-module with a free basis $\{(x_i - 1), 1 \leq i \leq n\}$ and the mappings ∂_i are projections to the corresponding free cyclic direct summands.* Thus any element $u \in \Delta$ can be uniquely written in the form $u = \sum \partial_i(u)(x_i - 1)$. Let ϕ be an endomorphism of the free group F given by

$$\phi : x_i \to y_i, 1 \leq i \leq n.$$

Then the matrix $J_\phi = ||\partial_j(y_i)||_{1 \leq i,j \leq n}$ is called the *Jacobian matrix* of ϕ. We can now formulate Birman's "inverse function theorem" which is the initial point for each of the three necessary conditions of tameness to follow.

Theorem 2.1. ([Bi]) *Let $\phi : x_i \to y_i, 1 \leq i \leq n$, be an endomorphism of the free group F. Then ϕ is an automorphism if and only if the Jacobian matrix J_ϕ is invertible in the ring $M_n(ZF)$ of $n \times n$ matrices over ZF.*

Denote by (Δ, Δ) the additive subgroup of ZF generated by all elements of the form $xy - yx, x, y \in \Delta$. We can now present the first necessary condition of tameness:

Theorem 2.2. ([BGLM]) *Let $m \geq 2$ and q be positive integers. Let $J \in M_q(ZF)$ be a matrix of the form $J = I + D$, where I is the identity matrix and $D \in M_q(\Delta^m)$. If J is invertible in $M_q(ZF)$ then the trace of D belongs to $\Delta^{m+1} + (\Delta, \Delta)$.*

As an application of this result (which becomes a necessary condition of tameness when applied to a Jacobian matrix) one obtains the existence of non-tame automorphisms of the free nilpotent groups $F/\gamma_{c+1}(F)$ of any rank $n \geq 2$ and class $c \geq 3$ (cf. Bachmuth [B2], Andreadakis [A1]). Another application shows the non-tameness of E. Stöhr's automorphism of the free centre-by-metabelian group mentioned in the Introduction. The necessary condition of Theorem 2.2 was later used by the first author and Levin [GL] to prove the existence of non-tame automorphisms of a large class of relatively free polynilpotent groups (more details are given in the next section).

We now give the next condition of tameness which is a particular case of a more general result due to Lustig and Moriah [LuMo2].

Theorem 2.3. *Let G be a group given by an n-generator presentation F/R with $n \geq 2$ minimal possible. Let I_R denote the ideal of ZF generated by all Fox derivatives of elements from R and let A be a commutative ring with a unit. Let $\rho : ZF \to A$ be a homomorphism such that $\rho(I_R) = 0$ and let it be naturally extended to the homomorphism $\rho : M_n(ZF) \to M_n(A)$ of matrix rings. Suppose $Y = \{y_1R, \ldots, y_nR\}$ is a generating system of the group G, and $J_Y = ||\partial_j(y_i)||_{1\leq i,j\leq n}$ the corresponding Jacobian matrix. A necessary condition for Y to be a tame generating system of G is that* $\det(\rho(J_Y))$ *belongs to the multiplicative subgroup of A generated by $\pm\rho(x_1), \ldots, \pm\rho(x_n)$.*

Some applications of this criterion have been given in [LuMo2] (see also [LuMo1], [LuMo3] for applications to topological groups). From the group-theoretic point of view, a systematic use of the above criterion is made in [MoSh]: the authors re-prove the existence of non-tame automorphisms of free nilpotent groups (this result has become a good test for new necessary conditions of tameness), and then they construct non-tame automorphisms in various extensions of the Burnside groups.

The second author has recently developed yet a third necessary condition of tameness. Unlike the previous two conditions, it uses the notion of "determinants" of matrices over non-commutative rings, extending the notion of such determinants of matrices over skew fields due to Dieudonné (see [Ar] for a very detailed exposition). A closer look at the argument in [Ar] leading to the definition of non-commutative determinant, shows that this argument applies to a more general situation as follows.

Let S be an associative ring with 1 satisfying the following property: every invertible square matrix with the entries from S has at least one invertible element in every row and in every column. Over such a ring S, any invertible matrix M can be written in a special canonical form $M = ED(\mu)$, where E is a product of *elementary matrices* (i.e. matrices possibly different from the identity matrix by a single element outside the diagonal), and $D(\mu) = [1, \ldots, 1, \mu]$ is the diagonal matrix with μ ($\neq 0$), as its lower right corner entry. Take now the group S^* of all invertible elements of the ring S, and let $S^a = S^*/[S^*, S^*]$ be the maximal abelian quotient group of S^*. Then the image of μ in S^a is defined uniquely and is called the determinant of the matrix M in the sense of Dieudonné. It has some usual properties of the determinant; in particular, the determinant of the product of two matrices from $GL_k(S)$ is equal to the product of their determinants.

An application of this construction to the study of the automorphisms of F is as follows. Given a matrix $M \in GL_n(ZF)$, we take a ring $S_m = QF/\Delta^m, m \geq 2$, Q the field of rationals, and come up with the determinant $\delta_m(M)$ in accordance with the construction described above using the fact that in the ring S_m, every element with non-zero augmentation is invertible. More specifically, given an element $u \in QF$ of the form $\alpha - w, 0 \neq \alpha \in Q; w \in$

Δ, one has modulo $\Delta^m : u^{-1} = \alpha^{-1} + \alpha^{-2}w + \ldots + \alpha^{-m}w^{m-1}$. Hence the commutator subgroup of the group S_m^* is generated as a group (actually as a semigroup) modulo Δ^m by the elements of the form

$$(1-v)(1-w)(1+v+\ldots+v^{m-1})(1+w+\ldots+w^{m-1}) \qquad (*)$$

with $v^\epsilon = w^\epsilon = 0$. Denote by P_m the (multiplicative) subsemigroup of ZF generated by all elements of the form (*).

We note that in case of the integral group ring ZF, the rings ZF/Δ^m don't satisfy the property of S, but still we can write any invertible matrix over such a ring in the form $ED(\mu)$ because the g.c.d. of the augmentations of the elements of every column must be 1, so we can get an element with augmentation 1 in every column and then in every row by multiplying the matrix by appropriate elementary matrices. Thus the determinant in this case exists, and we can prove that it is unique and satisfies all the necessary properties by embedding ZF into QF and using the standard argument.

Now the crucial observation is the following. Writing an automorphism ϕ of F as a product of some Nielsen automorphisms shows that the determinant $\delta_m(J_\phi)$ of the Jacobian matrix J_ϕ, is equal to the product of the determinants of elementary Jacobian matrices each of which is readily seen to be equal to $\pm g$ for some $g \in F$. This yields the following necessary condition for an endomorphism ϕ of the group F to be an automorphism.

Theorem 2.4. ([Sh2]) *Let ϕ be an automorphism of F and J_ϕ - the Jacobian matrix of ϕ. Then for an arbitrary $m \geq 2$, one has $\det_m(J_\phi) = \pm g p_m + w_m$, for some $g \in F$; $p_m \in P_m; w_m \in \Delta^m$, where $\det_m(J_\phi)$ is an arbitrary preimage (in ZF) of $\delta_m(J_\phi)$.*

In conclusion we simply remark that none of the three necessary conditions given in this section seem to follow from the other two except that, when restricted to the Jacobian matrices, Theorem 2.4 includes Theorem 2.2.

3. Lifting automorphisms of relatively free groups F/V

Let V be a fully invariant subgroup of a free group $F = \langle f_1, \ldots, f_n \rangle$ and consider the relatively free group $F/V = \langle x_1, \ldots, x_n \rangle$. If $\alpha = \{x_1 \to w_1, \ldots, x_n \to w_n\}$ defines an automorphism of F/V then α is induced by an endomorphism

$$\alpha^* = \{f_1 \to w_1^*, \ldots, f_n \to w_n^*\}$$

of $F(w_i^*(x_1, \ldots, x_n) = w_i)$ which may or may not define an automorphism of F. If, however, there exists a system $(v_1, \ldots, v_n)$ of elements of V such that the map $\{f_1 \to w_1^* v_1, \ldots, f_n \to w_n^* v_n\}$ defines an automorphism of F then we say that α lifts to an automorphism of F, and we call α a *tame* automorphism of F/V. If α does not lift to an automorphism of the free group F for any choice $(v_1, \ldots, v_n)$ of elements of V then we say that α is *wild* (or non-tame). If $\alpha \in \mathrm{Aut}(F/V)$ is tame then by a classical theorem of Nielsen it follows that α lies in the subgroup of $\mathrm{Aut}(F/V)$ generated by the following four elementary automorphisms:

$$\alpha_1 = \{x_1 \to x_2, x_2 \to x_3, \ldots, x_n \to x_1\};$$
$$\alpha_2 = \{x_1 \to x_2, x_2 \to x_1, x_i \to x_i, i \neq 1, 2\};$$
$$\alpha_3 = \{x_1 \to x_1^{-1}, x_i \to x_i, i \neq 1\};$$
$$\alpha_4 = \{x_1 \to x_1 x_2, x_i \to x_i, i \neq 1\}.$$

By definition, an IA-automorphism of F/V is an automorphism that fixes F/V modulo its commutator subgroup $(F/V)'$. Thus, if $\theta \in IA - \mathrm{Aut}(F/V)$ then it is an instance of an IA-endomorphism of the form: $\theta = \{x_1 \to x_1 d_1, x_2 \to x_2 d_2, \ldots, x_n \to x_n d_n\} (d_i \in (F/V)')$. We will limit our discussion to automorphisms of relatively free groups $F_n(\mathbf{V})$ of the varieties $\mathbf{V}$ which correspond to the fully invariant subgroups V generated by outer commutator words (e.g $[x, y, z], [[x, y, z], [u, v], w]$). These groups include, in particular, free polynilpotent groups. Denoting by T the subgroup of tame automorphisms of $F/V (= F_n(\mathbf{V}))$ shows that $\mathrm{Aut}(F/V) = \langle T, IA - \mathrm{Aut}(F/V) \rangle$, and the first important question to ask is if the group F/V can have non-tame IA-automorphisms: *Do there exist non-tame IA-automorphisms of F/V?* This question has a very satisfactory answer when V is the fully invariant closure of an outer commutator (i.e. a simple or complex commutator in all distinct variables). Let $\mathbf{A}, \mathbf{N}_2$ and $\mathbf{M}$ denote respectively the varieties of abelian groups, nilpotent class-2 groups and metabelian groups. It is known that the automorphisms of each of the following relatively free groups are tame: $F_n(\mathbf{A}) (n \geq 1), F_n(\mathbf{N}_2) (n \geq 2)$ [A1] and $F_n(\mathbf{M}) (n \geq 4)$ [BM2]. The following result answers the above question completely:

Theorem 3.1. ([GL]) *Let $\mathbf{V}$ be any variety defined by an outer-commutator word of weight m. Then except for the three known cases $F_n(\mathbf{A})$ $(n \geq 1)$, $F_n(\mathbf{N}_2)$ $(n \geq 2)$ and $Fn(\mathbf{M})$ $(n \geq 4)$, the group $F_n(\mathbf{V})$ $(n \geq m)$ has non-tame automorphisms.*

[The restriction $n \geq m$ can be relaxed in most cases (see [Sh2]). The second author and Narain Gupta [GSh] have proved that the two-generator free polynilpotent group $F/V, V \neq F$, has non-tame automorphisms except when $V = \gamma_2(F)$ or $V = \gamma_3(F)$, or when V is of the form $[\gamma_k(U), \gamma_k(U)], k \geq 2, U \leq F$.]

The proof of Theorem 3.1 uses an important invertibility criterion developed in [BGLM] (cf. Theorem 2.2). Recall that the k-th left partial derivative ∂_k is defined linearly on the free group ring $\mathbf{Z}F (= \mathbf{Z}F_n)$ by:

$\partial_k(f_k) = 1; \partial_k(f_i) = 0, i \neq k; \partial_k(uv) = \partial_k(u) + u\partial_k(v), u, v \in \mathbf{Z}F$. In particular, for any $w \in \gamma_m(F)$, the partial derivative $\partial_k(w)$ lies in Δ^{m-1} , and hence, modulo Δ^m, it can be represented as a polynomial $g(X_1, \ldots, X_n)$ in the non-commuting variables $X_i = f_i - 1, i = 1, \ldots, n$. For any $S_i, T_i \in \{X_1, \ldots, X_n\}$ we define an equivalence relation "$\approx$" on monomials by: $S_1 \ldots S_k \approx T_1 \ldots T_k$ if one is a cyclic permutation of the other. Finally, a polynomial $g(X_1, \ldots, X_n)$ is called *balanced* if $g(X_1, \ldots, X_n) \approx 0$, or, equivalently, *the sum of the coefficients of its cyclically equivalent terms is zero.* Then, through a technical analysis of the invertibility of the Jacobian matrix associated with an IA-endomorphism of F one has the following useful criterion:

Criterion 3.2. ([BGLM]) Let $w_i \in \gamma_m(F_n)$ for some $m \geq 3$ and let α be an endomorphism of F_n defined by: $\alpha(f_i) \equiv f_i w_i (\bmod\ \gamma_{m+1}(F_n)), i = 1, \ldots, n$. Let $\sum_i \partial_i(w_i) \equiv g(X_1, \ldots, X_n)(\bmod\ \Delta^m)$. If α defines an automorphism of F_n then $g(X_1, \ldots, X_n)$ must be balanced.

Since most free polynilpotent groups F/V have non-tame automorphisms, it would be most desirable to obtain presentations for $\mathrm{Aut}(F/V)$. The first step towards this goal is to recognize the generators of $IA-\mathrm{Aut}(F/V)$. This suggests to ask: *What endomorphisms* $\theta = \{x_1 \to x_1 d_1, \ldots, x_n \to x_n d_n\} (d_i \in (F/V)')$ *of* F/V *define its* IA*-automorphisms?*

We know very little towards an answer to this question except when V is one of the fully invariant subgroups $\gamma_{c+1}(F), F'', [F'', F]$ or $[F', F', F']$.

Automorphisms of free nilpotent groups

Let $F_{n,c} = \langle x_1, x_2, \ldots, x_n \rangle, n, c \geq 2$, denote the free nilpotent group of rank n and class c. Then $F_{n,c} \cong F_n/\gamma_{c+1}(F_n)$. Every automorphism of the free abelian group $F_{n,c}/\gamma_2(F_{n,c})$ is tame and the same is true for $F_{n,c}/\gamma_3(F_{n,c})$ [A1]. Thus $\mathrm{Aut}(F_{n,c}) = \langle T, IA^*-\mathrm{Aut}(F_{n,c}) \rangle, c \geq 3$, where $IA^*-\mathrm{Aut}(F_{n,c})$ consists of IA-automorphisms of the form $\{x_i \to x_i d_i, d_i \in \gamma_3(F_{n,c}), 1 \leq i \leq n\}$. The most satisfactory answer for free nilpotent groups comes from the following result (cf. [Go], [A2]).

Theorem 3.3. ([BG1]) *Let* $\theta = \{x_1 \to x_1[x_1, x_2, x_1], x_i \to x_i, i \neq 1\} \in \mathrm{Aut}(F_{n,c})$. *Then for* $n \geq c - 1$, $\mathrm{Aut}(F_{n,c}) = \langle T, \theta \rangle$.

For $n \leq c - 2$, more IA-automorphisms seem to be required to generate $\mathrm{Aut}(F_{n,c})$. For instance, for $n \geq c - 2 \geq 2$, $\mathrm{Aut}(F_{n,c}) =< T, \theta, \delta >$, where $\delta = \{x_1 \to x_1[x_1, x_2, x_1, x_1], x_i \to x_i, i \neq 1\}$. For a specific generating set when $n \geq c + 1/2$ we refer to [BG1].

Automorphisms of free metabelian nilpotent groups

Let $M_{n,c}$ denote the free metabelian nilpotent group of class c freely generated by $\{x_1, x_2, \ldots, x_n\}$. Then $M_{n,c} \cong F_n/\gamma_{c+1}(F_n)F_n''$. For $n = 2$, a complete description of $IA-\mathrm{Aut}(M_{2,c})$ in terms of generators and defining relations has been given by the first author [G].

In the general case, we have:

Theorem 3.4. ([AG])

(i) *If* $[c/2] < n \leq c$, *then* $\mathrm{Aut}(M_{n,c}) =< T, \delta_3, \ldots, \delta_c >$, *where* $\delta_k = \{x_1 \to x_1[x_1, x_2, x_1, \ldots, x_1]$ *($k-2$ repeats of x_1)*, $x_i \to x_i, i \neq 1\}$;

(ii) *If* $n \geq 2$ *and* $c \geq 3$ *then for each* $\alpha \in \mathrm{Aut}(M_{n,c})$ *there exists a positive integer* $a(\alpha)$ *such that* $\alpha^{a(\alpha)} \in < T, \delta_3, \ldots, \delta_c >$ *where, in addition, the prime factorization of* $a(\alpha)$ *uses primes dividing* $[c + 1/2]!$.

Automorphisms of free metabelian and centre-by-metabelian groups

Let M_n and C_n denote respectively the free metabelian and the free centre-by-metabelian group of rank $n \geq 2$. Then, by a result of Bachmuth [B1], $IA-\mathrm{Aut}(M_2) = \mathrm{Inner}-\mathrm{Aut}(M_2) = T$. For the free metabelian group $M_3, \mu = \{x \to x[y, z, x, x], y \to y, z \to z\}$ defines a non-tame automorphism [C] and $IA-\mathrm{Aut}(M_3)$ is infinitely generated [BM1]. However, for $n \geq 4$, it turns out that $IA-\mathrm{Aut}(M_n)$ is tame [BM2]. For free centre-by-metabelian groups C_n, Elena Stöhr [St] proved that

(i) $IA-\mathrm{Aut}(C_2) =< \mathrm{Inner}-\mathrm{Aut}(C_2), \alpha_p$ (all primes p) $>$, where $\alpha_p = \{x \to x[[x, y]^{x^p}, [x, y]], y \to y\}$;

(ii) When $n = 3$, $\mathrm{Aut}(C_3)$ is infinitely generated (follows easily from [BM1]);

(iii) For $n \geq 4, IA-\mathrm{Aut}(C_n)$ is the normal closure of a single automorphism defined by $\sigma = \{x \to x[[x, y], [u, v]], y \to y, \ldots, z \to z\}$.

[It is now known that σ is indeed a non-tame automorphism of C_n [BGLM].]

It is proved in [GL] that for F free of rank 4, the endomorphisms:

$$\sigma = \{x \to x[[x, y]^{x^m}, [u, v]], y \to y, u \to u, v \to v\}$$

define non-tame automorphisms of $F/[F', F', F']$.

4. Lifting primitive systems of relatively free groups F/V

Let $\mathbf{w} = (w_1, \ldots, w_m), m \leq n$, be a system of words in the free group $F = < f_1, \ldots, f_n > (= F_n)$ of rank n. The system $\mathbf{w}$ is said to be *primitive* if it can be included in some basis of F. Let V be a fully invariant subgroup of F. We say that a system $\mathbf{w} = (w_1, \ldots, w_m), m \leq n$, of words in F is *primitive mod* V if the system $(w_1 V, \ldots, w_m V)$ of cosets can be extended to some basis for F/V. Now let V, U be fully invariant subgroups of F with $V \geq U$, and let $\mathbf{w} = (w_1, \ldots, w_m), m \leq n$, be a *primitive system* mod V. Then we say that $\mathbf{w}$ can be *lifted to a primitive system* mod U if and only if there exist $v_i \in V$ such that the corresponding system $(w_1 v_1, \ldots, w_m v_m)$ is primitive mod U (if $U = 1$ then we simply say that $(w_1 v_1, \ldots, w_m v_m)$ is primitive). Since most groups F/V, where V is the fully invariant closure of an outer commutator, have non-tame automorphisms, it is natural to ask if a given primitive element of F/V can be lifted to a primitive element of F:

> *Does every primitive element of the relatively free group F/V lift to a primitive element of the free group F?*

The general answer is negative, since for instance, $x[x, y, y]$ is primitive mod $\gamma_4(F_2)$ but does not lift to a primitive element of F_2 (see remark below).

Lifting primitivity of free metabelian nilpotent groups

Let $\mathbf{w} = (w_1, \ldots, w_m), m \leq n$, be primitive mod $U = \gamma_{c+1}(F)F''$. We wish to lift this system to a primitive system of F. This is not always possible as for example if $F = < x, y, z >$, the system $(x[x, y, x], y)$ is primitive mod $(\gamma_4(F)F'')$ but the extended system $\mathbf{w} = (x[x, y, x]u, yv, zw)$ is not primitive in F for any choice of u, v in $\gamma_4(F)$ and w in F' (this can be seen using Bachmuth criteria by verifying that the Jacobian matrix $J(\mathbf{w})$ of the system is not invertible). However, for $n \geq 4$, we can take advantage of Bachmuth and Mochizuki's result which reduces the problem of lifting primitivity mod U to that of mod F''. Thus we can restrict to free metabelian nilpotent-of-class-c groups $M_{n,c}$ and need only to study the lifting primitivity mod $\gamma_{c+1}(M_n)$ to the free metabelian group M_n. In this direction, we have:

Theorem 4.1. ([GGR]) *If F is free of rank $n \geq 4$ then every primitive element mod $\gamma_{c+1}(M_n)$ can be lifted to a primitive element of F. Furthermore, for $n \geq 4$ and $m \leq n-2$, every primitive system mod $\gamma_{c+1}(M_n)$ can be lifted to a primitive system of F_n.*

In conjunction with [BM2] it follows from Theorem 4.1 that for $n \geq 4$ and $m \leq n-2$ every primitive system mod $\gamma_{c+1}(F)F''$ can be lifted to a primitive system of F.

Remarks.

(i) The restriction $m \leq n-2$ in the above theorem cannot be improved.

(ii) When rank of F is 3, the metabelian approach does not apply as $M_3 = < x, y, z >$ admits wild automorphisms [C]. The proof that every primitive element of $M_{3,c}, c \geq 3$, can be lifted (via $\gamma_{c+1}(M)F''$) to a primitive element of F_3 is quite technical and we refer to [GGR] for details.

(iii) Since every IA-automorphism of M_2 is inner, $g = x_1 u$ can be lifted to a primitive element of M_2 if and only if u is of the form $[x_1, v]$. Thus, for $c \geq 3$, not every primitive element of $M_{2,c}$ can be lifted to a basis of M_2.

(iv) The Chein automorphism $\{x \to x[y, z, x, x], y \to y, z \to z\}$ of M_3 cannot be lifted to an automorphism of the free group F_3, whereas the element $x[y, z, x, x]$ can be lifted to a primitive element of F_3 (see[GGR]).

Lifting primitivity of free nilpotent groups

Let $\mathbf{w} = (w_1, \ldots, w_m), m \leq n$, be primitive mod $\gamma_{c+1}(F_n)$. We wish to lift this system to a primitive system of F_n. Here we do not have the facility of working modulo F'' so certain further restrictions on m may be necessary. We have the following result.

Theorem 4.2. ([GG]) *For $n \geq m + c - 1$ every primitive system $\mathbf{w} = (w_1, \ldots, w_m), m \leq n$, mod $\gamma_{c+1}(F_n)$ can be lifted to a primitive system of F_n.*

Remarks.

(i) Let $w = w(f_1, \ldots, f_n)$ be such that the exponent sum of some f_i is ± 1. Then w is primitive modulo F' and hence also primitive modulo every term of the lower central series of F. As an application of Theorem 3.3 we deduce that for each $n \geq c \geq 2$ there exists an element $v = v(c) \in \gamma_{c+1}(F)$ such that wv is primitive in F.

(ii) For $n = 2, c \geq 3$, there is a primitive element mod $\gamma_{c+1}(F)$ which can not be lifted to a primitive element of F, and for $n = 3, c \geq 3$, there is a primitive system of two elements mod $\gamma_{c+1}(F)$ which cannot be lifted to a primitive system of F. For $c \geq 4, n = c - 1$, it would be of interest to know whether every primitive element mod $\gamma_{c+1}(F)$ can be lifted to a primitive element of F. The simplest case of the problem is to decide whether or not, for $n = 3, c = 4$, the element $x_1[x_1, x_2, x_2, x_3]$ can be lifted to a basis of F.

Lifting primitivity of relatively free nilpotent-by-abelian groups

Let V be a fully invariant subgroup of F_n such that $F_n'' \geq V \geq \gamma_{c+1}(F_n')$. Let $\mathbf{w} = \{w_1, \ldots, w_m\}, m \leq n$, be primitive mod V. We wish to lift this system to a primitive system of F_n. The next result solves this problem as follows.

Theorem 4.3. ([BG]) *For $n \geq m + 2c, n \geq 4$, every primitive system $\mathbf{w} = (w_1, \ldots, w_m), m \leq n$, mod V can be lifted to a primitive system of F_n.*

[Note that while Theorem 4.3 covers free centre-by-metabelian groups and the groups $F/[\gamma_3(F), \gamma_2(F)]$ we know practically nothing about free *abelian-by-nilpotent* groups.]

5. Automorphisms of $F_\omega(\mathbf{V})$

Consider the automorphism $\theta = \{x_1 \to x_1[x_1, x_2, x_1], x_i \to x_i, i \neq 1\}$ of the n-generator free nilpotent group $F_{n,3}$ of class 3. Then we know that θ is a non-tame automorphism for all $n \geq 2$. Now, let $G = F_{\omega,3} =< x_1, x_2, \ldots >$ be free nilpotent group of class 3 of countable infinite rank ω and extend $\theta = \{x_1 \to x_1[x_1, x_2, x_1], x_i \to x_i, i \neq 1\}$ to an automorphism of G. Then, surprisingly, θ defines a tame automorphism of G. This was proved by Gawron and Macedonska [GaMa]. In fact they proved that every automorphism of the free class-3 group of countable infinite rank is tame. This result has now been extended to free nilpotent groups (Bryant and Macedonska [BrMa]), free metabelian groups (Bryant and Groves [BrGr]) and free *nilpotent-by-abelian* groups (Bryant and the first author [BG2]). We do not know the status of the problem for free abelian-by-(nilpotent of class 2) groups of countable infinite rank. We add that there exists a variety $\mathbf{V}$ such that some automorphism of $F_\omega(\mathbf{V})$ does not lift to an automorphism of F_ω ([BrGr]).

6. Problems

We conclude our survey with a few problems about lifting the generating systems. Some of these problems seem to be new, others are well-known. We would like to attract the reader's attention to the survey article by Bachmuth and Mochizuki [BM3] published in 1989. It contained 7 problems; during the past 4 years 3 of these problems have been completely solved and much progress has been made on 2 more problems. We hope that the problems listed below will contribute towards keeping this subject popular.

Problem 1.

(a) (Well-known problem) Is there a generating system of the group $F_n \oplus F_n, n \geq 2$, that cannot be lifted to a generating system of F_{2n}?

The answer to this question is most probably "yes". Then one may ask:

(b) (R.I. Grigorchuk [K]) Is there an automorphism of the group $F_n \oplus F_n$, that cannot be lifted to an automorphism of F_{2n}?

[We note that while the free groups of finite rank are known to be Hopfian, the group $F_n \oplus F_n$ is not.]

Problem 2.

(a) ([BGLM]) Suppose we have an IA-automorphism ϕ of the free nilpotent group $F_{n,c}$ induced by $x_i \to x_i w_i, w_i \in \gamma_c(F), 1 \leq i \leq n$. Is the condition $\sum \partial_i(w_i) \in (\Delta, \Delta) + \Delta^c$ sufficient for ϕ to be tame?

When $n = 2$, Papistas [P] has proved that the answer to the above question is negative (see also [GSh] for another proof). However, the general case is open, and if the answer remains negative, one may ask:

(b) (R.M.Bryant) In the notation of the previous question, is the condition $\sum \partial_i(w_i) \in \Delta^c$ sufficient for ϕ to be tame? [Note that this condition is not necessary.]

Problem 3. (V.N.Remeslennikov [K]) Let $G_m = F_n/F_n^{(m)}$ be the free solvable group of length $m \geq 3$ and rank $n \geq 3$. Is AutG_m finitely generated?

[This question is not connected directly to the problem of lifting automorphisms, but we think it is very relevant in view of the fact that Aut$G_{m+1}, m \geq 3$, has more generators than AutG_m ([Sh2]).]

Problem 4.

(a) ([BM3]) Find an algorithm to determine whether or not a given automorphism of the free metabelian group M_3 of rank 3 is tame.

(b) Find a particular primitive element of M_3 that cannot be lifted to a primitive element of F_3.

[The existence of such an element has been proved by Roman'kov [R1]. See Evans [E] for an interesting application of Roman'kov's result.]

Problem 5. ([GG]) For $n \geq 4$, does every primitive element mod $\gamma_{n+1}(F_{n-1})$ lift to a primitive element of F_{n-1}? (see Theorem 4.2).

Problem 6. Let $G = F/R$ be a one-relator group with $R = < r >^F$ for some $r \in F$. For what words $r \in F$,

(a) does the group G have only tame generating systems?

(b) does the group G have only tame automorphisms?

(c) does the group G have non-tame generating system but has only tame autmorphisms?

[While there are many examples of non-Hopfian one-relator groups, the sets of elements satisfying (a) and (b) might as well coincide. There are examples of one-relator groups with non-tame automorphisms and of those having only tame generating systems (see [Ro]).]

Conjecture 7. Every group of the form F_n/R with $n \geq 4$ and $R \leq \gamma_7(F)$ has a non-tame generating system.

[This conjecture looks somewhat too bold but it is backed with some recent results [Sh3] on non-tameness in the matrix groups $GL_n(ZG)$.]

References

[A1] S. Andreadakis, On the automorphisms of free groups and free nilpotent groups, *Proc. London Math. Soc. (3)* **15**(1965), 239–268.

[A2] S. Andreadakis, Generators for Aut$G$, G free nilpotent, *Arch. Math.* **42**(1984), 296–300.

[AG] S. Andreadakis and C.K. Gupta, Automorphisms of free metabelian nilpotent groups, *Algebra i Logika* **29**(1990), 746–751.

[Ar] E. Artin, *Geometric algebra* (Interscience, New York, 1957).

[B1] S. Bachmuth, Automorphisms of free metabelian groups, *Trans. Amer. Math. Soc.* **118**(1965), 93–104,

[B2] S. Bachmuth, Induced automorphisms of free groups and free metabelian groups, *Trans. Amer. Math. Soc.*, **122**(1966), 1–17.

[BFM] S. Bachmuth, E. Formanek and H.Y. Mochizuki, IA-automorphisms of certain two-generator torsion-free groups, *J. Algebra* **40**(1976), 19–30.

[Bi] J. Birman, An inverse function theorem for free groups, *Proc. Amer. Math. Soc.* **41**(1973), 634–638.

[BG1] R.M. Bryant and C.K. Gupta, Automorphism groups of free nilpotent groups, *Arch. Math.* **52**(1989), 313–320.

[BG2] R.M. Bryant and C.K. Gupta, Automorphism of free nilpotent-by-abelian groups, *Math. Proc. Cambridge Philos. Soc.* **114**(1993), 143–147.

[BGLM] R.M. Bryant, C.K. Gupta, F. Levin, H.Y. Mochizuki, Non-tame automorphisms of free nilpotent groups, *Comm. Algebra* **18**(1990), 3619–3631.

[BM1] S. Bachmuth and H.Y. Mochizuki, The non-finite generation of Aut(G), G free metabelian of rank 3, *Trans. Amer. Math. Soc.* **270**(1982), 693–700.

[BM2] S. Bachmuth and H.Y. Mochizuki, Aut$(F) \to$ Aut(F/F'') is surjective for free group F of rank ≥ 4, *Trans. Amer. Math. Soc.* **292**(1985), 81–101.

[BM3] S. Bachmuth and H.Y. Mochizuki, The tame range of automorphism groups and GL_n, in *Group Theory (Singapore, 1987)* (De Gruyter, Berlin-New York, 1989), 241–251.

[BrMa] R.M. Bryant and O. Macedonska, Automorphisms of relatively free nilpotent groups of infinite rank, *J. Algebra* **121**(1989), 388–398.

[BrGr] R.M. Bryant, J.R.J. Groves, On automorphisms of relatively free groups, *J. Algebra* **137**(1991), 195-205.

[C] O. Chein, IA automorphisms of free and free metabelian groups, *Comm. Pure Appl. Math.* **21**(1968), 605–629.

[E] M.J. Evans, Presentations of free metabelian groups of rank 2, *Canad. Math. Bull.*, to appear.

[F] R.H. Fox, Free differential calculus, I. Derivations in the free group ring, *Ann. Math. (2)* **57**(1953), 547–560.

[G] C.K. Gupta, IA automorphisms of two generator metabelian groups, *Arch. Math.* **37**(1981), 106–112.

[GG] C.K. Gupta and N.D. Gupta, Lifting primitivity of free nilpotent groups, *Proc. Amer. Math. Soc.* **114**(1992), 617–621.

[GGR] C.K. Gupta, N.D. Gupta and V.Roman'kov, Primitivity in free groups and free metabelian groups, *Canadian J. Math.* **44**(1992), 516–523.

[GL1] C.K. Gupta and F. Levin, Tame range of automorphism groups of free polynilpotent groups, *Comm. Algebra* **19**(1991), 2497–2500.

[GL2] C.K. Gupta and F. Levin, Automorphisms of nilpotent by abelian groups, *Bull. Austral. Math. Soc.* **40**(1989), 207–213.

[Gu] N. Gupta, *Free group rings* (Contemp. Math., **66**, Amer. Math. Soc., 1987).

[GSh] N. Gupta and V. Shpilrain, Nielsen's commutator test for two-generator groups, *Math. Proc. Cambridge Philos. Soc.* **114**(1993), 295–301.

[GaMa] P.W. Gawron and O. Macedonska, All automorphisms of the 3-nilpotent free group of countably infinite rank can be lifted, *J. Algebra* **118**(1988), 120–128.

[Go] A.V. Goryaga, Generators of the automorphism group of a free nilpotent group, *Algebra i Logika* **15**(1976), 289–292.

[K] *Kourovka Notebook*, Novosibirsk, 1992.

[LS] R. Lyndon and P. Shupp, *Combinatorial Group Theory* (Series of Modern Studies in Math. **89**, Springer-Verlag).

[LuMo1] M. Lustig and Y. Moriah, Nielsen equivalence in Fuchsian groups and Seifert fibered spaces, *Topology* **30**(1991), 191–204.

[LuMo2] M. Lustig and Y. Moriah, Generating systems of groups and Reidemeister-Whitehead torsion, *J. Algebra* **157**(1993), 170–198.

[LuMo3] M. Lustig and Y. Moriah, Generalized Montesinos knots, tunnels and N-torsion, *Math. Ann.* **295**(1993), 167–189.

[MoSh] Y. Moriah and V. Shpilrain, Non-tame automorphisms of extensions of periodic groups, *Israel J. Math.*, **84**(1993), 17–31.

[N1] J. Nielsen, Die Isomorphismen der allgemeinen unendlichen Gruppe mit zwei Erzeugenden, *Math. Ann.* **78**(1918), 385–397.

[N2] J. Nielsen, Die Isomorphismengruppe der freien Gruppen, *Math. Ann.* **91**(1924), 169–209.

[P] A.I. Papistas, Non-tame automorphisms of free nilpotent groups of rank 2, *Comm. Algebra* **21**(1993), 1751–1759.

[R1] V.A . Roman'kov, Primitive elements of free groups of rank 3, *Math. Sb.* **182**(1991), 1074–1085.

[R2] V.A. Roman'kov, Automorphisms of groups, *Acta Appl. Math.* **29**(1992), 241–280.

[Ro] G. Rosenberger, Applications of Nielsen's reduction method to the solution of combinatorial problems in group theory: a survey, in *Homological group theory, Proceedings of the 1977 Durham symposium* (C.T.C. Wall (ed.), London Math. Soc. Lecture Notes Ser., **36**, 1979), 339–358.

[Sh1] V. Shpilrain, Automorphisms of F/R' groups, *Internat. J. Algebra Comput.* **1**(1991), 177–184.

[Sh2] V. Shpilrain, Automorphisms of groups and non-commutative determinants, *Bull. London Math. Soc.*, to appear.

[Sh3] V. Shpilrain, Non-tame elements of $GL_n(R)$, R a group ring, in preparation.

[St] E. Stöhr, On automorphisms of free centre-by-metabelian groups, *Arch. Math.* **48**(1987), 376–380.

(*MI*)-GROUPS ACTING UNISERIALLY ON A NORMAL SUBGROUP

PÉTER Z. HERMANN[1]

Department of Algebra and Number Theory, Eötvös University, Múzeum krt. 6–8., H–1088 Budapest, Hungary
E-mail: hermannzp@ludens.elte.hu

Abstract

Finite p–groups with all maximal subgroups isomorphic are considered acting uniserially on certain normal subgroups.

Introduction

We refer the reader to [2] where (MI)–groups are in some sense classified by means of their coclass. In this note some special types of (MI)–groups will be studied. Throughout the paper all groups are finite and of prime (p–) power order. The notation is standard and follows mainly that of [3].

Let us recall

Definition 1. A finite p–group $\mathfrak{P}$ is called an *(MI)–group* if all maximal subgroups of $\mathfrak{P}$ are isomorphic.

Definition 2. Let $\mathfrak{P}$ be a p–group and $\mathfrak{U} \lhd \mathfrak{P}$. The action of $\mathfrak{P}$ on $\mathfrak{U}$ is *uniserial* if for any $\mathfrak{P}$–invariant subgroup $\mathfrak{H} \neq 1$ in $\mathfrak{U}$

$$|\mathfrak{H} : [\mathfrak{H}, \mathfrak{P}]| = p$$

holds. $\mathfrak{P}$ is a *(CF)–group* if it acts uniserially on $\mathrm{K}_2(\mathfrak{P})$, the derived subgroup of $\mathfrak{P}$.

Remark. It is easy to see that in the case of $\mathfrak{P}$ acting uniserially on $\mathfrak{U}$, the $\mathfrak{P}$–invariant subgroups of $\mathfrak{U}$ are merely $\mathfrak{U}$, $[\mathfrak{U}, \mathfrak{P}]$, $[\mathfrak{U}, \mathfrak{P}, \mathfrak{P}]$, etc.

Let $cl(\mathfrak{X})$ denote the nilpotency class of a nilpotent group $\mathfrak{X}$. The following result from [2] will be needed.

Lemma 1. *If $\mathfrak{P}$ is an (MI)–group and $cl(\mathfrak{P}) = c$ then $\mathrm{Z}_k(\mathfrak{P}) \subseteq \Phi(\mathfrak{P})$ holds for all $k < c$.*

[1]This research was supported by the Hungarian National Foundation for Scientific Research, grant numbers T 7441 and T 7442.

PROOF. See Lemma 1 in [2]. □

It is proved in [2] that for any given positive integer r there are only finitely many (MI)-p-groups of coclass r; i.e. $\log_p |\mathfrak{P}|$ is bounded in terms of r for any such $\mathfrak{P}$. (Recently, a much better (exponential) upper bound was obtained by A. Mann in [4].) To get further information one certainly has to deal with (MI)-groups acting uniserially on at least one big section. In this paper we consider the simplest case, namely, when that section lies at the bottom.

Lemma 2. *Assume that $\mathfrak{P}$ is an $(MI)-p$-group and it acts uniserially on $\mathrm{K}_i(\mathfrak{P})$ for some $i \geq 2$. Then $cl(\mathfrak{P}) \leq 2i-1$.*

PROOF. Suppose that $cl(\mathfrak{P}) := c \geq 2i$. Set $\mathfrak{M}_0 = \mathrm{C}_{\mathfrak{P}}(\mathrm{K}_{c-1}(\mathfrak{P}))$. Since $|\mathrm{K}_{c-1}(\mathfrak{P})| = p^2$, $\mathfrak{M}_0$ is a maximal subgroup of $\mathfrak{P}$ and $\mathrm{K}_{c-1}(\mathfrak{P}) \subseteq \mathrm{Z}(\mathfrak{M}_0)$. Let $\mathfrak{M} \neq \mathfrak{M}_0$ denote a maximal subgroup in $\mathfrak{P}$, then $\mathrm{K}_i(\mathfrak{M}) \subseteq \mathrm{K}_i(\mathfrak{P})$ obviously implies that $\mathrm{K}_i(\mathfrak{M}) = \mathrm{K}_i(\mathfrak{M}_0)$. Thus $cl(\mathfrak{P}/\mathrm{K}_i(\mathfrak{M}_0) \leq 2i-2$ and $\mathrm{K}_i(\mathfrak{M}_0) \supseteq \mathrm{K}_{2i-1}(\mathfrak{P})$, consequently $\mathrm{K}_{c-1}(\mathfrak{P}) \subseteq \mathrm{K}_{2i-1}(\mathfrak{P}) \subseteq \mathrm{K}_i(\mathfrak{M}_0)$, so $\mathrm{K}_{c-1}(\mathfrak{P}) \subseteq \mathrm{K}_i(\mathfrak{M}_0) \cap \mathrm{Z}(\mathfrak{M}_0)$. We can conclude that (for any $\mathfrak{M}$ above), $|\mathrm{K}_i(\mathfrak{M}) \cap \mathrm{Z}(\mathfrak{M})| \geq p^2$, whence $\mathrm{K}_{c-1}(\mathfrak{P}) \subseteq \mathrm{K}_i(\mathfrak{M}) \cap \mathrm{Z}(\mathfrak{M})$. This gives

$$\mathrm{K}_{c-1}(\mathfrak{P}) \subseteq \bigcap_{\mathfrak{M}} \mathrm{Z}(\mathfrak{M}) = \mathrm{Z}(\mathfrak{P}),$$

a contradiction. □

Corollary. *Let $\mathfrak{P}$ be an $(MI)-p$-group of coclass $cc(\mathfrak{P})$, $\mathfrak{B} \triangleleft \mathfrak{P}$ and $|\mathfrak{P} : \mathfrak{B}| = p^{\ell}$. If $\mathfrak{P}$ acts uniserially on $\mathfrak{B}$ then*

(1) $cl(\mathfrak{P}) \leq 2\ell - 1$;

(2) $|\mathfrak{P}| \leq p^{3\ell-1}$;

(3) $|\mathfrak{P}| \leq p^{2\ell-1+cc(\mathfrak{P})}$.

PROOF. (1) This directly follows from Lemma 2.
Let $\mathfrak{B}_1 = \mathfrak{B}$, $\mathfrak{B}_2 = [\mathfrak{B}_1, \mathfrak{P}]$, $\mathfrak{B}_3 = [\mathfrak{B}_2, \mathfrak{P}]$, etc. Set $cl(\mathfrak{P}/\mathfrak{B}) = c$, then $\mathrm{K}_{c+1}(\mathfrak{P}) \subseteq \mathfrak{B}$, hence $\mathrm{K}_{c+1}(\mathfrak{P}) = \mathfrak{B}_v$ for some v. Thus

$$|\mathfrak{P}| = p^{\ell+(v-1)+(cl(\mathfrak{P})-c)}. \tag{$*$}$$

(2) Since $\mathfrak{B}_v \subseteq \mathrm{K}_v(\mathfrak{P})$ we have $v \leq c+1$, so $(*)$ and Lemma 2 give $|\mathfrak{P}| \leq p^{\ell+c+2\ell-1-c} = p^{3\ell-1}$.
(3) By Lemma 2 and the definition of coclass

$$|\mathfrak{P}| = p^{cc(\mathfrak{P})+cl(\mathfrak{P})} \leq p^{cc(\mathfrak{P})+2\ell-1},$$

which completes the proof. □

For (CF)-groups, i.e. in the $i=2$ case of Lemma 2, the nilpotency class is at most 3. Using this fact we can completely determine the "(MI) & (CF)" groups. (It is clear that the Abelian (MI)-groups are just the homocyclic ones.)

Proposition 1. *If the (MI)-group $\mathfrak{P}$ is (CF) and $cl(\mathfrak{P}) = 2$ then $\mathfrak{P}$ is isomorphic to one of the following:*

(1) the quaternion group (of order 8);

(2) $< a, b \mid a^{p^m} = b^{p^m} = 1,\ b^{-1}ab = a^{1+p^{m-1}} >$ $(m \geq 2)$;

(3) $< a_1, \ldots, a_k, b_1, \ldots, b_k \mid a_i^{p^m} = b_j^{p^m} = [a_i, b_i]^p = [a_i, b_i, a_j] = [a_i, b_i, b_j] = 1$ (for all i, j), $[a_i, a_j] = [a_i, b_j] = [b_i, b_j] = 1,\ [a_i, b_i] = [a_j, b_j]$ (for all $i \neq j$) $>$ $(p > 2$ or $p = 2$ and $m \geq 2)$.

PROOF. By assumption $|\mathrm{K}_2(\mathfrak{P})| = p$, hence $\Phi(\mathfrak{P}) \subseteq \mathrm{Z}(\mathfrak{P})$; this implies $\Phi(\mathfrak{P}) = \mathrm{Z}(\mathfrak{P})$ by Lemma 1. If the maximal subgroups of $\mathfrak{P}$ are Abelian then $\mathfrak{P}$ is obviously isomorphic to a group in (1) (2) or (3) (with $k = 1$). One can therefore assume that $\mathfrak{M}' = \mathfrak{P}'$ for all maximal subgroups $\mathfrak{M}$. Thus $\mathfrak{P}/\mathfrak{P}'$ is an Abelian (MI)-group of type $(p^m, \ldots, p^m)$, say. For $m = 1$ (i.e. $\mathfrak{P}$ extraspecial) $\mathfrak{P}$ is isomorphic to (1) or (3) (with $k > 1$, $m = 1$ and $p > 2$). From now on assume $m \geq 2$. Since $\Phi(\mathfrak{P}) = \mathrm{Z}(\mathfrak{P})$ and $|\mathfrak{P}'| = p$, every maximal subgroup is of the type $\mathfrak{M}_x = C_{\mathfrak{P}}(x)$ with $x \in \mathfrak{P} \setminus \Phi(\mathfrak{P})$. The exponent of $\Phi(\mathfrak{P})$ is at most p^m and that of $\mathfrak{P}$ is p^m or p^{m+1}. Suppose that there exists an element x_1 of order p^{m+1}. In this case, $\mathrm{Z}(\mathfrak{M}_{x_1}) =< x_1 > \Phi(\mathfrak{P})$ is of exponent p^{m+1}; hence (by the (MI)-property) every element outside $\Phi(\mathfrak{P})$ is of order p^{m+1}. Let $\mathfrak{M}_{x_2} \neq \mathfrak{M}_{x_1}$. We have $x_1^{\alpha p^m} = x_2^{\beta p^m}$ (with $\alpha\beta \not\equiv 0 \pmod p$). Set $y = x_1^{\alpha} \cdot x_2^{-\beta}$; then $y \notin \Phi(\mathfrak{P})$, and by $m \geq 2$

$$y^{p^m} = (x_1^{\alpha} \cdot x_2^{-\beta})^{p^m} = (x_1^{\alpha})^{p^m} \cdot (x_2^{-\beta})^{p^m} = 1,$$

which is a contradiction. Thus all elements in $\mathfrak{P} \setminus \Phi(\mathfrak{P})$ are of order p^m, so $\mathfrak{P}$ is isomorphic to some group in (3) (with $m \geq 2$). □

Proposition 2. *If the (MI)-group $\mathfrak{P}$ is (CF) and $cl(\mathfrak{P}) = 3$ then $\mathfrak{P}$ is isomorphic to one of the following:*

(1) $< a, b \mid a^4 = b^4 = [a, b]^2 = [a, b, b]^2 = [a, b, a] = [a, b, b, a] = [a, b, b, b] = 1 >$;

(2) $< a, b \mid a^{p^2} = b^{p^2} = [a, b]^{p^2} = [a, b, a] = [[a, b]^p, b] = 1,\ [a, b, b] = [a, b]^p >$ $(p \geq 3)$.

Let us note that the group in (1) is of order 2^6, while those in (2) are of order p^6.

PROOF. Let $\mathfrak{M}$ be a maximal subgroup in $\mathfrak{P}$. The inclusion $\mathfrak{M}' \subseteq \mathfrak{P}'$ allows three possibilities for $\mathfrak{M}'$: 1 , $\mathrm{K}_2(\mathfrak{P})$, $\mathrm{K}_3(\mathfrak{P})$. The first case can be excluded, since otherwise $\mathfrak{P}$ would be one step nonabelian, hence of class two. Suppose $\mathfrak{M}' = \mathrm{K}_2(\mathfrak{P})$ (for *all* $\mathfrak{M}$). Set $\mathfrak{M}_1 = C_{\mathfrak{P}}(\mathrm{K}_2(\mathfrak{P}))$; then $\mathfrak{M}_1$ is a maximal subgroup; therefore $\mathfrak{M}_1' = \mathrm{K}_2(\mathfrak{P}) \subseteq \mathrm{Z}(\mathfrak{M}_1)$. Thus $\mathrm{K}_2(\mathfrak{P})$ would coincide with the centre of any maximal subgroup, implying $\mathrm{K}_2(\mathfrak{P}) \subseteq \mathrm{Z}(\mathfrak{P})$, a contradiction. We have obtained:

(i) $\mathfrak{M}' = \mathrm{K}_3(\mathfrak{P})$; *hence* $\overline{\mathfrak{P}} := \mathfrak{P}/\mathrm{K}_3(\mathfrak{P})$ *is a one step nonabelian* (MI)-*group.*

By (i) $\mathfrak{P}$ can be generated by two elements, say $\mathfrak{P} = < x, y >$, $[x, y] = a$, $\mathrm{K}_3(\mathfrak{P}) = < c >$. By previous considerations we can assume that $x \in C_{\mathfrak{P}}(\mathrm{K}_2(\mathfrak{P}))$, i.e. $[a, x] = 1$. Set $[a, y] = c$. For any integers i, j

$$x^{y^i} = xa^ic^{\binom{i}{2}}, \quad (x^j)^{y^i} = x^ja^{ij}c^{\binom{i}{2}j},$$

and hence (by induction on n)

(ii) $(y^ix^j)^n = y^{ni}x^{nj}a^{\binom{n}{2}ij}c^{ij\Delta}$, where $\Delta = \frac{n(n-1)(i(2n-1)-3)}{12}$; in particular $[x^p, y] = a^p$, $[x, y^p] = a^pc^{\binom{p}{2}}$. Thus $[x^{p^2}, y] = [x^p, y^p] = [x, y^{p^2}] = 1$.

Case 1: $\mathfrak{P}'$ is cyclic.

Suppose that $p = 2$; then $[x, y^2] = a^2c = 1$, and hence $y^2 \in \mathrm{Z}(\mathfrak{P})$. Therefore the maximal subgroup $\mathfrak{M}_1 = < x, \Phi(\mathfrak{P}) > = < x, y^2, a >$ is Abelian, which yields that $\mathfrak{P}$ is one step nonabelian, in particular of class 2, which is impossible.

Thus $p \geq 3$, whence $[x, y^p] = a^p$. Assume that $\overline{\mathfrak{P}}$ is metacyclic; then $\mathfrak{P}$ itself is metacyclic (see 11.3 Hilfssatz, p. 336 in [3]) and the maximal subgroups of $\mathfrak{P}$ are one step nonabelian. Such types of groups does not exist by [1], and therefore

$$\overline{\mathfrak{P}} \cong < g, h \mid g^{p^m} = h^{p^m} = [g, h]^p = [g, h, g] = [g, h, h] = 1 > \ .$$

It is easy to see that $m \geq 2$. Consider the maximal subgroups $\mathfrak{M}_1 = < x, \Phi(\mathfrak{P}) >$ and $\mathfrak{M}_2 = < y, \Phi(\mathfrak{P}) >$. One has $\mathrm{Z}(\mathfrak{M}_1) = C_{\Phi(\mathfrak{P})}(x) = < x^p, y^{p^2}, a >$, $\mathrm{Z}(\mathfrak{M}_2) = C_{\Phi(\mathfrak{P})}(y) = < y^p, x^{-fp}a, a^p >$, where $a^{pf} = c$. These imply that $\mathrm{Z}(\mathfrak{M}_1)/\mathfrak{M}_1'$ is of type (p^{m-1}, p^{m-2}, p).
Similarly $\mathrm{Z}(\mathfrak{M}_2)/\mathfrak{M}_2'$ is of type (p^{m-1}, p^{m-1}). Equality of these types yields that $m = 2$, $|\mathfrak{P}| = p^6$. Take $x_1 = x^f$, $y_1 = y^i$ with $i \cdot f \equiv 1 \pmod p$. Then $a_1 := [x_1, y_1] \equiv a \pmod{\mathrm{K}_3(\mathfrak{P})}$, $[a_1, x_1] = 1$, $[a_1, y_1] = c^i = a_1^p$; therefore $f = 1$ can be assumed. One can easily check that $x^{p^2} = a^{\alpha p}$ and $y^{p^2} = a^{\beta p}$. We claim that $\alpha \equiv \beta \pmod p$. For any integers i, j (ii) gives $(y^ix^j)^p = y^{pi}x^{pj}a^{p\ell}$, $(y^ix^j)^{p^2} = y^{p^2i}x^{p^2j} = a^{(\alpha j + \beta i)p}$. Thus there exists an element $z = y^ix^j \in \mathfrak{P} \setminus \Phi(\mathfrak{P})$ such that $z^{p^2} = 1$. Then the maximal subgroup $\mathfrak{M}_z = < z, \Phi(\mathfrak{P}) >$ is of exponent p^2, hence the exponent of $\mathfrak{P}$ is also p^2. We conclude that $\mathfrak{P}$ is isomorphic to the group in (2).

Case 2: $\mathfrak{P}'$ is elementary Abelian.

Now we have $[x^p, y] = 1$, i.e. $x^p \in \mathrm{Z}(\mathfrak{P})$. Moreover, $[x, y^p] = c^{\binom{p}{2}}$. Set $\mathfrak{M}_1 = < x, \Phi(\mathfrak{P}) > = < x, y^p, a, c >$. Since $a \in \mathrm{Z}(\mathfrak{M}_1)$ and $\mathfrak{M}_1$ is nonabelian, then $p = 2$. For any maximal subgroup $\mathfrak{M}$, we have $\mathrm{K}_2(\mathfrak{P}) \subseteq \Omega_1(\mathfrak{M})$.
Since $\mathrm{K}_2(\mathfrak{P}) \not\subseteq \mathrm{Z}(\mathfrak{P})$, $\Omega_1(\mathfrak{M}) \not\subseteq \mathrm{Z}(\mathfrak{M})$, and in particular $\Omega_1(\mathfrak{M}_1) \not\subseteq \mathrm{Z}(\mathfrak{M}_1)$.
Using (i) we deduce that the order of y is 4. Also by (i), the elements $x\mathrm{K}_3(\mathfrak{P})$

and $y\mathrm{K}_3(\mathfrak{P})$ have the same order in $\overline{\mathfrak{P}}$, which implies the order of x is 4 or 8. Suppose that the order of x is 8; then every element of the maximal subgroup $\mathfrak{M}_2 := < y, \Phi(\mathfrak{P}) > = < y, x^2, a, c >$ can be written in the form $y^i x^{2j} a^k c^r$ and

$$(y^i x^{2j} a^k c^r)^4 = (y^i a^k)^4 (x^{2j} c^r)^4 = ((y^i a^k)^2)^2 = (y^{2i} c^{ik})^2 = 1,$$

which is a contradiction. Thus x also is of order 4, $|\mathfrak{P}| = 2^6$ and $\mathfrak{P}$ is isomorphic to the group in (1). □

One may have the feeling that the $\mathrm{K}_i(\mathfrak{P})^s$ (in Lemma 2) and $\mathfrak{B}$ (in the Corollary) are too big. It is certainly impossible to replace them by *any* normal subgroup, but, perhaps, some "sufficiently large" one of not necessarily small index may do. The following assertion is an attempt in this direction, although the result turns out to be very restrictive.

Proposition 3. *Assume that $\mathfrak{P}$ is an (MI)-group, $\mathfrak{U} \triangleleft \mathfrak{P}$ and $\mathfrak{P}$ acts uniserially on $\mathfrak{U}$. If $C_{\mathfrak{P}}(\mathfrak{U}) \subseteq \mathfrak{U}$ then $|\mathfrak{P}| = p^3$.*

PROOF. Let $\mathfrak{U}_0 = \mathfrak{U}$, $\mathfrak{U}_1 = [\mathfrak{U}_0, \mathfrak{P}]$, $\mathfrak{U}_2 = [\mathfrak{U}_1, \mathfrak{P}]$, ... The $\mathfrak{U}_i^s$ provide the only $\mathfrak{P}$–invariant subgroups in $\mathfrak{U}$. As $\mathrm{Z}(\mathfrak{P}) \subseteq \mathfrak{U}$, then $\mathrm{Z}(\mathfrak{P}) = \mathfrak{U}_{k-1}$ with $\mathfrak{U}_k = 1$. Set $\mathfrak{M}_0 = C_{\mathfrak{P}}(\mathfrak{U}_{k-2})$; then $\mathfrak{M}_0$ is a maximal subgroup of $\mathfrak{P}$. Let $\mathfrak{M}$ denote a maximal subgroup of $\mathfrak{P}$ containing $\mathfrak{U}$; then $\mathrm{Z}(\mathfrak{M}) \subseteq C_{\mathfrak{P}}(\mathfrak{U}) \subseteq \mathfrak{U}$, $\mathrm{Z}(\mathfrak{M}) \cap \mathfrak{U}_{k-2} \subseteq \mathrm{Z}(\mathfrak{M}) \cap \mathrm{Z}(\mathfrak{M}_0)$. Suppose that $\mathfrak{M} \neq \mathfrak{M}_0$, then $\mathrm{Z}(\mathfrak{M}) \cap \mathfrak{U}_{k-2} \subseteq \mathrm{Z}(\mathfrak{P})$, implying $|\mathrm{Z}(\mathfrak{M})| = p$, $|\mathfrak{P}| = p^2$, which is not the case. Thus $\mathfrak{M}_0$ is the unique maximal subgroup containing $\mathfrak{U}$, and so $\mathfrak{P}/\mathfrak{U}$ is cyclic. In particular $\mathfrak{U} \supseteq \mathrm{K}_2(\mathfrak{P})$, whence $\mathfrak{P}$ is a (CF)-group, and inspection of the groups in Propositions 1 and 2 implies the result. □

References

[1] Hermann, P.Z., On a class of finite groups having isomorphic maximal subgroups, *Ann. Univ. Sci. Budapest. Sect. Math.* **24**(1981), 87–92.

[2] Hermann, P.Z., On finite p-groups with isomorphic maximal subgroups, *J. Austral. Math. Soc. Ser. A* **48**(1990), 199–213.

[3] Huppert, B., *Endliche Gruppen I* (Springer, Berlin–Heidelberg–New York, 1967).

[4] Mann, A., On p-groups whose maximal subgroups are isomorphic, to appear.

REVISITING A THEOREM OF HIGMAN

E. JESPERS*[1], M.M. PARMENTER*[2] and P.F. SMITH†

*Department of Mathematics and Statistics, Memorial University of Newfoundland, St. John's, NF, Canada A1C 5S7

†Department of Mathematics, University of Glasgow, Glasgow G12 8QW, Scotland

Let $U(\boldsymbol{Z}G)$ denote the group of units of an integral group ring. The theorem of Higman we are interested in is the following.

Theorem. *For a finite group G, $U(\boldsymbol{Z}G) = \pm G$ if and only if G is abelian of exponent 2, 3, 4 or 6 or $G = E \times K_8$ where K_8 is the quaternion group of order 8 and E is an elementary abelian 2-group.*

Proofs using the important Dirichlet Unit Theorem or its equivalent can be found in [2] and [4]. We will give a self-contained elementary proof which will furthermore highlight the important role played by Bass cyclic and bicyclic units in integral group rings.

Terminology and notation will follow that of [4]. In particular, if $g \in G$ is of order n, we let $\hat{g}$ denote the sum $1 + g + \ldots + g^{n-1}$ in $\boldsymbol{Z}G$. Also a unit in $\boldsymbol{Z}G$ will be called trivial if it is of the form $\pm g$, and nontrivial otherwise. We will say that $U(\boldsymbol{Z}G)$ is trivial if all units in $\boldsymbol{Z}G$ are trivial. Finally, if $u = \Sigma \alpha_g g$ is a unit in $\boldsymbol{Z}G$, we recall that $u^* = \Sigma \alpha_g g^{-1}$ is also a unit in $\boldsymbol{Z}G$.

We first recall the definition of Bass cyclic and bicyclic units. The latter are the units of the form $u_{a,b} = 1 + (1-b)a\hat{b}$ where $a, b \in G$. Note that $u_{a,b}^{-1} = 1 - (1-b)a\hat{b}$, and that $u_{a,b}$ is nontrivial if and only if $a^{-1}ba$ is not a power of b.

The Bass cyclic units [1] are units in integral group rings of cyclic subgroups. If $g \in G$, and $o(g) = n$, choose an integer i such that gcd $(i, n) = 1$. Then if $m = \phi(n)$ (the Euler phi function), we define the Bass cyclic unit

$$u = (1 + g + g^2 + \ldots + g^{i-1})^m + \frac{1 - i^m}{n}\hat{g}.$$

Note that u is in $\boldsymbol{Z}G$ by Euler's Theorem and that if we choose $\alpha, \beta \in \boldsymbol{Z}$, $0 < \alpha < n$, with $i\alpha + n\beta = 1$, then

$$u^{-1} = (1 + g^i + g^{2i} + \ldots + g^{(\alpha-1)i})^m + \frac{1 - \alpha^m}{n}\hat{g}.$$

We will need to use the fact that if $i \not\equiv \pm 1 \pmod{n}$, then u is nontrivial. For completeness, we include a proof.

[1]First author supported in part by NSERC grant OGP0036631.

[2]Second author supported in part by NSERC grant A8775.

Assume $o(g) = n$, $\gcd(i, n) = 1$ and $g^i \neq g, g^{-1}$. Say to the contrary that $(1 + g + \ldots + g^{i-1})^m + \frac{1 - i^m}{n}\hat{g} = g^t$ for some t. Multiplying by $(1 - g)^m$, we get $(1 - g^i)^m = (1 - g)^m g^t$, so

$$1 - mg^i + \binom{m}{2} g^{2i} \ldots + g^{im} = g^t - mg^{t+1} + \binom{m}{2} g^{t+2} \ldots + g^{t+m}.$$

Note that the existence of such an i tells us that $m = \phi(n)$ is even and greater than 2 and also that the powers of g on the left hand side are distinct. It follows that $g^t = 1$ or $g^{t+m} = 1$.

Assume first that $g^t = 1$. Since $g^i \neq g$, we must have $g^i = g^{m-1}$. But then $g^{2i} = g^{2m-2} \neq g^{m-2}$ (since $n \nmid m$), so we must have $g^{2i} = g^2$. This says that $n | (2m - 4)$, forcing $n = 2m - 4$ since $2 < m < n$. Thus n is even. In that case, however, $m = \phi(n) \leq \frac{n}{2}$, giving a contradiction.

On the other hand, if $g^{t+m} = 1$, we must have $g^i = g^{1-m}$ (since $g^i \neq g^{-1}$). But then $g^{2i} = g^{2-2m} \neq g^{2-m}$, so $g^{2i} = g^{-2}$. This again says $n | (2m - 4)$, giving a contradiction.

We will now prove Higman's theorem. Some of the steps in the proof are standard ([2], [4]), but we include all details for completeness.

PROOF OF THEOREM. First we will show that if G is any group other than those stated, $\boldsymbol{Z}G$ contains a nontrivial unit. But this has essentially been done. For if G is not abelian and not Hamiltonian, there must exist $a, b \in G$ such that the bicyclic unit $u_{a,b}$ is nontrivial. On the other hand, if G is abelian or Hamiltonian, but not among those stated, then G contains an element g of order 5 or ≥ 7. In that case, $\phi(n) > 2$, and thus a nontrivial Bass cyclic unit exists.

The remainder of the proof consists of showing that $U(\boldsymbol{Z}G) = \pm G$ for all groups of the type listed. We will do this in four steps: combined together, these steps prove the result.

Step 1. $U(\boldsymbol{Z}G) = \pm G$ implies $U(\boldsymbol{Z}(G \times C_2)) = \pm(G \times C_2)$ where C_2 is the cyclic group of order 2.

To see this, assume $C_2 = \langle x \rangle$ and say $(\alpha + \beta x)(\theta + \delta x) = 1$ for $\alpha, \beta, \theta, \delta \in \boldsymbol{Z}G$. This means $\alpha\theta + \beta\delta = 1$ and $\alpha\delta + \beta\theta = 0$, so $(\alpha + \beta)(\theta + \delta) = 1$ and $(\alpha - \beta)(\theta - \delta) = 1$. Thus $\alpha + \beta = \pm g_1$, $\alpha - \beta = \pm g_2$ for some $g_1, g_2 \in G$. So $2\alpha = \pm g_1 \pm g_2$. It follows that $g_1 = g_2$ and either $\alpha = 0$ or $\beta = 0$, proving the statement.

Step 2. Let $K_8 = \langle x, y \mid x^4 = y^4 = 1, x^2 = y^2, yx = x^{-1}y \rangle$ be the quaternion group of order 8. Then $U(\boldsymbol{Z}K_8) = \pm K_8$.

Let $u \in U(\boldsymbol{Z}K_8)$. Since $K_8/\langle x^2 \rangle \cong C_2 \times C_2$, the above result says that $U(\boldsymbol{Z}(K_8/\langle x^2 \rangle))$ is trivial, so multiplying by $\pm g$ for some g in K_8, we may assume that

$$u = 1 + (1 - x^2)(\alpha_0 + \alpha_1 x + \beta_0 y + \beta_1 xy)$$

for some α_0, α_1, β_0, β_1 in $\boldsymbol{Z}$. Note that uu^* equals

$$\begin{aligned}
& (1 + (1 - x^2)(\alpha_0 + \alpha_1 x + \beta_0 y + \beta_1 xy)) \\
& \qquad (1 + (1 - x^2)(\alpha_0 + \alpha_1 x^3 + \beta_0 y^3 + \beta_1 x^3 y)) \\
= & (1 + (1 - x^2)(\alpha_0 + \alpha_1 x + \beta_0 y + \beta_1 xy)) \\
& \qquad (1 + (1 - x^2)(\alpha_0 - \alpha_1 x - \beta_0 y - \beta_1 xy)) \\
= & 1 + 2(1 - x^2)(\alpha_0 + \alpha_0^2 + \alpha_1^2 + \beta_0^2 + \beta_1^2).
\end{aligned}$$

Since $\boldsymbol{Z}\langle x^2 \rangle$ has only trivial units by the earlier result, we conclude that $\alpha_0 + \alpha_0^2 + \alpha_1^2 + \beta_0^2 + \beta_1^2 = 0$. But $\alpha_0 + \alpha_0^2 \geq 0$, so this forces $\alpha_1 = \beta_0 = \beta_1 = 0$ and $\alpha_0 = 0$ or -1. So $u = 1$ or $u = x^2$, and we're done.

Step 3. If $G = C_4 \times C_4 \times \ldots \times C_4$, then $U(\boldsymbol{Z}G) = \pm G$.

We proceed by induction on the rank of G. If rank $G = 1$, i.e. $G = \langle a \rangle$ where $a^4 = 1$, then G is a subgroup of K_8 so the previous result tells us $U(\boldsymbol{Z}G) = \pm G$.

Assume then that rank $G = n > 1$, and the result is true for all such groups of rank $\leq n - 1$. We have $G = A \times \langle a \rangle \times \langle b \rangle$ where $a^4 = b^4 = 1$ and rank $A = n - 2$.

Let $u \in U(\boldsymbol{Z}G)$. The inductive hypothesis, together with result 1, tells us that $U(\boldsymbol{Z}(G/\langle b^2 \rangle))$ is trivial, so (multiplying by a trivial unit) we can assume

$$u = 1 + (1 - b^2)(\alpha_0 + \alpha_1 a + \alpha_2 a^2 + \alpha_3 a^3 + \beta_0 b + \beta_1 ab + \beta_2 a^2 b + \beta_3 a^3 b)$$

for some $\alpha_i, \beta_i \in \boldsymbol{Z}A$.

Now $U(\boldsymbol{Z}(G/\langle a^2 \rangle))$ is also trivial. Applying this to u, we obtain $\alpha_1 + \alpha_3 = 0$, $\beta_0 + \beta_2 = 0$, $\beta_1 + \beta_3 = 0$ and $\alpha_0 + \alpha_2 = 0$ or -1.

Similarly $U(\boldsymbol{Z}(G/\langle a^2 b^2 \rangle))$ is trivial. Applying this to u, we obtain $\alpha_1 - \alpha_3 = 0$, $\beta_0 - \beta_2 = 0$, $\beta_1 - \beta_3 = 0$ and $\alpha_0 - \alpha_2 = 0$ or -1.

Combining the last two facts, we obtain $\alpha_1 = \alpha_3 = \beta_0 = \beta_2 = \beta_1 = \beta_3 = \alpha_2 = 0$, and $\alpha_0 = 0$ or -1. This means $u = 1$ or $u = b^2$, so u is certainly trivial.

Step 4. If $G = C_3 \times C_3 \times \ldots \times C_3$, then $U(\boldsymbol{Z}G) = \pm G$.

Again, we proceed by induction on the rank of G. First assume $G = \langle a \rangle$ where $a^3 = 1$. Let $u \in U(\boldsymbol{Z}G)$. Using the fact that $(1 - a)(1 + a + a^2) = 0$, we may assume (multiplying by a trivial unit) that $u = 1 + (1 - a)(\beta_0 + \beta_1 a)$ for some $\beta_0, \beta_1 \in \boldsymbol{Z}$.

The quotient ring $\boldsymbol{Z}G/\langle 1 + a + a^2 \rangle$ is isomorphic to $\boldsymbol{Z}[\xi]$ where ξ is a third root of 1, and thus has only $\pm\xi^i$ as units. In this quotient, u becomes $\bar{u} = 1 + (1 - \xi)(\beta_0 + \beta_1 \xi) = 1 + \beta_0 + \beta_1 + (2\beta_1 - \beta_0)\xi$. It is easily seen that all possibilities for $\bar{u} = \pm\xi^i$ yield trivial units for u.

Now assume rank $G = n > 1$, and the result is true for all such groups of rank $\leq n-1$. We have $G = A \times \langle a \rangle \times \langle b \rangle$ where $a^3 = b^3 = 1$ and rank $A = n-2$.

Let $u \in U(\mathbf{Z}G)$. The inductive hypothesis tells us that $U(\mathbf{Z}(G/\langle b \rangle))$ is trivial so (multiplying by a trivial unit) we can assume $u = 1 + (1-b)[(\gamma_0 + \gamma_1 a + \gamma_2 a^2) + (\theta_0 + \theta_1 a + \theta_2 a^2)b]$ for some $\gamma_i, \theta_i \in \mathbf{Z}A$.

Now $G/\langle ab \rangle$ also has rank $n-1$, so the image of u in $\mathbf{Z}(G/\langle ab \rangle)$ must be trivial. This image is

$$\begin{aligned}\bar{u} &= 1 + (1-a^2)(\gamma_0 + \gamma_1 a + \gamma_2 a^2 + (\theta_0 + \theta_1 a + \theta_2 a^2)a^2) \\ &= 1 + \gamma_0 - \gamma_1 + \theta_1 - \theta_2 + (\gamma_1 - \gamma_2 + \theta_2 - \theta_0)a + (\gamma_2 - \gamma_0 + \theta_0 - \theta_1)a^2\end{aligned}$$

It follows that precisely 2 of the above coefficients (in $\mathbf{Z}A$) must equal 0, and this leads to three cases.

Case I. Assume $\gamma_1 - \gamma_2 + \theta_2 - \theta_0 = 0$ and $\gamma_2 - \gamma_0 + \theta_0 - \theta_1 = 0$. Again, the inductive hypothesis tells us that the image of u in $\mathbf{Z}(G/\langle a^2 b \rangle)$ is trivial. This image is

$$\begin{aligned}\bar{\bar{u}} &= 1 + (1-a)(\gamma_0 + \gamma_1 a + \gamma_2 a^2 + (\theta_0 + \theta_1 a + \theta_2 a^2)a) \\ &= 1 + \gamma_0 - \gamma_2 + \theta_2 - \theta_1 + (\gamma_1 - \gamma_0 + \theta_0 - \theta_2)a + (\gamma_2 - \gamma_1 + \theta_1 - \theta_0)a^2 \\ &= 1 + \theta_0 - 2\theta_1 + \theta_2 + (\theta_0 + \theta_1 - 2\theta_2)a + (-2\theta_0 + \theta_1 + \theta_2)a^2\end{aligned}$$

Either the second or third coefficient must equal zero, and therefore the first is congruent to 1 modulo 3. It follows that $\theta_0 + \theta_1 - 2\theta_2 = 0 = -2\theta_0 + \theta_1 + \theta_2$, forcing $\theta_0 = \theta_1 = \theta_2$ and $\gamma_0 = \gamma_1 = \gamma_2$.

Finally, we note that the image of u in $\mathbf{Z}(G/\langle a \rangle)$ is also trivial, and this image is $1 + (1-b)(3\gamma_0 + 3\theta_0 b)$.

It follows that $\gamma_0 = \theta_0 = 0$, and we conclude that $u = 1$.

Case II. Assume $1 + \gamma_0 - \gamma_1 + \theta_1 - \theta_2 = 0$ and $\gamma_1 - \gamma_2 + \theta_2 - \theta_0 = 0$. In this case, the image of u in $\mathbf{Z}(G/\langle a^2 b \rangle)$ is

$$\bar{\bar{u}} = \theta_0 - 2\theta_1 + \theta_2 + (1 + \theta_0 + \theta_1 - 2\theta_2)a + (-2\theta_0 + \theta_1 + \theta_2)a^2.$$

As in Case I, we conclude that $\theta_0 - 2\theta_1 + \theta_2 = 0 = -2\theta_0 + \theta_1 + \theta_2$. It follows that $\theta_0 = \theta_1 = \theta_2$ and $\gamma_1 = \gamma_2 = 1 + \gamma_0$. The image of u in $\mathbf{Z}(G/\langle a \rangle)$ is now $1 + (1-b)(2 + 3\gamma_0 + 3\theta_0 b)$, so we conclude that $\gamma_0 = -1$, $\theta_0 = 0$ and $u = b$.

Case III. Assume $1 + \gamma_0 - \gamma_1 + \theta_1 - \theta_2 = 0$ and $\gamma_2 - \gamma_0 + \theta_0 - \theta_1 = 0$. Here the image of u in $\mathbf{Z}(G/\langle a^2 b \rangle)$ is

$$\bar{\bar{u}} = 1 + \theta_0 - 2\theta_1 + \theta_2 + (1 + \theta_0 + \theta_1 - 2\theta_2)a + (-1 - 2\theta_0 + \theta_1 + \theta_2)a^2.$$

Again as in Case I, $1 + \theta_0 - 2\theta_1 + \theta_2 = 0 = 1 + \theta_0 + \theta_1 - 2\theta_2$, so $\theta_2 = \theta_1 = 1 + \theta_0$ and $\gamma_2 = \gamma_1 = 1 + \gamma_0$. Finally, the image of u in $\mathbf{Z}(G/\langle a \rangle)$ is $1 + (1-b)(2 + 3\gamma_0 + (2 + 3\theta_0)b)$, so we conclude that $\gamma_0 = \theta_0 = -1$ and $u = b^2$.

In all cases, u has been shown to be trivial and our proof is complete. □

We remark in closing that a different elementary argument for the special cases $G = C_3$ or C_4 was given by Moore in [3].

References

[1] H. Bass, The Dirichlet Unit Theorem, induced characters and Whitehead groups of finite groups, *Topology* 4(1966), 391–410.

[2] G. Higman, The units of group rings, *Proc. London Math. Soc.* **46**(1940), 231–248.

[3] P. Moore, Units of integral group rings for finite groups (Research report M/CS 91-94, Mount Allison University).

[4] S.K. Sehgal, *Topics in Group Rings* (Marcel Dekker, New York, 1978).

COHOMOLOGICAL FINITENESS CONDITIONS

PETER H. KROPHOLLER[1]

School of Mathematical Sciences, Queen Mary and Westfield College, Mile End Road, London E1 4NS
E-mail: P.H.Kropholler@qmw.ac.uk

Abstract

This article is based on four lectures delivered at the conference. In the months following the conference Jonathan Cornick and I made progress with some aspects of the theory, and I have included a survey of these results here. The starting point is the theorem that if G is a group of type $(\mathrm{FP})_\infty$ which belongs to a reasonably large class of groups, $\mathbf{H}\mathfrak{F}$, and which is torsion-free, then G has finite cohomological dimension. After the conference we discovered proofs of various more general results. In particular if G is an $\mathbf{H}\mathfrak{F}$-group of type $(\mathrm{FP})_\infty$ then every torsion-free subgroup has finite cohomological dimension and if G is a residually finite $\mathbf{H}\mathfrak{F}$-group of type $(\mathrm{FP})_\infty$ then G has finite virtual cohomological dimension (vcd). In particular, all linear groups of type $(\mathrm{FP})_\infty$ have finite vcd regardless of the characteristic. The proofs make use of *complete cohomology*, a theory developed by Vogel and later but independently by Mislin. We use only a few facts about this theory, the crucial one being that it provides a simple cohomological criterion for a module over a ring to have finite projective dimension. The simplicity of the arguments allows one to generalize results from group algebras to strongly graded rings. The typical result states that under suitable conditions a module of type $(\mathrm{FP})_\infty$ over a G-graded ring has finite projective dimension if and only if its restrictions to finite subgroups of G have finite projective dimension.

Contents

---

[1]This research was partially supported by SERC grant GR/F80616.

1. Introduction

These notes are based on a series of four lectures delivered at the conference. In the months following the conference, Jonathan Cornick and I made several improvements to the theory of groups and modules of type (FP)$_\infty$, and I have included a summary of the new results here. I hope this does not obscure the underlying simplicity. There are two principle methods for studying cohomological finiteness conditions for groups:

- Cohomology
- Group actions on spaces

In this article we only need to use these methods in their very simplest forms. We only need a few simple properties of cohomology theories and we only work with nice group actions on nice spaces.

1.1. Cohomology

Let G be a group. Cohomology of G can be computed with coefficients in any G-module, and one gets a sequence of cohomology groups for each module. A few things make this appear more mysterious than it really is. One problem is that there usually seems to be a large gap between the definition of the cohomology theory and the problem of computing any cohomology groups. This problem is partly addressed by axioms: certain basic properties that the cohomology theory satisfies which can be used to begin calculations. The principle axiom which we need is the Long Exact Sequence Axiom. It says that associated to any short exact sequence

$$A \rightarrowtail B \twoheadrightarrow C$$

of G-modules there is a corresponding long exact sequence of cohomology groups involving cohomology with coefficients in the modules A, B, C. One point of view is that this sequence relates the cohomology with coefficients in C to that with coefficients in A and B, and the simplest way of using it is as follows:

Lemma 1.1.1. *If all cohomology groups with coefficients in A and B are zero, then all cohomology groups with coefficients in C are zero.*

There is a more general form of this, which follows from this easily and which we shall use repeatedly:

Lemma 1.1.2. *Let*

$$0 \to N_r \to \cdots \to N_1 \to N_0 \to M \to 0$$

be an exact sequence of G-modules. If all cohomology groups with coefficients in the N_i are zero, then all cohomology groups with coefficients in M are zero.

A second problem is that you can have various cohomology theories. In this article we make a lot of use of a complete cohomology theory generalizing Tate cohomology of finite groups. It just happens that the axioms which this theory satisfies are ideally suited to studying cohomological finiteness conditions.

If you like, you can think of the cohomology theory as a black box, having axioms which help to predict what will come out when certain things are put in. There is often a great deal of room for manoeuvre: it may be possible to formulate a problem in cohomological terms in several different ways. When this happens, it is usually advantageous to try to choose a formulation where solving the problem is equivalent to showing that a certain cohomology group is zero. It is simply the fact that it is generally much easier to prove that a cohomology group is zero (if it is zero) than to understand a cohomology group when it is non-zero. For this reason, we shall discuss various *vanishing theorems* for cohomology. Lemma 1.1.2 is the simplest and it is also the building block for others.

1.2. Group actions on spaces

All the spaces we work with are cell complexes. These are built up starting with a discrete set of 0-cells, or vertices. The vertices are joined by 1-cells, or edges to form a graph (the 1-skeleton) and then 2-cells are attached to the 1-skeleton to form a 2-complex, (the 2-skeleton), and so on. All the cell complexes we work with are finite dimensional, and so this process stops once one reaches this dimension. Usually, the cell complexes will involve infinitely many cells.

Associated to an r-dimensional cell complex X, there is the cellular chain complex

$$0 \to C_r(X) \to \cdots \to C_1(X) \to C_0(X) \to 0.$$

Essentially, $C_i(X)$ is the free abelian group on the set of i-dimensional cells. The way the i-cells are attached to the $i-1$-skeleton is used to define the map $C_i(X) \to C_{i-1}(X)$. The result is a chain complex: the composite of two consecutive arrows is zero. In each dimension i, the kernel of the outgoing map modulo the image of the incoming map is the ith cellular homology group $H_i(X)$. Ostensibly, the cellular homology groups depend on the precise way in which X is carved up into cells. In fact, the cellular homology group do not depend on the way the space is carved up into cells, and they are even invariants of the homotopy type of the space. We only need this fact in a very special case:

Lemma 1.2.1 *If X is contractible then $H_i(X) = 0$ for $i \neq 0$ and $H_0(X) = \mathbb{Z}$.*

An equivalent way of putting this is to say that if X is homotopy equivalent to a point then X has the homology of a point. There is a convenient way of phrasing this in terms of the *augmented* cellular chain complex.

Lemma 1.2.2. *If X is contractible then the augmented cellular chain complex*

$$0 \to C_r(X) \to \cdots \to C_1(X) \to C_0(X) \to \mathbb{Z} \to 0$$

is an exact sequence.

Here, the augmentation map $\varepsilon : C_0(X) \to \mathbb{Z}$ is defined by sending each vertex to 1.

Now let G be a group. We shall consider cellular actions of G on X. This means that as well as acting by self-homeomorphisms of X, elements of G carry cells to cells and the setwise stabilizer of each cell is equal to its pointwise stabilizer. Thus G permutes the set of i-dimensional cells and the free abelian group $C_i(X)$ naturally inherits the structure of a permutation module. In this situation, the augmented cellular chain complex becomes a sequence of G-modules and G-maps. (It is the fact that G acts by self-homeomorphisms which ensures that the connecting maps $C_i(X) \to C_{i-1}(X)$ are G-maps.) Let Σ_i be a set of G-orbit representatives of the i-dimensional cells, and let G_σ be the stabilizer of the cell σ. The ith chain group $C_i(X)$ breaks up as the direct sum of the the free abelian groups on each orbit, and so we have

$$C_i(X) \cong \bigoplus_{\sigma \in \Sigma_i} \mathbb{Z} \otimes_{\mathbb{Z}G_\sigma} \mathbb{Z}G.$$

This description of $C_i(X)$ is crucial throughout this article. In practice, there are often infinitely many orbits of i-cells in which case the above describes $C_i(X)$ as an infinite direct sum. This creates subtle problems, because cohomology theories do not always behave well in relation to infinite direct sums, even though they always behave well with respect to finite direct sums. The point is that, by allowing group actions with infinitely many orbits of cells, we will be able to work with a vastly larger class of groups.

2. Cohomological finiteness conditions

The idea of a *finiteness condition* or a *cohomological finiteness condition* in group theory is often intuitively clear in context. In this section, we give precise definitions of what these terms mean. Then it becomes easier to understand why cohomological finiteness conditions are sometimes very different from finiteness conditions, and why they are sometimes very similar.

2.1. Finiteness conditions versus cohomological finiteness conditions

A *finiteness condition* in group theory is a property of groups which holds for all *finite* groups. Examples include the properties of being

- finitely generated,
- finitely presented,
- residually finite.

Finiteness conditions have played an important role in infinite groups theory for many years, beginning with the pioneering work of Philip Hall [12] and [13] in which he studied finiteness conditions for certain classes of finitely generated soluble group. His theorem that every finitely generated abelian-by-nilpotent group is residually finite is typical of this approach: one restricts to a special class, in this case abelian-by-nilpotent groups, and there, one finds that one rather weak finiteness condition *finite generation* implies another stronger one *residual finiteness.*

In this paper we shall be concerned with *cohomological finiteness conditions*. By a cohomological finiteness condition we mean a property of groups that holds for all those groups which are fundamental groups of a finite aspherical CW-complex, that is to say, groups G for which there is a finite model for the Eilenberg-Mac Lane space $K(G,1)$. Groups G with a finite $K(G,1)$ are said to be of type (F). Examples of cohomological finiteness conditions include the properties of being

- finitely generated,
- finitely presented,
- torsion-free.

Notice that some cohomological finiteness conditions are also finiteness conditions and some are not. Torsion-freeness is a particularly interesting case because it is so far away from being a finiteness condition, but as we shall see it is remarkably strong amongst *cohomological* finiteness conditions. This leads to the philosophy: if you want to prove that a group is torsion-free you can expect that cohomological methods, or methods from combinatorial topology, will be helpful.

The fundamental theorem of Wall identifies what cohomological finiteness conditions are in algebraic terms. We need one definition in order to state this.

Definition 2.1.1. A group G is said to be of type (FL) if and only if the trivial module admits a finite free resolution: that is, an exact sequence

$$0 \to F_r \to \cdots \to F_1 \to F_0 \to \mathbb{Z} \to 0$$

of finite length and in which each F_i is a free $\mathbb{Z}G$-module of finite rank.

2.2. The types $(\mathbf{FP})_*$ and $(\mathbf{F})_*$

Now, the following result reveals what *cohomologically finite* means in purely algebraic terms, and it leads quickly to a whole series of important cohomological finiteness conditions.

Theorem 2.2.1. *Let G be a group. The following are equivalent:*

(1) G is of type (F)*;*

(2) G is finitely presented and of type (FL).

We next summarize some of the important cohomological finiteness conditions which can be defined in terms of projective resolutions. Projective resolutions are exact sequences like the one in (2.1.1) but in which one allows any projective modules, not just free modules. In homological algebra it is much more natural to work with projective modules rather than free modules, and definitions made in terms of projective resolutions instead of free resolutions are often easier to work with.

Definition 2.2.2. Let G be a group. If it is possible to choose a projective resolution $P_* \twoheadrightarrow \mathbb{Z}$ of $\mathbb{Z}$ over $\mathbb{Z}G$ such that

(1) P_i is finitely generated for $0 \leq i \leq n$ then G is said to be of type $(\mathrm{FP})_n$:
(2) P_i is finitely generated for all i then G is said to be of type $(\mathrm{FP})_\infty$:
(3) P_* is finite (i.e. the resolution has finite length and finite type) then G is said to be of type (FP).

We also have the following definitions:

Definition 2.2.3.

(1) G is of type $(\mathrm{F})_n$ if and only if there is a $K(G,1)$ with finite n-skeleton;
(2) G is of type $(\mathrm{F})_\infty$ if and only if there is a $K(G,1)$ with finite n-skeleton for all n.
(3) G is of type (FD) if and only if there is a $K(G,1)$ which is finitely dominated, (i.e. there is a finite complex X and continuous maps $K(G,1) \to X$ and $X \to K(G,1)$ such that the composite $K(G,1) \to X \to K(G,1)$ is homotopic to the identity on $K(G,1)$).

The relationships between the various properties are illustrated in Figure 1. Discounting the first line of the diagram, the right hand property on each line is equivalent to the left hand property plus finite presentation. For example, a group is of type (FD) if and only if it is finitely presented and of type (FP). However, in none of these cases is it known that the left hand and right hand properties actually differ: it remains conceivable that every group of type $(\mathrm{FP})_2$ is finitely presented and if this were proved to be the case then equivalence of the left hand and right hand columns would follow automatically. In addition to this mystery, there is also no known example of a group of type (FP) which is not of type (FL), and so in fact there is no example to distinguish the four types (FP), (FL), (FD), (F).

It is easy to prove that a group is of type $(\mathrm{FP})_\infty$ if and only if it is of type $(\mathrm{FP})_n$ for all n, and similarly; of type $(\mathrm{F})_\infty$ if and only if of type $(\mathrm{F})_n$ for all n.

2.3. Dimension

There is one further important definition.

Definition 2.3.1. Let n be a non-negative integer. Then G has cohomological dimension at most n if and only if there is a projective resolution $P_* \twoheadrightarrow \mathbb{Z}$ in which $P_i = 0$ for all $i > n$.

$$\begin{array}{ccccc}
(\mathrm{FP})_1 & \Leftrightarrow & (\mathrm{F})_1 & \Leftrightarrow & \text{finitely generated} \\
\Uparrow & & \Uparrow & & \\
(\mathrm{FP})_2 & \Leftarrow & (\mathrm{F})_2 & \Leftrightarrow & \text{finitely presented} \\
\Uparrow & & \Uparrow & & \\
(\mathrm{FP})_3 & \Leftarrow & (\mathrm{F})_3 & & \\
\Uparrow & & \Uparrow & & \\
\vdots & & \vdots & & \\
\Uparrow & & \Uparrow & & \\
(\mathrm{FP})_\infty & \Leftarrow & (\mathrm{F})_\infty & & \\
\Uparrow & & \Uparrow & & \\
(\mathrm{FP}) & \Leftarrow & (\mathrm{FD}) & & \\
\Uparrow & & \Uparrow & & \\
(\mathrm{FL}) & \Leftarrow & (\mathrm{F}) & &
\end{array}$$

Figure 1.

We denote the cohomological dimension of G by $\mathrm{cd}(G)$. One can also define the geometric dimension of G to be the least n such that G admits an n-dimensional Eilenberg–MacLane space of dimension n. Both the property of having finite geometric dimension and of having finite cohomological dimension are cohomological finiteness conditions. Denoting the geometric dimension by $\mathrm{gd}(G)$, it is known that

Lemma 2.3.2. *For any group G, either*

(1) $\mathrm{cd}(G) = \mathrm{gd}(G)$, *or*

(2) $\mathrm{cd}(G) = 2$ *and* $\mathrm{gd}(G) = 3$.

The Eilenberg–Ganea conjecture asserts that (1) always holds.

In this article we shall be greatly concerned with the question of when $\mathrm{cd}(G)$ is finite. If H is a subgroup of G then $\mathrm{cd}(H) \leq \mathrm{cd}(G)$ and if H is a non-trivial finite group then $\mathrm{cd}(H) = \infty$. From this we see that groups of finite cohomological dimension are always torsion-free, and in particular *torsion-freeness* is a cohomological finiteness condition. It turns out that many naturally occuring groups which are not torsion-free nevertheless contain a subgroup of finite index which has finite cohomological dimension. This property holds for all arithmetic groups, polycyclic-by-finite groups, Coxeter groups. Groups with this property are said to have finite virtual cohomological dimension, or finite vcd.

The problem of computing the exact value of $\mathrm{cd}(G)$ when it is known to be finite can often be very subtle, and we shall not have much to say about this. Soluble groups provide an intriguing example. It is relatively easy to prove that a soluble group G has finite cohomological dimension if and only if it is torsion-free of finite rank, (see [19]), and that it in this case $\mathrm{cd}(G)$ is equal to either $h(G)$ or $h(G)+1$, where $h(G)$ is the Hirsch length of G. Determining which of the two cases is a deeper question, although this is now understood:

Theorem 2.3.3. *Let G be a soluble group of finite cohomological dimension. Then $\mathrm{cd}(G) = h(G)$ if and only if G is of type* (F).

This result was conjectured by Gildenhuys and Strebel, and important steps towards the result were taken in [9] and [10]. In [14], the result was proved together with the more detailed information (also conjectured by Gildenhuys and Strebel) that these groups are constructible in the sense of Baumslag and Bieri: that is, they are built up from the trivial group by a sequence of HNN-extensions and finite extensions.

There are many other cases where computation of cohomological dimension has led to important advances in group theory, such as the classical Stallings-Swan theorem that groups of cohomological dimension 1 are free, [18, 20] and more recent work of Vogtmann computing the (virtual) cohomological dimension of outer automorphism groups of free groups.

In the remarkable paper [1], Alperin and Shalen find a criterion for a linear group (in characteristic zero) to have finite cohomological dimension. There arguments revolve around some ingenious valuation theoretic techniques which are used to get actions of linear groups on buildings, and have had a strong influence on our study of the class $\mathbf{H}\mathfrak{F}$, reviewed below: see [15] for further details. The Alperin–Shalen argument can in fact also be applied to linear groups in characteristic p, as pointed out in [7].

2.4. Cohomology and continuity

The properties $(\mathrm{FP})_n$ have a useful interpretation in terms of continuity of cohomology functors. This point of view was first taken by Bieri and Eckmann in their study of groups satisfying homological duality, [3]. In their work they used the fact that because *homology* functors are always continuous, therefore groups satisfying duality relating homology and cohomology have continuous cohomology functors and so must satisfy strong cohomological finiteness conditions. To make all this precise, let $\mathcal{C}$ and $\mathcal{D}$ be abelian categories which admit direct limits (i.e. colimits over directed posets). The principal example is the category of modules over a ring, and so we shall refer to *module* rather than object of $\mathcal{C}$. Let $F : \mathcal{C} \to \mathcal{D}$ be an additive functor. We say that F is *continuous* if and only if the natural map

$$\varinjlim_{\lambda} F(M_\lambda) \to F(\varinjlim_{\lambda} M_\lambda)$$

is an isomorphism for all direct limit systems $(M_\lambda \mid \lambda \in \Lambda)$ of modules. We say that F is continuous at zero if and only if $\varinjlim_{\lambda} F(M_\lambda) = 0$ whenever $\varinjlim_{\lambda} M_\lambda = 0$. The following facts are explained in [2]. They relate cohomological finiteness conditions to properties of the cohomology functors associated with a group. For the moment, it is not too important to know what these cohomology functors are. One of the reasons why they provide a good way of looking at cohomological finiteness conditions is that they are invariants: independent of a particular choice of projective resolution and independent of a choice of model for the Eilenberg–MacLane space.

Lemma 2.4.1. *Let G be a group. Then*

(1) *G is of type $(\mathrm{FP})_n$ if and only if the cohomology functors $H^i(G, \ )$ are continuous at zero for $i \leq n$; and*

(2) *if G is of type $(\mathrm{FP})_n$ then the functors $H^i(G, \ )$ are continuous for $i < n$.*

In practice, one of the most useful consequences of this is that for G of type $(\mathrm{FP})_n$, the functors $H^i(G, \ )$ commute with direct sums for $i \leq n$. In this article we shall make very considerable use of this principle for groups of type $(\mathrm{FP})_\infty$.

3. H$\mathfrak{F}$-groups

In this section we review the class of **H$\mathfrak{F}$**-groups. This is a very large class of groups: most everyday countable groups belong to it, and some uncountable groups belong to it. At present, very few examples are known of non-**H$\mathfrak{F}$**-groups. The Brown–Geoghegan example

$$G := \langle x_0, x_1, x_2, \dots \mid x_n^{x_i} = x_{n+1} \forall i < n \rangle$$

is essentially the only known example of a group not in the class. There is a family of related groups also not in **H$\mathfrak{F}$**, (see [6]) and since **H$\mathfrak{F}$** is subgroup closed, any group containing an isomorphic copy of one of these is also not in the class.

It seems likely that any genuinely different examples of non-$\mathbf{H}\mathfrak{F}$-groups would be of considerable interest.

3.1. The definition of $\mathbf{H}\mathfrak{F}$

We recall the following definitions from [15]. Let $\mathfrak{X}$ be a class of groups. We follow the convention which goes back to Philip Hall's work, that $\mathfrak{X}$ always contains the trivial group and that if it contains G then it contains every group isomorphic to G. Here are two recipes for defining larger classes from $\mathfrak{X}$:

Definition 3.1.1.

(1) $\mathbf{H}_1\mathfrak{X}$ denotes the class of all groups G which admit an action on a finite dimensional contractible cell complex with cell stabilizers in $\mathfrak{X}$.

(2) $\mathbf{H}\mathfrak{X}$ denotes the smallest class of groups containing $\mathfrak{X}$ with the property that a group G belongs to $\mathbf{H}\mathfrak{X}$ whenever it admits an action on a finite dimensional contractible complex with cell stabilizers already in $\mathbf{H}\mathfrak{X}$.

Let $\mathfrak{F}$ be the class of finite groups. Then $\mathbf{H}_1\mathfrak{F}$ is already a very large class. It contains all groups of finite virtual cohomological dimension, Gromov hyperbolic groups, and the known Burnside groups of odd exponent, amongst others.

But, by comparison, $\mathbf{H}\mathfrak{F}$ is absolutely enormous. It contains all countable soluble groups, all countable linear groups, it is extension closed and subgroup closed, and it is closed under forming fundamental groups of graphs of groups. We can think of $\mathbf{H}\mathfrak{F}$ as built up hierarchically:

Let $\mathbf{H}_0\mathfrak{F} := \mathfrak{F}$. For each ordinal α, we define the class $\mathbf{H}_\alpha\mathfrak{X}$ inductively as follows. If α is a successor ordinal then $\mathbf{H}_\alpha\mathfrak{F} := \mathbf{H}_1\mathbf{H}_{\alpha-1}\mathfrak{F}$ and if α is a limit ordinal then $\mathbf{H}_\alpha\mathfrak{F} := \bigcup_{\beta<\alpha}\mathbf{H}_\beta\mathfrak{F}$. A group G belongs to $\mathbf{H}\mathfrak{F}$ if and only if there is an ordinal α such that $G \in \mathbf{H}_\alpha\mathfrak{F}$, and the least such ordinal is called the *height* of G. Results about $\mathbf{H}\mathfrak{F}$-groups are usually proved by induction on the height. The classes $\mathbf{H}_\alpha\mathfrak{F}$ grow very rapidly with α. In practice, most $\mathbf{H}\mathfrak{F}$-groups that one meets ordinarily have turned up by the time $\alpha = 5$, and possibly even earlier: it is usually difficult to determine the height of an $\mathbf{H}\mathfrak{F}$-group. For example, all finitely generated linear groups belong to $\mathbf{H}_5\mathfrak{F}$, but probably many, and possibly all, already belong to $\mathbf{H}_4\mathfrak{F}$. It is fairly easy to construct groups with height exactly 2 because of the following fact:

Lemma 3.1.2. *If G is a torsion-free group then G belongs to $\mathbf{H}_1\mathfrak{F}$ if and only if G has finite cohomological dimension.*

PROOF. The result is well known, but we include the proof in one direction because it illustrates an idea which is used repeatedly later on. Suppose that G is a torsion-free $\mathbf{H}_1\mathfrak{F}$-group. Then G acts on a finite dimensional contractible cell complex X with finite cell stabilizers. Since G is torsion-free the stabilizers are trivial, and G is acting freely. Now, the augmented cellular chain complex of X is an exact sequence of finite length:

$$0 \to C_r \to \cdots \to C_1 \to C_0 \to \mathbb{Z} \to 0.$$

The length r is equal to the dimension of X, and exactness holds because X is contractible. Each C_i is essentially the free abelian group on the set of i-dimensional cells, and because G is acting freely, each C_i becomes a free $\mathbb{Z}G$-module. Then the chain complex is a free resolution of $\mathbb{Z}$ of finite length and G has finite cohomological dimension ($\leq r$). □

As a consequence, we have

Lemma 3.1.3. *If G is a countable group of infinite cohomological dimension such that every finitely generated subgroup has finite cohomological dimension, then G is an $\mathbf{H}\mathfrak{F}$-group of height 2.*

PROOF. This follows using the fact that any countable group admits an action on a tree with finitely generated stabilizers. □

For example, a free abelian group of countably infinite rank is an $\mathbf{H}\mathfrak{F}$-group of height 2.

3.2. Some $\mathbf{H}\mathfrak{F}$-groups of height > 2

The fact that the classes $\mathbf{H}_\alpha\mathfrak{F}$ grow very fast with α makes it hard to establish that $\mathbf{H}_\alpha\mathfrak{F} < \mathbf{H}_{\alpha+1}$ for any particular value of α. We know that $\mathbf{H}_0\mathfrak{F} < \mathbf{H}_1\mathfrak{F} < \mathbf{H}_2\mathfrak{F}$. In this section we sketch an argument to prove that $\mathbf{H}_2\mathfrak{F} < \mathbf{H}_3\mathfrak{F}$. We begin with a result which can be used to construct groups outside the class $\mathbf{H}_2\mathfrak{F}$.

Lemma 3.2.1. *Let G be a torsion-free group with a sequence of subgroups H_n, $n \geq 1$ such that*

(1) H_n is of type $(\mathrm{FP})_n$, and

(2) the cohomology functors $H^*(H_n, \quad)$ vanish on all free $\mathbb{Z}H_n$-modules.

Then G does not belong to $\mathbf{H}_2\mathfrak{F}$.

PROOF. Suppose that $G \in \mathbf{H}_2\mathfrak{F}$. Then G admits an action on a finite dimensional contractible cell complex X so that the stabilizers are in $\mathbf{H}_1\mathfrak{F}$. Let n be the dimension of X, and consider the action of H_n on X. The augmented cellular chain complex of X can be regarded as an exact sequence of $\mathbb{Z}H_n$-modules:

$$0 \to C_n \to \cdots \to C_1 \to C_0 \to \mathbb{Z} \to 0.$$

Since $H^0(H_n, \mathbb{Z}) = \mathbb{Z}$ is non-zero, it can be shown that for some i, $H^i(H_n, C_i)$ is non-zero. This follows from a precise form of Lemma 1.1.2. Now C_i is a permutation module, and is isomorphic to

$$\bigoplus_{\sigma} \mathbb{Z} \otimes_{\mathbb{Z}K_\sigma} \mathbb{Z}H_n$$

where K_σ is the stabilizer in H_n of the i-cell σ and σ runs through a set of H_n-orbit representatives. Each K_σ has finite cohomological dimension. Since $i \leq n$ and G is of type $(\mathrm{FP})_n$, $H^i(H_n, \quad)$ commutes with direct sums, and hence there is a choice of σ such that $H^i(H_n, \mathbb{Z} \otimes_{\mathbb{Z}K_\sigma} \mathbb{Z}H_n)$ is non-zero. Let $K := K_\sigma$. Now, K has finite cohomological dimension, and hence there is a projective resolution

$$0 \to P_r \to \cdots \to P_1 \to P_0 \to \mathbb{Z} \to 0$$

of finite length over $\mathbb{Z}K$. Applying induction from K to H_n, we obtain the exact sequence

$$0 \to P_r \otimes_{\mathbb{Z}K} \mathbb{Z}H_n \to \cdots \to P_1 \otimes_{\mathbb{Z}K} \mathbb{Z}H_n \to P_0 \otimes_{\mathbb{Z}K} \mathbb{Z}H_n \to \mathbb{Z} \otimes_{\mathbb{Z}K} \mathbb{Z}H_n \to 0.$$

Since the ith cohomology is non-zero on the module at the right hand end here, one can deduce that for some j, $H^{i+j}(H_n, P_j \otimes_{\mathbb{Z}K} \mathbb{Z}H_n)$ is non-zero. But this is a contradiction because the cohomology of H_n is supposed to vanish on all projective modules. Hence G cannot belong to $\mathbf{H}_2\mathfrak{F}$. □

Lemma 3.2.2. *Every countable metabelian-by-finite group belongs to* $\mathbf{H}_4\mathfrak{F}$.

PROOF. First, the classical theory of crytallographic groups shows that every finitely generated abelian-by-finite group acts discretely and properly discontinuously as a group of affine transformations of a Euclidean space. Therefore these groups belong to $\mathbf{H}_1\mathfrak{F}$, and it follows that all countable abelian-by-finite groups belong to $\mathbf{H}_2\mathfrak{F}$. Now let G be a finitely generated metabelian-by-finite group and let A be a normal abelian subgroup such that $Q := G/A$ is abelian-by-finite. A crystallographic action of Q on a Euclidean space can be regarded as an action of G for which all isotropy groups are finite extensions of A and hence are countable and abelian-by-finite. Thus every such G belongs to $\mathbf{H}_3\mathfrak{F}$. It follows that all countable metabelian-by-finite groups belong to $\mathbf{H}_4\mathfrak{F}$. □

Corollary 3.2.3. *There exists a metabelian* $\mathbf{H}\mathfrak{F}$*-group of height* > 2.

PROOF. Choose finitely generated torsion-free metabelian groups H_n of infinite rank and of type $(\mathrm{FP})_n$ for each $n \geq 1$. It can be shown that the cohomology of the H_n vanishes on projective modules. Let G be the direct product of the H_n. It follows from Lemma 3.2.1 that G does not belong to $\mathbf{H}_2\mathfrak{F}$. On the other hand, Lemma 3.2.2 shows that G belongs to $\mathbf{H}_4\mathfrak{F}$. Thus G is an $\mathbf{H}\mathfrak{F}$-group of height 3 or 4. □

It is conceivable that for some ordinal α, $\mathbf{H}_\alpha\mathfrak{F} = \mathbf{H}\mathfrak{F}$. Stunningly little is known about this question. The Corollary shows that if there is such an ordinal then it is at least 3. This is the best known result at the time of writing.

4. Finite dimensionality of certain groups and modules of type $(\mathrm{FP})_\infty$

In this section we sketch the proof of the remarkable fact that if G is a torsion-free $\mathbf{H}\mathfrak{F}$-group of type $(\mathrm{FP})_\infty$ then G has finite cohomological dimension. The proof is very simple and it can be generalized considerably both to give information about non-torsion-free groups of type $(\mathrm{FP})_\infty$ and to yield certain ring theoretic results. The arguments make use of a *complete cohomology* theory invented first by Vogel and then later using a different approach by Mislin. An account of Vogel's approach has now been published by Goichot [11]. Mislin's approach appears in [16]. Our first step is to describe some properties of this theory.

4.1. Complete cohomology and the first application

Let G be a group and let M be a $\mathbb{Z}G$-module. For each integer n we then have a cohomology group $H^n(G, M)$. Each $H^n(G, \ )$ is a functor from $\mathbb{Z}G$-modules to abelian groups. When n is negative, the functors are zero. For $n = 0$, one has the fixed point functor. When n is positive, the functors can be complicated but can also provide useful information about G.

Complete cohomology is similar in spirit to ordinary cohomology. We shall denote the complete cohomology groups by $\widehat{H^n}(G, M)$. These can be non-zero even if n is negative. They are functorial in the same way as ordinary cohomology groups and they satisfy certain axioms. To illustrate how complete cohomology can be applied we begin by outlining the proof of the following theorem.

Theorem 4.1.1. *Let G be a torsion-free $\mathbf{H}\mathfrak{F}$-group of type $(\mathrm{FP})_\infty$. Then G has finite cohomological dimension.*

To prove this we need just four properties of complete cohomology:

(1) There is a long exact sequence axiom for complete cohomology. That is, for any short exact sequence $A \rightarrowtail B \twoheadrightarrow C$ of $\mathbb{Z}G$-modules there are natural connecting homomorphisms $\widehat{H^n}(G,C) \to \widehat{H^{n+1}}(G,A)$ so that together with functorially induced maps, you get a long exact sequence:

$$\cdots \to \widehat{H^n}(G,A) \to \widehat{H^n}(G,B) \to \widehat{H^n}(G,C) \to \widehat{H^{n+1}}(G,A) \to \cdots$$

(2) If G is of type $(\mathrm{FP})_\infty$, all the complete cohomology functors $\widehat{H^n}(G, \ )$ commute with direct sums. That is, for any family $(M_\lambda \mid \lambda \in \Lambda)$ of $\mathbb{Z}G$-modules, the natural map

$$\bigoplus_{\lambda\in\Lambda} \widehat{H^n}(G,M_\lambda) \to \widehat{H^n}(G, \textstyle\bigoplus_{\lambda\in\Lambda} M_\lambda)$$

is an isomorphism.

(3) For all integers n, $\widehat{H^n}(G,\mathbb{Z}G) = 0$.

(4) G has finite cohomological dimension if and only if $\widehat{H^0}(G,\mathbb{Z}) = 0$.

The first property is also a property of ordinary cohomology. It is a very familiar axiom for any cohomology theory, and it is surely the least we should expect of a decent theory. Property (2) is also a property of ordinary cohomology, and so it should not be regarded as too surprising. As a matter of fact, it follows from property (1) that complete cohomology always commutes with *finite* direct sums, regardless of whether or not G is $(\mathrm{FP})_\infty$. The point is that (2) also allows for infinite direct sums.

Property (3) does not always hold for ordinary cohomology. In fact, for many groups G there is at least one integer n for which $H^n(G,\mathbb{Z}G)$ is non-zero. Usually these cohomology groups are hard to compute. When $n = 0$ the situation is simple: $H^0(G,\mathbb{Z}G)$ is non-zero if and only if G is finite. When $n = 1$, $H^1(G,\mathbb{Z}G)$ is related to the number of ends of G, and a great deal is understood about this case from the work of Stallings, Swan, Dunwoody and others. For higher n, one imagines that these cohomology groups are related to higher dimensional end invariants of the universal cover of an Eilenberg MacLane space, but calculation is usually very hard. On the other hand, it is often the case that $H^n(G,\mathbb{Z}G)$ is zero for many n, perhaps all but one n. For complete cohomology, the situation is much simpler: with hats on, all these cohomology groups vanish. This property is built in to complete cohomology at an early stage when it is set up and it is very convenient.

Property (4) is quite startling. It reduces the problem of showing that G has finite cohomological dimension to the problem of calculating $\widehat{H^0}(G,\mathbb{Z})$. Now, in general $\widehat{H^0}(G,\mathbb{Z})$ is extremely difficult to compute, and remarkably little is known about it. But, as in any cohomological situation, things are a lot happier when you need to prove a cohomology group is zero.

The strategy of proof. The strategy of proof is very simple. You start with an $\mathbf{H}\mathfrak{F}$-group of type $(\mathrm{FP})_\infty$. We want to prove that $\widehat{H^0}(G,\mathbb{Z})$ vanishes. To get started we should like to know that at least some cohomology *do* vanish, and for this we can use property (3). To draw the conclusion, we can use properties (1) and (2).

PROOF. Let G be as in the theorem. We shall prove that for all subgroups H of G and all integers i,

$$\widehat{H^i}(G,\mathbb{Z}\otimes_{\mathbb{Z}H}\mathbb{Z}G)=0.$$

This is more than enough, because in the special case when $H=G$ and $i=0$, it reduces to the statement $\widehat{H^0}(G,\mathbb{Z})=0$ which implies that G has finite cohomological dimension by property (4).

The advantage of proving the more general statement is that you can use induction on the least α such that H belongs to $\mathbf{H}_\alpha\mathfrak{F}$. If $\alpha=0$ then H is finite and since G is torsion-free, this means that H is the trivial subgroup. In this case the claim reduces to the statement that $\widehat{H^i}(G,\mathbb{Z}G)=0$ for all integers i and this is property (3).

Now suppose that $\alpha>0$. Then there is an action of H on a finite dimensional contractible cell complex X such that each isotropy group belongs to $\mathbf{H}_\beta\mathfrak{F}$ for some $\beta<\alpha$. The augmented cellular chain complex of X is an exact sequence of $\mathbb{Z}H$-modules:

$$0\to C_r\to\cdots\to C_1\to C_0\to\mathbb{Z}\to 0.$$

This is an exact sequence because X is contractible. Each C_i is a permutation module: in effect, it is the free abelian group on the set of i-dimensional cells of X, and since H is permuting these cells there is an induced action of H on C_i. If Σ_i is a set of H-orbit representatives of the i-dimensional cells then we can identify C_i as the direct sum of induced modules:

$$C_i=\bigoplus_{\sigma\in\Sigma_i}\mathbb{Z}\otimes_{\mathbb{Z}H_\sigma}\mathbb{Z}H.$$

Now we can apply induction from H to G to the cellular chain complex. The induction functor $-\otimes_{\mathbb{Z}H}\mathbb{Z}G$ preserves exactness and so the new sequence,

$$0\to C_r\otimes_{\mathbb{Z}H}\mathbb{Z}G\to\cdots\to C_1\otimes_{\mathbb{Z}H}\mathbb{Z}G\to C_0\otimes_{\mathbb{Z}H}\mathbb{Z}G\to\mathbb{Z}\otimes_{\mathbb{Z}H}\mathbb{Z}G\to 0,$$

is still exact. Moreover, each $C_i\otimes_{\mathbb{Z}H}\mathbb{Z}G$ is isomorphic to

$$\bigoplus_{\sigma\in\Sigma_i}\mathbb{Z}\otimes_{\mathbb{Z}H_\sigma}\mathbb{Z}G$$

and since the H_σ have smaller height than H in the hierarchical decomposition, we know by induction that complete cohomology vanishes on each of these modules. Property (2) now shows that the complete cohomology also vanishes on each $C_i \otimes_{\mathbb{Z}H} \mathbb{Z}G$. Property (1), the long exact sequence axiom can now be used to deduce that the complete cohomology must also vanish on the module $\mathbb{Z} \otimes_{\mathbb{Z}H} \mathbb{Z}G$ at the right-hand end of the exact sequence. This completes the inductive step, and the Theorem follows. □

The fact that this proof is so simple suggests that more should be true, and indeed there are several important generalizations which can be made. We describe these next.

4.2. Complete cohomology for arbitrary rings

Let k be a commutative ring and let R be a k-algebra. An R-module M has *finite projective dimension* if and only if it has a projective resolution of finite length (i.e. eventually the projectives are all zero), and we say that M has type $(\mathrm{FP})_\infty$ if and only if it has a projective resolution of finite type, (i.e. all the projectives are finitely generated). These notions generalize the corresponding notions for groups. If G is a group, then G has finite cohomological dimension if and only if the trivial module $\mathbb{Z}$ has finite projective dimension over the group ring $\mathbb{Z}G$, and G has type $(\mathrm{FP})_\infty$ if and only if the trivial module $\mathbb{Z}$ is a module of type $(\mathrm{FP})_\infty$ over $\mathbb{Z}G$. Theorem 4.1.1 can also be generalized by using a more general version of complete cohomology. For any algebra R, complete cohomology groups $\widehat{\mathrm{Ext}}^*_R(A, B)$ are defined for any pair of right R-modules A, B. These groups sit in the same relation to ordinary Ext groups as complete cohomology of groups sits in relation to ordinary cohomology of groups. The properties (1)—(4) have analogues in this general setting:

(1) $\widehat{\mathrm{Ext}}^*_R(-,-)$ satisfies long exact sequence axioms for both the left hand variable and the right hand variable.

(2) If M is an R-module of type $(\mathrm{FP})_\infty$ then all the functors $\widehat{\mathrm{Ext}}^i_R(M, \quad)$ commute with direct sums.

(3) If either A or B is a projective module then $\widehat{\mathrm{Ext}}^i_R(A, B) = 0$ for all integers i.

(4) An R-module M has finite projective dimension if and only if

$$\widehat{\mathrm{Ext}}^0_R(M, M) = 0$$

Property (4) provides a technique for proving that certain modules have finite projective dimension, and to apply it, one need criteria for complete cohomology to vanish. Using the long exact sequence axioms together with property (3) it is easy to see that the following holds:

Lemma 4.2.1. *If either A or B has finite projective dimension then $\widehat{\mathrm{Ext}}^i_R(A, B) = 0$ for all integers i.*

PROOF. Suppose that B has finite projective dimension. Let

$$0 \to P_r \to \cdots \to P_1 \to P_0 \to B \to 0$$

be a projective resolution of finite length. Now the cohomology functors $\widehat{\mathrm{Ext}}^*_R(A, \ )$ vanish on all the P_j, therefore, by the basic vanishing lemma Lemma 1.1.2, they vanish on B. A similar argument applies if A has finite projective dimension. □

Using a variation on the proof of Theorem 4.1.1 one has the following vanishing theorem:

Theorem 4.2.2. *Let k be a commutative ring and let G be an $\mathbf{H}\mathfrak{F}$-group. Let M and N be kG-modules where M is of type $(\mathrm{FP})_\infty$ and N has finite projective dimension as kH-module for all finite subgroups H of G. Then $\widehat{\mathrm{Ext}}^i_{kG}(M, N) = 0$ for all integers i.*

This has many corollaries. The simplest follows by applying it with $M = N$:

Corollary 4.2.3. *Let G be an $\mathbf{H}\mathfrak{F}$-group and let M be a kG-module of type $(\mathrm{FP})_\infty$. Then M has finite projective dimension if and only if it has finite projective dimension as a kH-module for all finite subgroups H of G.*

The consequences are particularly decisive if G is a torsion-free $\mathbf{H}\mathfrak{F}$-group and k has finite global dimension. In this case the corollary shows that every kG-module of type $(\mathrm{FP})_\infty$ has finite projective dimension.

The proof of the vanishing theorem (4.2.2) depends on the tensor identity trick. Let M be a kG-module and let H be a subgroup of G. Let $k[H\backslash G]$ denote the permutation module on the cosets Hg of H. Then $M \otimes_k k[H\backslash G]$ can be made into a kG-module via the diagonal action of G. The tensor identity asserts that

Lemma 4.2.3. *$M \otimes_k k[H\backslash G]$ is isomorphic to the induced module $M \otimes_{kH} kG$.*

PROOF. It is easy to check that the two maps

$$M \otimes_{kH} kG \to M \otimes_k k[H\backslash G]$$

given by

$$m \otimes g \mapsto mg \otimes g,$$

and

$$M \otimes_k k[H\backslash G] \to M \otimes_{kH} kG$$

given by

$$m \otimes Hg \mapsto mg^{-1} \otimes g,$$

are well-defined and are mutually inverse isomorphisms. □

We can now sketch the proof of the vanishing theorem (4.2.2). The idea is to prove the more general vanishing result that

$$\widehat{\mathrm{Ext}}^*_{kG}(M, N \otimes_k k[H\backslash G]) = 0$$

for all subgroups H of G. This can be proved by induction on the hierarchical height of H, and the special case $H = G$ gives the desired result. When H is finite then N has finite projective dimension over kH and so $N \otimes_{kH} kG$ has finite projective dimension over kG. In view of the tensor identity and the basic vanishing result 4.2.1 it follows that $\widehat{\mathrm{Ext}}^*_{kG}(M, N \otimes_k k[H\backslash G]) = 0$ in this case. When H is infinite, it admits an action on a finite dimensional contractible cell complex so that the cell stabilizers are lower down in the hierarchy. Just as in the proof of (4.1.1) we work with the augmented cellular chain complex induced from H to G. But here, in addition, we apply $N \otimes_k -$ to it as well. The inductive argument using properties (1) and (2) then works in the same way as before.

4.3. Further applications of the vanishing theorem

Theorem 4.2.2 can be applied in various ways. Here are two striking applications, both coming out of joint work with Jonathan Cornick [7].

Theorem 4.3.1. *If G is an $\mathbf{H}\mathfrak{F}$-group of type $(\mathrm{FP})_\infty$ then every torsion-free subgroup of G has finite cohomological dimension.*

Theorem 4.3.2. *If G is a residually finite $\mathbf{H}\mathfrak{F}$-group of type $(\mathrm{FP})_\infty$ then G has a subgroup of finite index which has finite cohomological dimension. (That is, G has finite virtual cohomological dimension.)*

Both are special cases of assertions which can be made about arbitrary modules of type $(\mathrm{FP})_\infty$. To understand how these are proved we need to know a little more about complete cohomology. First of all, there is a natural map from ordinary cohomology to complete cohomology. That is, for all i, there are natural maps

$$\mathrm{Ext}^i_R(A, B) \to \widehat{\mathrm{Ext}}^i_R(A, B).$$

When $i = 0$ and $A = B$, $\mathrm{Ext}^0_R(A, A)$ is just the endomorphism ring $\hom_R(A, A)$ of A and in this case, $\widehat{\mathrm{Ext}}^0_R(A, A)$ also has a ring structure and the natural map

$$\hom_R(A, A) \to \widehat{\mathrm{Ext}}^0_R(A, A)$$

is a ring homomorphism. Secondly, there are restriction maps for complete cohomology, just as for ordinary cohomology. That is, if S is a subring of R such that R is projective as an S-module, then there are restriction maps

$$\widehat{\operatorname{Ext}}^i_R(A,B) \to \widehat{\operatorname{Ext}}^i_S(A,B)$$

defined for each i. When $i = 0$ and $A = B$, the restriction map

$$\widehat{\operatorname{Ext}}^0_R(A,A) \to \widehat{\operatorname{Ext}}^0_S(A,A)$$

is a ring homomorphism. These facts lead to a simple proof of the following lemma:

Lemma 4.3.3. *Let G be a group and let H be a subgroup. Let $\iota : M \to N$ be a kH-split monomorphism of kG-modules. If $\widehat{\operatorname{Ext}}^0_{kG}(M,N) = 0$ then M has finite projective dimension as a kH-module.*

PROOF. There is a commutative diagram:

$$\begin{array}{ccc}
\hom_{kG}(M,N) & \longrightarrow & \widehat{\operatorname{Ext}}^0_{kG}(M,N) \\
\downarrow & & \downarrow \\
\hom_{kH}(M,N) & \longrightarrow & \widehat{\operatorname{Ext}}^0_{kH}(M,N) \\
\downarrow & & \downarrow \\
\hom_{kH}(M,M) & \longrightarrow & \widehat{\operatorname{Ext}}^0_{kH}(M,M).
\end{array}$$

The vertical maps are restriction maps followed by maps induced by a kH-splitting. We can follow the inclusion ι around this diagram in two ways. Following anti-clockwise, round the left and the bottom of the diagram, we see that it maps to the identity endomorphism in $\hom_{kH}(M,M)$ and hence to the identity element of the ring $\widehat{\operatorname{Ext}}^0_{kH}(M,M)$. Following it around clockwise, we see that it maps to zero in $\widehat{\operatorname{Ext}}^0_{kH}(M,M)$ because the route passes through $\widehat{\operatorname{Ext}}^0_{kG}(M,N) = 0$. Thus the identity element of $\widehat{\operatorname{Ext}}^0_{kH}(M,M)$ is zero, the ring must be zero, and M has finite projective dimension over kH by property (4). □

In order to apply this in conjunction with the vanishing theorem (4.2.2) we can use the following device:

Lemma 4.3.4. *Let G be a group and let H be a subgroup. Suppose that M is a kG-module which has finite projective dimension as a kF-module for all finite subgroups F of H. Then there is a kH-split monomorphism of M into a kG-module N which has finite projective dimension over kF for all finite subgroups F of G.*

PROOF. We outline one way of doing this. Let B denote the set of bounded function from G/H to $\mathbb{Z}$. Here G/H denotes the set of left cosets gH of H. Then B can be made into a right $\mathbb{Z}G$-module using the recipe $\phi^g(g'H) = \phi(gg'H)$, for $\phi \in B$. Now B contains a copy of the trivial module $\mathbb{Z}$ in the form of the constant functions, and the inclusion $\mathbb{Z} \to B$ is $\mathbb{Z}H$-split by the map $B \to \mathbb{Z}$, *evaluation at* H. Set N equal to $M \otimes B$ with the diagonal action. Then we get a kH-split inclusion of M into N. It is somewhat miraculous that N inherits the property of having finite projective dimension over all finite subgroups of G from the fact that M has this property for all finite subgroups of H. But it is true, and details can be found in [7]. □

As a consequence of all this we have

Corollary 4.3.5. *Let G be an $\mathbf{H}\mathfrak{F}$-group, let H be a subgroup of G and let M be a kG-module of type $(\mathrm{FP})_\infty$. Then M has finite projective dimension over kH if and only if M has finite projective dimension over kF for all finite subgroups F of H.*

PROOF. This is easy in view of the results above. Suppose that M has finite projective dimension over all finite subgroups of H. Using (4.3.4) we can find a kH-split monomorphism into a kG-module N which has finite projective dimension over all finite subgroups of G. Now the vanishing theorem (4.2.2) shows that $\widehat{\mathrm{Ext}}^0_{kG}(M, N) = 0$. The result now follows from the finite dimensionality criterion (4.3.3). □

Theorem 4.3.1 is immediate: Just apply this with $k = M = \mathbb{Z}$ and with H a torsion-free subgroup of G.

Similar methods yield Theorem 4.3.2. One needs the following variation on the trick (4.3.4): suppose that G is a residually finite group. For each subgroup H of finite index, there is the coinduced module $\hom_{\mathbb{Z}H}(\mathbb{Z}G, \mathbb{Z})$ and if $K \subseteq H$ then there is a natural inclusion

$$\hom_{\mathbb{Z}H}(\mathbb{Z}G, \mathbb{Z}) \hookrightarrow \hom_{\mathbb{Z}K}(\mathbb{Z}G, \mathbb{Z}).$$

Let C denote the direct limit $\varinjlim \hom_{\mathbb{Z}H}(\mathbb{Z}G, \mathbb{Z})$ as H runs through the subgroups of finite index. Since G is residually finite, it can be shown that C is free as a $\mathbb{Z}F$-module for all finite F. If M is any kG-module which has finite projective dimension as k-module then $M \otimes C$ has finite projective dimension as kF-module for all finite F. In particular, if M is of type $(\mathrm{FP})_\infty$ then the vanishing theorem (4.2.2) guarantees that $\widehat{\mathrm{Ext}}^0_{kG}(M, M \otimes C) = 0$. A small strengthening of property (2) of complete cohomology allows one to take the direct limit outside, so that we have

$$\varinjlim \widehat{\mathrm{Ext}}^0_{kG}(M, M \otimes \hom_{\mathbb{Z}H}(\mathbb{Z}G, \mathbb{Z})) = 0.$$

Now, there is a natural map

$$\hom_{kG}(M, M) \to \varinjlim \widehat{\mathrm{Ext}}^0_{kG}(M, M \otimes \hom_{\mathbb{Z}H}(\mathbb{Z}G, \mathbb{Z})),$$

and since the right-hand group is zero, there must be an H of finite index in G such that the identity endomorphism of M becomes zero in $\widehat{\mathrm{Ext}}^0_{kG}(M, M \otimes \hom_{\mathbb{Z}G}(\mathbb{Z}H, \mathbb{Z}))$. Using the same argument that proves (4.3.3) it can now be shown that M has finite projective dimension as a kH-module. Thus we have sketched a proof of the following:

Corollary 4.3.6. *Let G be a residually finite* $\mathbf{H}\mathfrak{F}$*-group and let M be a kG-module of type* $(\mathrm{FP})_\infty$ *which has finite projective dimension over k. Then there is a subgroup H of finite index in G such that M has finite projective dimension over kH.*

Theorem 4.3.2 follows at once, by applying this with $k = M = \mathbb{Z}$.

4.4. Crossed products and strongly graded rings

In fact, all these results can be set in a far more general ring-theoretic context. Let k be a commutative ring and let G be a monoid. Naively, a G-graded k-algebra is a k-algebra R which has a k-module decomposition

$$R = \bigoplus_{g \in G} R_g$$

such that $R_g R_h \subseteq R_{gh}$ for all $g, h \in G$. We should also require that the unit map $k \to R$ carries k into the degree one part R_1. The monoid algebra kG is the simplest example of a G-graded k-algebra, and this has a natural coalgebra structure as well: the comultiplication is defined by

$$c : kG \to kG \otimes_k kG$$
$$c(g) = g \otimes g.$$

In this way kG becomes a bialgebra (various axioms are satisfied, the comultiplication is an algebra map, the ring multiplication is a coalgebra map *et cetera*). It is really through the bialgebra structure that one can define diagonal actions of G on a tensor product of G-modules. This technique works rather more generally, because there is another way of formulating the definition of a G-graded k-algebra:

Lemma 4.4.1. *Let R be a G-graded k-algebra. Then there is a k-algebra map $\gamma : R \mapsto R \otimes_k kG$ which makes R into a kG-comodule.*

PROOF. We define $\gamma(r) = r \otimes g$ for $r \in R_g$. □

We do not need to know it, but the lemma provides an alternative definition of G-graded k-algebra which is equivalent to the naive one. What is important is that it allows us to define diagonal actions of R in certain circumstances. Let M be an R-module and let V be a kG-module. Then $M \otimes_k V$ is an $R \otimes_k kG$-module. When R is G-graded there is a k-algebra map γ as in the lemma, and through this, any $R \otimes_k kG$-module can be viewed as an R-module. In particular, we get an induced "diagonal" action of R on $M \otimes_k V$. It is even true that part of the proof of the tensor identity (4.2.4) holds in this generality. Let H be a submonoid of G, and let R_H be the subring supported on H. That is,

$$R_H = \bigoplus_{h \in H} R_h.$$

The compatibility of multiplication with the grading ensures that this is a subring. Now, taking V to be the permutation module $k[H \backslash G]$ on the right cosets of H, we have a natural map of R-modules

$$M \otimes_{R_H} R \to M \otimes_k k[H \backslash G]$$

given by

$$m \otimes r \mapsto mr \otimes g,$$

for r of degree g. It is easy to check that this map is a well defined R-module homomorphism. If it were an isomorphism then the tensor identity would hold and it would be possible to carry through much of the $(\mathrm{FP})_\infty$ theory we have developed so far for groups and group algebras. It turns out that it is an isomorphism in a fairly general setting. In the following definition, we use the convention that if X and Y are additive subgroups of a ring R then XY denotes the additive subgroup generated by the set of products xy.

Definition 4.4.2. Let G be a group.

(1) A strongly G-graded k-algebra is a G-graded k-algebra R with the property that $R_g R_h = R_{gh}$ for all g, h.

(2) A kG-crossed product is a G-graded k-algebra R such that every R_g contains a unit.

Every crossed product is strongly graded, but there are also strongly graded rings which are not crossed products. For example the 3×3 matrix ring $R := M_3(k)$ can be regarded as a strongly C_2-graded k-algebra with

$$R_1 = \begin{pmatrix} * & * & 0 \\ * & * & 0 \\ 0 & 0 & * \end{pmatrix}, \; R_{-1} = \begin{pmatrix} 0 & 0 & * \\ 0 & 0 & * \\ * & * & 0 \end{pmatrix}.$$

This is not a crossed product because R_{-1} does not contain a unit, but it is strongly graded. For our purposes the key property of strongly graded rings is that a version of the tensor identity (4.2.4) holds:

Lemma 4.4.3. *Let G be a group, let H be a subgroup and let R be a strongly G-graded k-algebra. Then*

(1) the tensor identity holds: that is, if M is any R-module then the natural map $M \otimes_{R_H} R \to M \otimes_k k[H\backslash G]$ is an isomorphism;

(2) R is projective as an R_H-module.

PROOF. Here we just sketch a proof of (1). We prove this by showing that the natural map $M \otimes_{R_H} R \to M \otimes_k k[H\backslash G]$ is injective and surjective. First, for surjectivity it suffices to show that given $m \in M$ and $g \in G$, then $m \otimes Hg$ lies in the image. Since R is strongly graded, $R_{g^{-1}}R_g = R_1$ and in particular, there is an expression of the form

$$1 = \sum_i x_i y_i$$

for the identity element as a finite sum with $x_i \in R_{g^{-1}}$ and $y_i \in R_g$. Now we have

$$\sum_i m x_i \otimes y_i \mapsto \sum_i m x_i y_i \otimes Hg = m \otimes Hg.$$

To establish surjectivity, first note that the left-hand side breaks up as a direct sum

$$M \otimes_{R_H} R = \bigoplus_g M \otimes_{R_H} R_{Hg},$$

where g runs through a transversal to H in G. The summand $M \otimes_{R_H} R_{Hg}$ maps into the summand $M \otimes Hg$ of $M \otimes_k k[H\backslash G]$. Thus it suffices to prove that for each g, the map

$$M \otimes_{R_H} R_{Hg} \to M$$

given by $m \otimes r \mapsto mr$ is injective. Fix $g \in G$. Suppose that $\sum_j m_j \otimes r_j$ lies in the kernel, where r_j belongs to R_{Hg} for all j. We may assume that each r_j is homogeneous. Then $\sum_j m_j \otimes r_j \mapsto \sum_j m_j r_j$ and hence $\sum_j m_j r_j = 0$ As before, write $1 = \sum_i x_i y_i$ with $x_i \in R_{g^{-1}}$ and $y_i \in R_g$. Then

$$\sum_i m_j \otimes r_j = \sum_{i,j} m_j \otimes r_j x_i y_i = \sum_{i,j} m_j r_j x_i \otimes y_i = 0,$$

as required. □

Using this one can establish the following version of the vanishing theorem (4.2.2):

Theorem 4.4.4. *Let G be an $\mathbf{H}\mathfrak{F}$-group and let R be a strongly G-graded k-algebra. Let M and N be R-modules where M is of type $(\mathrm{FP})_\infty$ and N has finite projective dimension as an R_F-module for all finite subgroups F of G. Then $\widehat{\mathrm{Ext}}^i_0(M,N) = 0$ for all i.*

By combining this with the other methods we can, for example, deduce the following:

Corollary 4.4.5. *Let G be an $\mathbf{H}\mathfrak{F}$-group, let R be a strongly G-graded k-algebra, let M be an R-module of type $(\mathrm{FP})_\infty$ and let H be a subgroup of G. Then M has finite projective dimension as an R_H-module if and only if it has finite projective dimension as an R_F-module for all finite subgroups F of H.*

4.5. Some random remarks

Throughout this section, we have used complete cohomology to get a handle on cohomological or projective dimension. However, large chunks of the argument can be carried out using ordinary cohomology. Just to draw attention to the contrast, here are the properties of ordinary cohomology which correspond to properties (1)—(4) of complete cohomology:

(1) There is a long exact sequence axiom for cohomology. That is, for any short exact sequence $A \rightarrowtail B \twoheadrightarrow C$ of $\mathbb{Z}G$-modules there are natural connecting homomorphisms $H^n(G,C) \to H^{n+1}(G,A)$ so that together with functorially induced maps, you get a long exact sequence:

$$\cdots \to H^n(G,A) \to H^n(G,B) \to H^n(G,C) \to H^{n+1}(G,A) \to \cdots.$$

(2) If G is of type $(\mathrm{FP})_\infty$ all of the cohomology functors $H^n(G,\ \ )$ commute with direct sums. That is, for any family $(M_\lambda \mid \lambda \in \Lambda)$ of $\mathbb{Z}G$-modules, the natural map

$$\bigoplus_{\lambda\in\Lambda} H^n(G,M_\lambda) \to H^n(G, \textstyle\bigoplus_{\lambda\in\Lambda} M_\lambda)$$

is an isomorphism.

(3) $H^n(G,\mathbb{Z}G)$ is often zero, but not always.

(4) $H^0(G,\mathbb{Z}) = \mathbb{Z}$.

The first two properties are the same. The third and fourth are dramatically different. If one carries through the same techniques with ordinary cohomology the results are different. They are in fact rather less satisfactory, probably because property (3) is indecisive and (4) is almost vacuous. Nevertheless one can draw the following conclusion which may have useful applications:

Theorem 4.5.1. *Let G be an $\mathbf{H}\mathfrak{F}$-group of type* $(\mathrm{FP})_\infty$. *Then there is an integer n such that $H^n(G, \mathbb{Z}G)$ is non-zero.*

Here is a sketch of the proof. Let α be an ordinal such that G belongs to $\mathbf{H}_\alpha\mathfrak{F}$. (In practice, bear in mind that α is probably about 3.) Now look at the set $\mathcal{O}$ of ordinals β such that G has an $\mathbf{H}_\beta\mathfrak{F}$-subgroup H for which $H^i(G, \mathbb{Z} \otimes_{\mathbb{Z}H} \mathbb{Z}G)$ is non-zero for some i. The first step is to show that 0 belongs to $\mathcal{O}$. Now α surely belongs to $\mathcal{O}$, for we can take $H := G$ and $i = 0$. Now the process of hierarchical decomposition can be used to show that if $\beta > 0$ belongs to $\mathcal{O}$ then $\mathcal{O}$ must contain an ordinal smaller than β. It follows by transfinite induction that $0 \in \mathcal{O}$ and hence there is a finite subgroup H and an i such that $H^i(G, \mathbb{Z} \otimes_{\mathbb{Z}H} \mathbb{Z}G)$ is non-zero. Since H is finite, it is possible to choose an exact sequence

$$0 \to \mathbb{Z} \to P_0 \to P_1 \to P_2 \to \cdots$$

of $\mathbb{Z}H$-modules in which each P_i is free. (This is a kind of "reverse" projective resolution, and its existence depends heavily on the fact that H is finite.) Applying induction from H to G to this sequence, we obtain the exact sequence

$$0 \to \mathbb{Z} \otimes_{\mathbb{Z}H} \mathbb{Z}G \to P_0 \otimes_{\mathbb{Z}H} \mathbb{Z}G \to P_1 \otimes_{\mathbb{Z}H} \mathbb{Z}G \to P_2 \otimes_{\mathbb{Z}H} \mathbb{Z}G \to \cdots,$$

which is a "reverse" projective resolution of $\mathbb{Z} \otimes_{\mathbb{Z}H} \mathbb{Z}G$ by free $\mathbb{Z}G$-modules. Note that the ith cohomology of G is non-zero on the left-hand module here. A *dimension-shifting* argument can be used to deduce that there is a $j \geq 0$ such that $H^{i-j}(G, P_j \otimes_{\mathbb{Z}H} \mathbb{Z}G)$ is non-zero and the result follows with $n := i - j$.

It is an open problem whether $H^n(G, \mathbb{Z}G)$ could be non-zero for infinitely many n. It seems natural to conjecture that this cannot happen, and it even seems natural to make further closely related conjectures which links up with work of Gedrich and Gruenberg [8]:

Conjecture 4.5.2. Let G be an $\mathbf{H}\mathfrak{F}$-group of type $(\mathrm{FP})_\infty$. Then

(1) every projective $\mathbb{Z}G$-module has finite injective dimension;
(2) every injective $\mathbb{Z}G$-module has finite projective dimension;
(3) $\mathbb{Z}G$ has finite finitistic dimension (that is, there is a finite bound on the projective dimensions of the $\mathbb{Z}G$-modules of finite projective dimension);
(4) if M is any $\mathbb{Z}G$-module then M has finite projective dimension if and only if it has finite projective dimension as a $\mathbb{Z}F$-module for all finite subgroups F of G.

The conjecture is valid if G is torsion-free or even virtually torsion-free. But it is natural to ask for a result like this in general. (There are known examples of $\mathbf{H}\mathfrak{F}$-groups of type $(\mathrm{FP})_\infty$ which are not virtually torsion-free.) The different parts of the conjecture are very closely related: if one had a handle on one of them then they would almost certainly all succumb.

5. Topological motivation

In this section we review some of the background material which motivated the arguments and proofs in §§3 and 4.

5.1. The Bieri-Strebel Lemma

The study of finitely presented groups took a dramatic turn when Bieri and Strebel established the following lemma. It is the starting point for their theory of finitely presented metabelian groups [4, 5] and it has had many other applications.

Lemma 5.1.1. (Bieri–Strebel) *Let G be a finitely presented group and let $\theta : G \twoheadrightarrow \mathbb{Z}$ be a surjective homomorphism. Then G is an HNN-extension $G = B*_{H,t}$ where B and H are finitely generated subgroups of* $\ker\theta$ *and* $\theta(t) = 1$.

The reason why this result had such a big impact on the study of finitely presented metabelian groups is that HNN-extensions usually contain non-cyclic free subgroups whereas metabelian groups never contain non-cyclic free subgroups. Bieri and Strebel succeeded in making use of the fact that finitely presented metabelian groups admit many homomorphisms to $\mathbb{Z}$ and for each of these the lemma provides an HNN-extension which must be of a very special type.

Moreover, Bieri and Strebel also observed that the assumption that G is finitely presented could be weakened to the assumption that G is of type $(\mathrm{FP})_2$. The starting point for the research described in this article was the search, as yet unsuccessful, for a higher dimensional version of the lemma which would give useful information about groups of type $(\mathrm{FP})_3$ or of type $(\mathrm{FP})_n$ for higher values of n. We shall suggest some plausible generalizations of the lemma which would be powerful if they could be proved. But first, we shall study some weakened forms of the Bieri–Strebel lemma which are easier to generalize to higher dimensions.

To understand these weaker versions we need the notion of ends of a pair of groups. Let G be a group and H a subgroup. Then there is an invariant $e(G, H)$, called the number of ends of the pair G, H, introduced in [17], which is 0 if H has finite index in G, which is usually 1 if H has infinite index in G, but which can under special conditions take values greater than 1. *HNN*-extensions provide one example of these special circumstances:

Lemma 5.1.2. *Suppose that $G = B*_{H,t}$ is an HNN-extension. Then $e(G, H) > 1$.*

In view of this we can state our first weak form of the Bieri–Strebel Lemma:

Lemma 5.1.3. *Let G be a finitely presented group and let $\theta : G \twoheadrightarrow \mathbb{Z}$ be a surjective homomorphism. Then there is a finitely generated subgroup H of $\ker\theta$ such that $e(G, H) > 1$.*

Clearly this follows immediately from the two previous lemmas. I shall not describe precisely how $e(G, H)$ is defined: it can be defined in a geometric way, and it can also be defined in an algebraic way using cohomology groups. From the algebraic point of view, the following lemma is significant:

Lemma 5.1.4. *Let $H \leq G$ be groups with $e(G, H) > 1$. Then it follows that $H^1(G, \mathbb{Z} \otimes_{\mathbb{Z}H} \mathbb{Z}G)$ is non-zero.*

In view of this, we can state a further weakening of the Bieri-Strebel Lemma:

Lemma 5.1.5. *Let G be a finitely presented group and let $\theta : G \twoheadrightarrow \mathbb{Z}$ be a surjective homomorphism. Then there is a finitely generated subgroup H of G such that $H^1(G, \mathbb{Z} \otimes_{\mathbb{Z}H} \mathbb{Z}G)$ is non-zero.*

This is just a faint shadow of the original Lemma, and it is rather easy to give a direct proof of this form.

PROOF. Let K be the kernel of θ and let M be any G/K-module. There is an inflation map $H^i(G/K, M) \to H^i(G, M)$ for each i and for $i = 1$ it is always injective. (When $i = 1$ it arises at the beginning of the so-called inflation-restriction exact sequence.) Taking M to be the free module $\mathbb{Z}[G/K]$, one has $H^1(G/K, \mathbb{Z}[G/K]) \cong \mathbb{Z}$ because G/K is an infinite cyclic group, and hence, from the inflation map, $H^1(G, \mathbb{Z}[G/K])$ is non-zero. Now $\mathbb{Z}[G/K]$ is isomorphic to the induced module $\mathbb{Z} \otimes_{\mathbb{Z}K} \mathbb{Z}G$. Let (K_λ) be the family of all finitely generated subgroups of K. Then K is the directed union of the K_λ and $\mathbb{Z} \otimes_{\mathbb{Z}K} \mathbb{Z}G$ can in turn be viewed as the direct limit of the induced modules $\mathbb{Z} \otimes_{\mathbb{Z}K_\lambda} \mathbb{Z}G$. Since G is finitely presented, $H^1(G, \quad)$ commutes with direct limits, so we have

$$\varinjlim_{\lambda} H^1(G, \mathbb{Z} \otimes_{\mathbb{Z}K_\lambda} \mathbb{Z}G) \cong H^1(G, \mathbb{Z} \otimes_{\mathbb{Z}K} \mathbb{Z}G).$$

Given that the right-hand group is non-zero, it follows that for some λ, $H^1(G, \mathbb{Z} \otimes_{\mathbb{Z}K_\lambda} \mathbb{Z}G)$ is non-zero and we can take $H := K_\lambda$ for such a λ. □

Essentially, it is this argument which we have generalized to higher dimensions throughout this article. The focus of attention has always been the groups of type $(\mathrm{FP})_\infty$, but one can also apply this argument to give some information about groups of type $(\mathrm{FP})_n$. For example, it can be shown that the following holds:

Proposition 5.1.6. *Let G be a group of type $(\mathrm{FP})_n$ and let $\theta : G \twoheadrightarrow \mathbb{Z}^{n-1}$ be a surjective homomorphism onto the free abelian group of rank $n-1$. Then there is a finitely generated subgroup H of $\ker\theta$ such that $H^i(G, \mathbb{Z} \otimes_{\mathbb{Z}H} \mathbb{Z}G)$ is non-zero for some $i < n$.*

Conceivably, this Proposition might give a new way of studying Bieri and Groves' conjectured characterization of metabelian groups of type $(\mathrm{FP})_n$. However, it would be much better to have a higher dimensional analogue of the real Bieri–Strebel Lemma rather than a higher dimensional analogue of our poor shadow of it. To this end, we need to reinterpret the original Lemma in a more geometric way.

5.2. Group actions on trees

For this purpose we need to study group actions on trees. We define a graph Γ to be a quadruple (V, E, ι, τ) where V and E are sets and ι, τ are functions from E to V. Think of V and E as the sets of vertices and edges of Γ. The maps ι and τ specify the initial and terminal vertices of each edge. Associated to a graph Γ there is a sequence

$$0 \to \mathbb{Z}E \to \mathbb{Z}V \to \mathbb{Z} \to 0$$

where $\mathbb{Z}E$ and $\mathbb{Z}V$ denote the free abelian groups on E and V respectively, and the maps are defined by $e \mapsto \tau e - \iota e$ and $v \mapsto 1$. In effect, the graph is a one dimensional cell complex, and this sequence is its augmented cellular chain complex. Intuitively, a tree is a connected graph without loops. If Γ is a tree then its chain complex is a short exact sequence.

Let G be a group. A G-graph is a graph Γ in which E and V are G-sets and ι, τ are G-maps. A G-tree is a G-graph which is a tree. The simplest example of a G-tree arises with $G = \langle g \rangle$ being infinite cyclic, with $V = E = G$ and with $\iota(g^i) = g^i$, $\tau(g^i) = g^{i+1}$. Geometrically, the infinite cyclic group acts as translations along the real line. More generally, if G is any group which admits a surjective homomorphism to an infinite cyclic group then we can make G act on the line via this homomorphism.

Now, the infinite cyclic group is an example, the simplest example, of an HNN-extension, and in fact any HNN-extension $G = B*_{H,t}$ admits an action on a tree in the following way. Let V be the set of cosets Bg, let E be the set of cosets Hg, let $\iota(Hg) = Bg$ and let $\tau(Hg) = Btg$. Then V and E become G-sets through right multiplication, and we obtain a G-graph. Special properties of the HNN-extension ensure that it is in fact a G-tree.

This point of view sheds a new light on the Bieri–Strebel Lemma. For notice that both the hypothesis that G maps onto $\mathbb{Z}$ and the conclusion that G is an HNN-extension with finitely generated base and associated subgroup (B and H) can be interpreted in terms of certain group actions on trees. There is a generalized form of the Bieri–Strebel Lemma due to Dicks and Dunwoody which can be stated as follows:

Theorem 5.2.1. (Dicks–Dunwoody) *Let G be a finitely presented group (or more generally any group of type $(\mathrm{FP})_2$) and let T be a G-tree. Then there is a G-tree T' together with a G-map $T' \to T$ such that every vertex and edge stabilizer of T' is finitely generated and such that the quotient graph T'/G is finite.*

This is now in a form which looks as though it might be a special case of some higher dimensional fact. Notice that trees are contractible 1-dimensional complexes and that the property $(\mathrm{FP})_2$ is a two dimensional finiteness condition.

5.3. Wild Conjectures

In view of this, the strongest conjecture one could make is as follows:

Conjecture 5.3.1. Let G be a group of type $(\mathrm{FP})_n$ and suppose we are given an action of G on an $(n-1)$-dimensional contractible cell complex X. Then there is a cocompact action of G on a contractible $(n-1)$-dimensional complex X', together with a G-map $X' \to X$, such that every cell stabilizer for X' is finitely generated.

This conjecture is far too strong for there to be any hope of proving it. Apart from anything else, it already implies that every group of type (FP) is of type (F). If some weakened form could be proved it would undoubtedly be very useful. Serious problems arise already when $n = 3$. While one can imitate the Dicks–Dunwoody argument to produce an action of G on a new 2-complex X' with finitely generated cell stabilizers, it seems almost hopeless to prove that this new complex is contractible. Indeed, the most naive construction might well lead to a non-contractible 2-complex and one would have to envisage surgery techniques for replacing X' by something better.

References

[1] R.C. Alperin and P.B. Shalen, Linear groups of finite cohomoloical dimension, *Invent. Math.* **66**(1982), 89–98.

[2] R. Bieri, *Homological Dimension of Discrete Groups, Mathematics Notes* (Queen Mary College, London, 2nd ed., 1981).

[3] R. Bieri and B. Eckmann, Groups with homological duality generalizing Poincaré duality, *Invent. Math.* **20**(1973), 103–124.

[4] R. Bieri and R. Strebel, Valuations and finitely presented metabelian groups, *Proc. London Math. Soc. (3)* **41**(1980), 439–464.

[5] R. Bieri and R. Strebel, A geometric invariant for modules over a finitely generated abelian group, *J. Reine Angew. Math.* **322**(1981), 170–189.

[6] R. Bieri and R. Strebel, Finiteness properties of groups, *J. Pure Appl. Algebra* **44**(1987), 45–75.

[7] J. Cornick and P.H. Kropholler, Some cohomological properties of a new class of groups, in preparation.

[8] T.V. Gedrich and K.W. Gruenberg, Complete cohomological functors on groups, *Topology Appl.* **25**(1987), 203–223.

[9] D. Gildenhuys, Classification of soluble groups of cohomological dimension 2, *Math. Z.* **2**(1979), 21–25.

[10] D. Gildenhuys and R. Strebel, On the cohomological dimension of soluble groups, *Canad. Math. Bull.* **24**(1981), 385–392.

[11] F. Goichot, Homologie de Tate-Vogel équivariante, *J. Pure Appl. Algebra* **82**(1992), 39–64.

[12] P. Hall, Finiteness conditions for soluble groups, *Proc. London Math. Soc. (3)* **4**(1954), 419–436.

[13] P. Hall, On the finiteness of certain soluble groups, *Proc. London Math. Soc. (3)* **9**(1959), 595–622.

[14] P.H. Kropholler, Cohomological dimension of soluble groups, *J. Pure Appl. Algebra* **43**(1986), 281–287.

[15] P.H. Kropholler, On groups of type $(FP)_\infty$, *J. Pure Appl. Algebra* **90**(1993), 55–67.

[16] G. Mislin, Tate cohomology for arbitrary groups via satellites, *Topology Appl.*, to appear.

[17] P. Scott, Ends of pairs of groups, *J. Pure Appl. Algebra* **11**(1977), 179–198.

[18] J.R. Stallings, On torsion-free groups with infinitely many ends, *Ann. of Math.* **88**(1968), 312–334.

[19] U. Stammbach, On the weak (homological) dimension of the group algebra of solvable groups, *J. London Math. Soc. (2)* **2**(1970), 567–570.

[20] R.G. Swan, Groups of cohomological dimension one, *J. Algebra* **12**(1969), 585–610.

Printed in the United States
By Bookmasters